T0255194

Basic Concepts in Computational Physics

Benjamin A. Stickler • Ewald Schachinger

Basic Concepts in Computational Physics

Second Edition

Benjamin A. Stickler
Faculty of Physics
University of Duisburg-Essen
Duisburg
Germany

Ewald Schachinger
Institute of Theoretical and Computational Physics
Graz University of Technology
Graz, Austria

Supplementary material and data can be found on extras.springer.com

ISBN 978-3-319-80103-2 ISBN 978-3-319-27265-8 (eBook)
DOI 10.1007/978-3-319-27265-8

Softcover reprint of the hardcover 2nd edition 2016

Printed on acid-free paper

This Springer imprint is published by Springer Nature
The registered company is Springer International Publishing AG Switzerland

Preface

Traditionally physics is divided into two fields of activities: theoretical and experimental. As a consequence of the stunning increase in computer power and of the development of more powerful numerical techniques, a new branch of physics was established over the last decades: Computational Physics. This new branch was introduced as a spin-off of what nowadays is commonly called *computer simulations*. They play an increasingly important role in physics and in related sciences as well as in industrial applications and serve two purposes, namely:

- Direct simulation of physical processes such as
  - Molecular dynamics or
  - Monte Carlo simulation of physical processes
- Solution of complex mathematical problems such as
  - Differential equations
  - Minimization problems
  - High-dimensional integrals or sums

This book addresses all these scenarios on a very basic level. It is addressed to lecturers who will have to teach a basic course/basic courses in Computational Physics or numerical methods and to students as a companion in their first steps into the realm of this fascinating field of modern research. Following these intentions this book was divided into two parts. Part I deals with deterministic methods in Computational Physics. We discuss, in particular, numerical differentiation and integration, the treatment of ordinary differential equations, and we present some notes on the numerics of partial differential equations. Each section within this part of the book is complemented by numerous applications. Part II of this book provides an introduction to stochastic methods in Computational Physics. In particular, we will examine how to generate random numbers following a given distribution, summarize the basics of stochastics in order to establish the necessary background to understand techniques like MARKOV-Chain Monte Carlo. Finally, algorithms of stochastic optimization are discussed. Again, numerous examples out of physics like diffusion processes or the POTTS model are investigated exhaustively. Finally, this book contains an appendix that augments the main parts of the book with a detailed discussion of supplementary topics.

This book is not meant to be just a collection of algorithms which can immediately be applied to various problems which may arise in Computational Physics. On the contrary, the scope of this book is to provide the reader with a mathematically well-founded glance behind the scene of Computational Physics. Thus, particular emphasis is on a clear analysis of the various topics and to even provide in some cases the necessary means to understand the very background of these methods. Although there is a barely comprehensible amount of excellent literature on Computational Physics, most of these books seem to concentrate either on deterministic methods or on stochastic methods. It is not our goal to compete with these rather specific works. On the contrary, it is the particular focus of this book to discuss deterministic methods on par with stochastic methods and to motivate these methods by concrete examples out of physics and/or engineering.

Nevertheless, a certain overlap with existing literature was unavoidable and we apologize if we were not able to cite appropriately all existing works which are of importance and which influenced this book. However, we believe that by putting the emphasis on an exact mathematical analysis of both, deterministic and stochastic methods, we created a stimulating presentation of the basic concepts applied in Computational Physics.

If we assume two basic courses in Computational Physics to be part of the curriculum, nicknamed here *Computational Physics 101* and *Computational Physics 102*, then we would like to suggest to present/study the various topics of this book according to the following syllabus:

- Computational Physics 101:
  - Chapter 1: Some Basic Remarks
  - Chapter 2: Numerical Differentiation
  - Chapter 3: Numerical Integration
  - Chapter 4: The KEPLER Problem
  - Chapter 5: Ordinary Differential Equations: Initial Value Problems
  - Chapter 6: The Double Pendulum
  - Chapter 7: Molecular Dynamics
  - Chapter 8: Numerics of Ordinary Differential Equations: Boundary Value Problems
  - Chapter 9: The One-Dimensional Stationary Heat Equation
  - Chapter 10: The One-Dimensional Stationary SCHRÖDINGER Equation
  - Chapter 12: Pseudo-random Number Generators
- Computational Physics 102:
  - Chapter 11: Partial Differential Equations
  - Chapter 13: Random Sampling Methods
  - Chapter 14: A Brief Introduction to Monte Carlo Methods
  - Chapter 15: The ISING Model

– Chapter 16: Some Basics of Stochastic Processes
– Chapter 17: The Random Walk and Diffusion Theory
– Chapter 18: MARKOV-Chain Monte Carlo and the POTTS Model
– Chapter 19: Data Analysis
– Chapter 20: Stochastic Optimization

The various chapters are augmented by problems of medium complexity which help to understand better the numerical part of the topics discussed within this book.

Although thc manuscript has been carefully checked several times, we cannot exclude that some errors escaped our scrutiny. We apologize in advance and would highly appreciate reports of potential mistakes or typos.

Throughout the book SI-units are used except stated otherwise.

Graz, Austria
July 2015

Benjamin A. Stickler
Ewald Schachinger

Acknowledgments

The authors are grateful to Profs. Dr. C. Lang and Dr. C. Gattringer (Karl-Franzens Universität, Graz, Austria) and to Profs. Dr. W. von der Linden, Dr. H.-G. Evertz, and Dr. H. Sormann (Graz University of Technology, Austria). They inspired this book with their lectures on various topics of Computational Physics, computer simulation, and numerical analysis. Last but not least, the authors thank Dr. Chris Theis (CERN) for meticulously reading the manuscript, for pointing out inconsistencies, and for suggestions to improve the text.

Contents

Chapter 1
Some Basic Remarks

1.1 Motivation

Computational Physics aims at solving physical problems by means of numerical methods developed in the field of numerical analysis [1, 2]. According to I. JACQUES and C. JUDD [3], it is defined as:

> Numerical analysis is concerned with the development and analysis of methods for the numerical solution of practical problems.

Although the term *practical problems* remained unspecified in this definition, it is certainly necessary to reflect on ways to find approximate solutions to complex problems which occur regularly in natural sciences. In fact, in most cases it is not possible to find analytic solutions and one must rely on good approximations. Let us give some examples.

Consider the definite integral

$$\int_a^b \mathrm{d}x \exp\left(-x^2\right) , \tag{1.1}$$

which, for instance, may occur when it is required to calculate the probability that an event following a normal distribution takes on a value within the interval $[a, b]$, where $a, b \in \mathbb{R}$. In contrast to the much simpler integral

$$\int_a^b \mathrm{d}x \exp(x) = \exp(b) - \exp(a) , \tag{1.2}$$

the integral (1.1) cannot be solved analytically because there is no elementary function which differentiates to $\exp\left(-x^2\right)$. Hence, we have to approximate this integral in such a way that the approximation is accurate enough for our purpose. This example illustrates that even mathematical expressions which appear quite

B.A. Stickler, E. Schachinger, *Basic Concepts in Computational Physics*,
DOI 10.1007/978-3-319-27265-8_1

simple at first glance may need a closer inspection when a numerical estimate for the expression is required. In fact, most numerical methods we will encounter within this book have been designed before the invention of modern computers or calculators. However, the applicability of these methods has increased and is still increasing drastically with the development of even more powerful machines. We give another example, namely the oscillation of a pendulum. We know from basic mechanics [4–8] that the time evolution of a frictionless pendulum of mass m and length ℓ in a gravitational field is modeled by the differential equation

$$\ddot{\theta} + \frac{g}{\ell} \sin(\theta) = 0 . \tag{1.3}$$

The solution of this equation describes the oscillatory motion of the pendulum around the origin O within a two-dimensional plane (Fig. 1.1). Here θ is the angular displacement and g is the acceleration due to gravity. Furthermore, a common situation is described by initial conditions of the form:

$$\begin{cases} \theta(0) = \theta_0 , \\ \dot{\theta}(0) = 0 . \end{cases} \tag{1.4}$$

For small initial angular displacements, $\theta_0 \ll 1$, we set in Eq. (1.3) $\sin(\theta) \approx \theta$ and obtain the differential equation of the harmonic oscillator:

$$\ddot{\theta} + \frac{g}{\ell} \theta = 0 . \tag{1.5}$$

Together with the initial conditions (1.4) we arrive at the solution

$$\theta(t) = \theta_0 \cos(\omega t) , \tag{1.6}$$

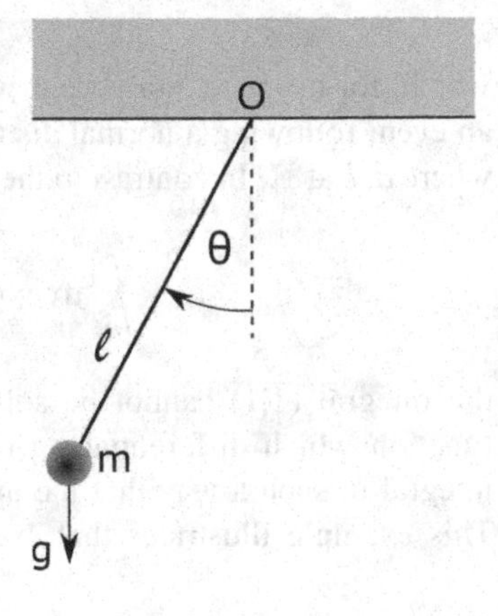

Fig. 1.1 Schematic illustration of the pendulum

with $\omega = \sqrt{g/\ell}$. The period τ of the pendulum follows immediately:

$$\tau = 2\pi \sqrt{\frac{\ell}{g}} . \tag{1.7}$$

However, if the approximation of a small angular displacement $\theta_0 \ll 1$ is not applicable, expressions (1.6) and (1.7) will not be valid. Thus, it is advisable to apply energy conservation in order to arrive at analytic results. The total energy of the pendulum is given by:

$$E = \frac{1}{2}mv^2 + mg\ell\,[1 - \cos(\theta)] = \frac{1}{2}mv_0^2 + mg\ell\,[1 - \cos(\theta_0)] \ . \tag{1.8}$$

Here v is the velocity of the point mass m and v_0 and θ_0 are defined by the initial conditions (1.4). Since $\dot{\theta}(0) = 0$ we have

$$\begin{aligned} E &= mg\ell\,[1 - \cos(\theta_0)] \\ &= 2mg\ell \sin^2\left(\frac{\theta_0}{2}\right) , \end{aligned} \tag{1.9}$$

where we made use of the relation: $1 - \cos(x) = 2\sin^2(x/2)$. We use this result in Eq. (1.8) and arrive at:

$$\frac{1}{2}v^2 = 2g\ell\left[\sin^2\left(\frac{\theta_0}{2}\right) - \sin^2\left(\frac{\theta}{2}\right)\right] . \tag{1.10}$$

Since $v^2 = \ell^2\dot{\theta}^2$ we have

$$\dot{\theta} = 2\sqrt{\frac{g}{\ell}}\sqrt{\sin^2\left(\frac{\theta_0}{2}\right) - \sin^2\left(\frac{\theta}{2}\right)} . \tag{1.11}$$

Separation of variables yields

$$\sqrt{\frac{g}{\ell}}t = \frac{1}{2k}\int_0^{\theta} \frac{d\varphi}{\sqrt{1 - \frac{1}{k^2}\sin^2\left(\frac{\varphi}{2}\right)}} , \tag{1.12}$$

with $k = \sin(\theta_0/2)$. For $t = \tau$ we have $\theta = \theta_0$ and we obtain for the period

$$\tau = \frac{2}{k}\sqrt{\frac{\ell}{g}}\int_0^{\theta_0} \frac{d\varphi}{\sqrt{1 - \frac{1}{k^2}\sin^2\left(\frac{\varphi}{2}\right)}} . \tag{1.13}$$

Let us transform the above integral into a more convenient form with help of the substitution $k \sin(\alpha) = \sin(\varphi/2)$. Thus, $\alpha \in [0, \pi/2]$ and a straightforward calculation yields:

$$\begin{aligned} \tau &= 4\sqrt{\frac{\ell}{g}} \int_0^{\frac{\pi}{2}} \frac{d\alpha}{\sqrt{1 - k^2 \sin^2(\alpha)}} \\ &= 4\sqrt{\frac{\ell}{g}} K_1(k) \, . \end{aligned} \tag{1.14}$$

The function $K_1(k)$ introduced in (1.14) for $k \in \mathbb{R}$ is referred to as the complete elliptic integral of the first kind [9–12]. All these manipulations did not really result in a simplification of the problem at hand because we are still confronted with the integral in Eq. (1.14) which cannot be evaluated without the use of additional approximations which will, in the end, result in a numerical solution of the problem. A natural way to proceed would be to expand the complete elliptic integral in a power series up to order N, where N is chosen in such a way that the *truncation error* $R_N(k)$ becomes negligible. We can find the desired expression in any text on special functions [9, 11, 12]. It reads

$$\begin{aligned} K_1(k) &= \frac{\pi}{2} \sum_{n=0}^{\infty} \left[\frac{(2n)!}{2^{2n}(n!)^2} \right]^2 k^{2n} \\ &= \frac{\pi}{2} \sum_{n=0}^{N} \left[\frac{(2n)!}{2^{2n}(n!)^2} \right]^2 k^{2n} + R_N(k) \, . \end{aligned} \tag{1.15}$$

Imagine now the inverse problem: the period τ is given and the initial angle θ_0 is unknown. Again, we could expand the integrand in a power series and solve the corresponding polynomial for θ_0. However, such an approach would be very inefficient due to two reasons: first of all, we are confronted with the impossibility of finding analytically the roots of a polynomial of order $N > 4$[1] and, secondly, at which value of N should we truncate the power series if θ_0 is unknown? A glance in a book on special functions might give us a better, i.e. more convenient, alternative. Indeed, the inverse function of the elliptic integral $K_1(k)$ with respect to k can be given explicitly in terms of JACOBI elliptic functions [9–12]. Series expansions of these functions have been developed such that we can approximate θ_0 by truncating the respective series.

This example helped to illustrate that we depend on numerical approximations of definite expressions in a multitude of cases. Even if an numerically approximate solution has been found for a particular problem it will be adamant to check quite

[1]The roots of a real valued polynomial of order $N = 3$ or 4 are referred to as CARDANO's or FERRARI's solutions [13], respectively.

carefully if the approach was (i) justified within the required accuracy, and (ii) if it allowed to improve the induced error of the result. The second point is known as the *stability* of a routine. We will discuss this topic in more detail in Sect. 1.4.

Throughout this book we will be confronted with numerous methods which will allow approximate solutions of problems similar to the two examples illustrated above. First of all, we would like to specify the properties we expect these methods to have. Primarily, the method is to be formulated as an unambiguous mathematical recipe which can be applied to the set of problems it was designed for. Its applicability should be well defined and it should allow to determine an estimate for the error. Moreover, infinite repetition of the procedure should approximate the exact result to arbitrary accuracy. In other words, we want the method to be well defined in *algorithmic* form. Consequently, let us define an algorithm as *a sequence of logical and arithmetic operations (addition, subtraction, multiplication or division) which allows to approximate the solution of the problem under consideration within any accuracy desired.* This implies, of course, that numerical errors will be unavoidable.

Let us classify the occurring errors based on the structure every numerical routine follows: We have input-errors, algorithmic-errors, and output-errors as indicated schematically in Fig. 1.2. This structural classification can be refined: input-errors are divided into roundoff errors and measurement errors contained in the input data; algorithmic-errors consist of roundoff errors during evaluation and of methodological errors due to mathematical approximations; finally, output errors are, in fact, roundoff errors. In Sects. 1.2 and 1.3 we will concentrate on roundoff errors and methodological errors. Since in most cases measurement errors cannot be influenced by the theoretical physicist concerned with numerical modeling, this particular part will not be discussed in this book. However, we will discuss the stability of numerical routines, i.e. the influence of slight modifications of the input parameters on the outcome of a particular algorithm in Sect. 1.4.

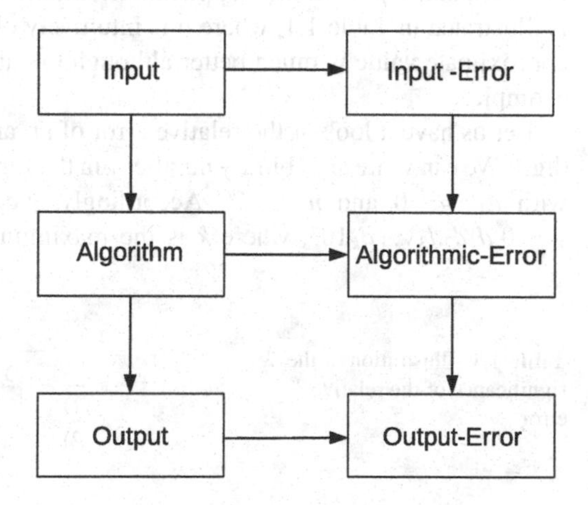

Fig. 1.2 Schematic classification of the errors occurring within a numerical procedure

1.2 Roundoff Errors

In fact, since every number is stored in a computer using a finite number of digits, we have to truncate every non-terminating number at some point. For instance, consider $\frac{2}{3} = 0.666666666666\ldots$ which will be stored as 0.6666666667 if the machine allows only ten digits. Actually, computers use binary arithmetic (for which even $0.1_{10} = 0.000110011001100\ldots_2$ is problematic[2]) but for the moment we shall ignore this fact since the above example suffices to illustrate the crucial point. Let $\mathrm{Fl}(x)$ denote the floating-point form of a number x within the numerical range of the machine. For the above example, i.e. a ten digit storage, we have

$$\mathrm{Fl}\left(\frac{2}{3}\right) = 0.6666666667\,. \tag{1.16}$$

This has the consequence that, for instance, $\mathrm{Fl}(\sqrt{3}) \cdot \mathrm{Fl}(\sqrt{3}) \neq \mathrm{Fl}(\sqrt{3} \cdot \sqrt{3}) = 3$. However, $\mathrm{Fl}(\sqrt{3}) \cdot \mathrm{Fl}(\sqrt{3}) \approx 3$ within the defined range. Before we continue our discussion on roundoff errors we have to introduce the concepts of the absolute and the relative error. We denote the true value of a quantity by y and its approximate value by $\bar{y}$. Then the absolute error ϵ_a is defined as

$$\epsilon_a = |y - \bar{y}|\,, \tag{1.17}$$

while the relative error ϵ_r is given by

$$\epsilon_r = \left|\frac{y - \bar{y}}{y}\right| = \frac{\epsilon_a}{|y|}\,, \tag{1.18}$$

provided that $y \neq 0$. In most applications, the relative error is more significant. This is illustrated in Table 1.1, where it is intuitively obvious that in the second case the approximate value is much better although the absolute error is the same for both examples.

Let us have a look at the relative error of an arbitrary number stored to the k-th digit: We can write an arbitrary number y in the form $y = 0.d_1 d_2 d_3 \ldots d_k d_{k+1} \ldots 10^n$ with $d_1 \neq 0$ and $n \in \mathbb{Z}$. Accordingly, we write its approximate value as $\bar{y} = 0.d_1 d_2 d_3 \ldots d_k 10^n$, where k is the maximum number of digits stored by the

Table 1.1 Illustration of the significance of the relative error

	y	$\bar{y}$	ϵ_a	ϵ_r
(1)	0.1	0.09	0.01	0.1
(2)	1000.0	999.99	0.01	0.00001

[2] A disastrous effect of this binary approximation of 0.1 was discussed by T. Chartier [14].

machine. Hence we obtain for the relative error

$$\begin{aligned}
\epsilon_r &= \left| \frac{0.d_1 d_2 d_3 \ldots d_k d_{k+1} \ldots 10^n - 0.d_1 d_2 d_3 \ldots d_k 10^n}{0.d_1 d_2 d_3 \ldots d_k d_{k+1} \ldots 10^n} \right| \\
&= \left| \frac{0.d_{k+1} d_{k+2} \ldots 10^{n-k}}{0.d_1 d_2 d_3 \ldots 10^n} \right| \\
&= \left| \frac{0.d_{k+1} d_{k+2} \ldots}{0.d_1 d_2 d_3 \ldots} \right| 10^{-k} \\
&\leq \frac{1}{0.1} 10^{-k} \\
&= 10^{-k+1} \,.
\end{aligned} \tag{1.19}$$

In the last steps we employed that, since $d_1 \neq 0$, we have $0.d_1 d_2 d_3 \ldots \geq 0.1$ and accordingly $0.d_{k+1} d_{k+2} \ldots < 1$. If the last digit would have been rounded (for $d_{k+1} \geq 5$ we set $d_k = d_k + 1$ otherwise d_k remains unchanged) instead of a simple truncation, the relative error of a variable y would be $\epsilon_r = 0.5 \cdot 10^{-k+1}$.

Whenever an arithmetic operation is performed, the errors of the variables involved is transferred to the result [15]. This can occur in an advantageous or disadvantageous way, where we understand disadvantageous as an increase in the relative error. Particular care is required when two nearly identical numbers are subtracted (*subtractive cancellation*) or when a large number is divided by a, in comparison, small number. In such cases the roundoff error will increase dramatically. We note that it might be necessary to avoid such operations in our aim to design an algorithm which is required to produce reasonable results. An illustrative example and its remedy will be discussed in Sect. 1.3. However, before proceeding to the next section we introduce a lower bound to the accuracy which is achievable with a non-ideal computer, the *machine-number*. The machine-number is smallest positive number η which can be added to another number, such that a change in the result is observed. In particular,

$$\eta = \min_{\delta} \left\{ \delta > 0 \middle| 1 + \delta > 1 \right\} . \tag{1.20}$$

For a (nonexistent) super-computer, which is capable of saving as much digits as desired, η would be arbitrarily small. A typical value for double-precision in FORTRAN or C is $\eta \approx 10^{-16}$.

1.3 Methodological Errors

A methodological error is introduced into the routine whenever a complex mathematical expression is replaced by an approximate, simpler one. We already came across an example when we regarded the series representation of the elliptic integral (1.12) in Sect. 1.1. Although we could evaluate the series up to an arbitrary order N, we are definitely not able to sum up the coefficients to infinite order. Hence, it is not possible to get rid of methodological errors whenever we have to deal with expressions we cannot evaluate analytically. Another intriguing example is the numerical differentiation of a given function. The standard approximation of a derivative reads

$$f'(x_0) = \frac{\mathrm{d}}{\mathrm{d}x} f(x) \bigg|_{x=x_0} \approx \frac{f(x_0 + h) - f(x_0)}{h} . \tag{1.21}$$

This approximation is referred to as *finite difference* and will be discussed in more detail in Chap. 2. One would, in a first guess, expect that the obtained value gets closer to the true value of the derivative $f'(x_0)$ with decreasing values of h. From a calculus point of view, this is correct since by definition

$$\frac{\mathrm{d}}{\mathrm{d}x} f(x) \bigg|_{x=x_0} = \lim_{h \to 0} \frac{f(x_0 + h) - f(x_0)}{h} . \tag{1.22}$$

However, this is not the case numerically. In particular, one can find a value $\hat{h}$ for which the relative error is minimal, while for values $h < \hat{h}$ and $h > \hat{h}$ the approximation obtained is worse in comparison. The reason is that for small values of h the roundoff errors dominate the result since $f(x_0 + h)$ and $f(x_0)$ almost cancel while $1/h$ is very small. For $h > \hat{h}$, the methodological error, i.e. the replacement of a derivative by a finite difference, controls the result.

We give one further example [16] in order to illustrate the interplay between methodological errors and roundoff errors. We regard the, apparently nonhazardous, numerical solution of a quadratic equation

$$ax^2 + bx + c = 0 , \tag{1.23}$$

where $a, b, c \in \mathbb{R}$, $a \neq 0$. The well known solutions read

$$x_1 = \frac{-b + \sqrt{b^2 - 4ac}}{2a} \quad \text{and} \quad x_2 = \frac{-b - \sqrt{b^2 - 4ac}}{2a} . \tag{1.24}$$

Cautious because of the above examples, we immediately diagnose the danger of a subtractive cancellation in the expression of x_1 for $b > 0$ or in x_2 for $b < 0$, and rewrite the above expression for x_1:

$$x_1 = \frac{(-b + \sqrt{b^2 - 4ac})}{2a} \frac{(-b - \sqrt{b^2 - 4ac})}{(-b - \sqrt{b^2 - 4ac})} = \frac{2c}{-b - \sqrt{b^2 - 4ac}} . \tag{1.25}$$

For x_2 we obtain

$$x_2 = \frac{2c}{-b + \sqrt{b^2 - 4ac}} \,. \tag{1.26}$$

Consequently, if $b > 0$ x_1 should be calculated using Eq. (1.25) and if $b < 0$ Eq. (1.26) should be used to calculate x_2. Moreover, the above expressions can be cast into one expression by setting

$$x_1 = \frac{q}{a} \quad \text{and} \quad x_2 = \frac{c}{q} \,, \tag{1.27}$$

with

$$q = -\frac{1}{2}\left[b + \mathrm{sgn}(b)\sqrt{b^2 - 4ac}\right] \,. \tag{1.28}$$

Thus, Eqs. (1.27) and (1.28) can be used to calculate x_1 and x_2 for any sign of b.

1.4 Stability

When a new numerical method is designed *stability* is the third crucial point after roundoff errors and methodological errors [17]. We give an introductory definition:

> An algorithm, equation or, even more general, a problem is referred to as unstable or ill-conditioned if small changes in the input cause a large change in the output.

It will be followed by a couple of elucidating examples [3].[3] To be more specific, let us now, for instance, consider the following system of equations

$$\begin{aligned} x + y &= 2.0, \\ x + 1.01y &= 2.01 \,. \end{aligned} \tag{1.29}$$

These equations are easily solved and give $x = 1.0$ and $y = 1.0$. To make our point we consider now the case in which the right hand side of the second equation of (1.29) is subjected to a small perturbation, i.e. we consider in particular the following system of equations

$$\begin{aligned} x + y &= 2.0, \\ x + 1.01y &= 2.02 \,. \end{aligned} \tag{1.30}$$

[3] Although unstable behavior is not desirable in the first place the discovery of unstable systems was the birth of a specific branch in physics called *Chaos Theory*. We briefly comment on this point at the end of this section.

The corresponding solution is $x = 0.0$ and $y = 2.0$. We observe that a relative change of 0.05 % on the right hand side of the second equation in (1.29) resulted in a 100 % relative change of the solution. Moreover, if the coefficient of y in the second equation of (1.29) were 1.0 instead of 1.01, which corresponds to a relative change of 1 %, the equations would be unsolvable. This is a behavior typical for ill-conditioned problems which, for obvious reasons, should be avoided whenever possible.

We give a second example: We consider the following initial value problem

$$\begin{cases} \ddot{y} - 10\dot{y} - 11y = 0 , \\ y(0) = 1, \qquad \dot{y}(0) = -1 . \end{cases} \tag{1.31}$$

The general solution is readily obtained to be of the form

$$y = A \exp(-x) + B \exp(11x) , \tag{1.32}$$

with numerical constants A and B. The initial conditions yield the unique solution

$$y = \exp(-x) . \tag{1.33}$$

The initial conditions are now changed by two small parameters $\delta, \epsilon > 0$ to give:

$$y(0) = 1 + \delta \qquad \text{and} \qquad \dot{y}(0) = -1 + \epsilon . \tag{1.34}$$

The unique solution which satisfies these initial conditions is:

$$\overline{y} = \left(1 + \frac{11\delta}{12} - \frac{\epsilon}{12}\right) \exp(-x) + \left(\frac{\delta}{12} + \frac{\epsilon}{12}\right) \exp(11x) . \tag{1.35}$$

We calculate the relative error

$$\begin{aligned} \epsilon_r &= \left|\frac{y - \overline{y}}{y}\right| \\ &= \left(\frac{11\delta}{12} - \frac{\epsilon}{12}\right) + \left(\frac{\delta}{12} + \frac{\epsilon}{12}\right) \exp(12x) , \end{aligned} \tag{1.36}$$

which indicates that the problem is ill-conditioned since for large values of x the second term definitely overrules the first one.

Another, but not less serious kind of problem is *induced instability*:

A method is referred to as induced unstable if a small error at one point of the calculation induces a large error at some subsequent point.

Induced instability is particularly dangerous since small roundoff errors are unavoidable in most calculations. Hence, if some part of the whole algorithm is ill-conditioned, the final output will be dominated by the error induced in such a way. Again, an example will help to illustrate such behavior. The definite integral

$$I_n = \int_0^1 \mathrm{d}x\, x^n \exp(x-1) , \tag{1.37}$$

is considered. Integration by parts yields

$$I_n = 1 - nI_{n-1} . \tag{1.38}$$

This expression can be used to recursively calculate I_n from I_0, where

$$I_0 = 1 - \exp(-1) . \tag{1.39}$$

Although the recursion formula (1.38) is exact we will run into massive problems using it. The reason is easily illustrated:

$$\begin{aligned} I_n &= 1 - nI_{n-1} \\ &= 1 - n + n(n-1)I_{n-2} \\ &= 1 - n + n(n-1) - n(n-1)(n-2)I_{n-3} \\ &\vdots \\ &= 1 + \sum_{k=1}^{n-1} (-1)^k \frac{n!}{(n-k)!} + (-1)^{n-1} n! I_0 . \end{aligned} \tag{1.40}$$

Thus, the initial roundoff error included in the numerical value of I_0 is multiplied with $n!$. Note that for large n we have according to STIRLING's approximation

$$n! \approx \sqrt{2\pi} n^{n+\frac{1}{2}} \exp(-n) , \tag{1.41}$$

i.e. an initial error increases almost as n^n.

However, Eq. (1.38) can be reformulated to give

$$I_n = \frac{1}{n+1} (1 - I_{n+1}) , \tag{1.42}$$

and this opens an alternative method for a recursive calculation of I_n. We can start with some value $N \gg n$ and simply set $I_N = 0$. The error introduced in such a way may in the end not be acceptable, nevertheless, it decreases with every iteration step due to the division by n in Eq. (1.42).

Having discussed some basic features of stability in numerical algorithms we would like to add a few remarks on *Chaos Theory*. Chaos theory investigates dynamical processes which are very sensitive to initial conditions. One of the best known examples for such a behavior is the weather prediction. Although, POINCARÉ already observed chaotic behavior while working on the three body problem, one of the pioneers of chaos theory was E.N. LORENZ [18] (not to be confused with H. LORENTZ, who introduced the LORENTZ transformation). In 1961 he ran weather simulations on a computer of restricted capacity. However, when he tried to reproduce one particular result by restarting the calculation with new parameters calculated the days before, he observed that the outcome was completely different [19]. The reason was that the equations he dealt with were ill-conditioned, and the roundoff error he introduced by simply typing in the numbers of the graphical output, increased drastically, and, hence, produced a completely different result. Nowadays, various physical systems are known which indeed behave in such a way. Further examples are turbulences in fluids, oscillations in electrical circuits, oscillating chemical reactions, population growth in ecology, the time evolution of the magnetic field of celestial bodies, ...

It is important to note, that chaotic behavior induced in such systems is *deterministic*, yet *unpredictable*. This is due to the impossibility of an exact knowledge of the initial conditions required to predict, for instance, the weather over a reasonably long period. A feature which is referred to as the *butterfly effect*: a hurricane can form because a butterfly flapped its wings several weeks before. However, these effects have nothing to do with intrinsically probabilistic properties which are solely a feature of quantum mechanics. In contrast to this, in chaos theory, the future is uniquely determined by initial conditions, however, still unpredictable. This is often referred to as *deterministic chaos*.

It has to be emphasized that chaos in physical systems is a consequence of the equations describing the processes and not a consequence of the numerical method used for modeling. Therefore, it is important to distinguish between the stability of a numerical method and the stability of a physical system in general.

We will come across chaotic behavior again in Sect. 6.3 where we discuss chaotic behavior in the dynamics of the double pendulum [4–8].

1.5 Concluding Remarks

In this chapter we dealt with the basic features of numerical errors one is always confronted with when developing an algorithm. One point we neglected in our discussion is the *computational cost*, i.e. the time a program needs to be executed. Although this is a very important point, it is beyond the scope of this book. However, one has to find a balance between the need of achieving the most accurate result and the computing time required to achieve it. The most accurate result is useless if the programmer does not get the result within his lifetime. D. ADAMS [20] put in a nutshell: the super-computer *Deep Thought* was asked to compute the answer to

"The Ultimate Question of Life, the Universe and Everything", quote:

> "How long?" he said.
> "Seven and a half million years."

Another quite crucial point, which we neglected so far, is the error analysis of a computational method which is based on random numbers (in fact it is pseudo-random numbers and this point will be discussed in the second part of this book). In this case, the situation changes completely, because, similar to experimental results, the observed values are distributed around a mean with a certain variance. Such results have to be interpreted within a statistical context. However, it turns out that for many problems the computational efficiency can be significantly increased using such methods. Typical applications are estimates of integrals or solutions to optimization problems. Such topics will be treated in the second part of this book.

References

1. Süli, E., Mayers, D.: An Introduction to Numerical Analysis. Cambridge University Press, Cambridge (2003)
2. Gautschi, W.: Numerical Analysis. Springer, Berlin/Heidelberg (2012)
3. Jacques, I., Judd, C.: Numerical Analysis. Chapman and Hall, London (1987)
4. Arnol'd, V.I.: Mathematical Methods of Classical Mechanics, 2nd edn. Graduate Texts in Mathematics, vol. 60. Springer, Berlin/Heidelberg (1989)
5. Fetter, A.L., Walecka, J.D.: Theoretical Mechanics of Particles and Continua. Dover, New York (2004)
6. Scheck, F.: Mechanics, 5th edn. Springer, Berlin/Heidelberg (2010)
7. Goldstein, H., Poole, C., Safko, J.: Classical Mechanics, 3rd edn. Addison-Wesley, Menlo Park (2013)
8. Fließbach, T.: Mechanik, 7th edn. Lehrbuch zur Theoretischen Physik I. Springer, Berlin/Heidelberg (2015)
9. Abramovitz, M., Stegun, I.A. (eds.): Handbook of Mathemathical Functions. Dover, New York (1965)
10. Olver, F.W.J., Lozier, D.W., Boisvert, R.F., Clark, C.W.: NIST Handbook of Mathematical Functions. Cambridge University Press, Cambridge (2010)
11. Mathai, A.M., Haubold, H.J.: Special Functions for Applied Scientists. Springer, Berlin/Heidelberg (2008)
12. Beals, R., Wong, R.: Special Functions. Cambridge Studies in Advanced Mathematics. Cambridge University Press, Cambridge (2010)
13. Clark, A.: Elements of Abstract Algebra. Dover, New York (1971)
14. Chartier, T.: Devastating roundoff error. Math. Horiz. **13**, 11 (2006). http://www.jstor.org/stable/25678616
15. Ueberhuber, C.W.: Numerical Computation 1: Methods, Software and Analysis. Springer, Berlin/Heidelberg (1997)
16. Burden, R.L., Faires, J.D.: Numerical Analysis. PWS-Kent Publishing Comp., Boston (1993)
17. Higham, N.J.: Accuracy and Stability of Numerical Algorithms, 2nd edn. Society for Industrial and Applied Mathematics (SIAM), Philadelphia (2002)
18. Lorenz, E.N.: Deterministic nonperiodic flow. J. Atmos. Sci. **20**, 130–141 (1963)
19. Roulstone, I., Norbury, J.: Invisible in the Storm: The Role of Mathematics in Understanding Weather. Prinecton University Press, Princeton (2013)
20. Adams, D.: The Hitchhiker's Guide to the Galaxy. Pan Books, London (1979)

Part I
Deterministic Methods

Chapter 2
Numerical Differentiation

2.1 Introduction

This chapter is the first of two systematic introductions to the numerical treatment of differential equations. Differential equations and, thus, derivatives and integrals are of eminent importance in the modern formulation of natural sciences and in particular of physics. Very often the complexity of the expressions involved does not allow an analytical approach, although modern *symbolic* software can ease a physicists life significantly. Thus, in many cases a numerical treatment is unavoidable and one should be prepared.

We introduce here the notion of finite differences as a basic concept of numerical differentiation [1–3]. In contrast, the next chapter will deal with the concepts of numerical quadrature. Together, these two chapters will set the stage for a comprehensive discussion of algorithms designed to solve numerically differential equations. In particular, the solution of ordinary differential equations will always be based on an integration.

This chapter is composed of four sections. The first repeats some basic concepts of calculus and introduces formally finite differences. The second formulates approximates to derivatives based on finite differences, while the third section includes a more systematic approach based on an operator technique. It allows an arbitrarily close approximation of derivatives with the advantage that the expressions discussed in this section can immediately be applied to the problems at hand. The chapter is concluded with a discussion of some additional aspects.

B.A. Stickler, E. Schachinger, *Basic Concepts in Computational Physics*,
DOI 10.1007/978-3-319-27265-8_2

2.2 Finite Differences

Let us consider a smooth function $f(x)$ on the finite interval $[a, b] \subset \mathbb{R}$ of the real axis. The interval $[a, b]$ is divided into $N - 1 \in \mathbb{N}$ equally spaced sub-intervals of the form $[x_i, x_{i+1}]$ where $x_1 = a$, $x_N = b$. Obviously, x_i is then given by

$$x_i = a + (i - 1)\frac{b - a}{N - 1}, \quad i = 1, \ldots, N\,. \tag{2.1}$$

We introduce the distance h between two grid-points x_i by:

$$h = x_{i+1} - x_i = \frac{b - a}{N - 1}, \quad \forall i = 1, \ldots, N - 1\,. \tag{2.2}$$

For the sake of a more compact notation we restrict our discussion to equally spaced grid-points keeping in mind that the extension to arbitrarily spaced grid-points by replacing h by h_i is straight forward and leaves the discussion essentially unchanged.

Note that the number of grid-points and, thus, their distance h, has to be chosen in such a way that the function $f(x)$ can be sufficiently well approximated by its function values $f(x_i)$ as indicated in Fig. 2.1. We understand by *sufficiently well approximated* that some interpolation scheme in the interval $[x_i, x_{i+1}]$ will reproduce the function $f(x)$ within a required accuracy. In cases where the function is strongly varying within some sub-interval $[c, d] \subset [a, b]$ and is slowly varying within

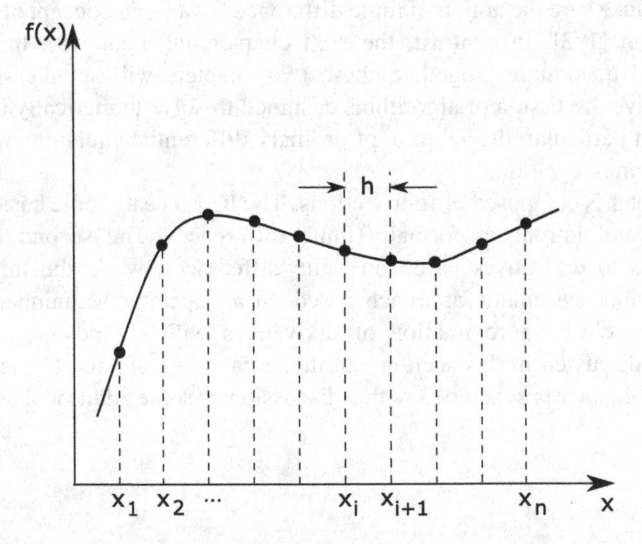

Fig. 2.1 We define equally spaced grid-points x_i on a finite interval on the real axis in such a way that the function $f(x)$ is sufficiently well approximated by its functional values $f(x_i)$ at these grid-points

$[a, b] \setminus [c, d]$ it might be advisable to use variable grid-spacing in order to reduce the computational cost of the procedure.

We introduce the following notation: The function value of $f(x)$ at the grid-point x_i will be denoted by $f_i \equiv f(x_i)$ and its n-th derivative:

$$f_i^{(n)} \equiv f^{(n)}(x_i) = \left. \frac{d^n f(x)}{dx^n} \right|_{x=x_i} . \tag{2.3}$$

Furthermore, we define for arbitrary $\xi \in [x_i, x_{i+1})$

$$f_{i+\epsilon}^{(n)} = f^{(n)}(\xi) , \tag{2.4}$$

where $f_{i+\epsilon}^{(0)} \equiv f_{i+\epsilon}$ and ϵ is chosen to give:

$$\xi = x_i + \epsilon h , \qquad \epsilon \in [0, 1) . \tag{2.5}$$

Let us remember some basics from calculus: The first derivative, denoted $f'(x)$ of a function $f(x)$ which is smooth within the interval $[a, b]$, i.e. $f(x) \in \mathscr{C}^\infty[a, b]$ for arbitrary $x \in [a, b]$, is defined as

$$\begin{aligned} f'(x) &:= \lim_{h \to 0} \frac{f(x+h) - f(x)}{h} \\ &= \lim_{h \to 0} \frac{f(x) - f(x-h)}{h} \\ &= \lim_{h \to 0} \frac{f(x+h) - f(x-h)}{2h} . \end{aligned} \tag{2.6}$$

However, it is impossible to draw numerically the limit $h \to 0$ as discussed in Sect. 1.3, Eq. (1.22). This manifests itself in a non-negligible error due to subtractive cancellation.

This problem is circumvented by the use of TAYLOR's theorem. It states that if there is a function which is $(n + 1)$-times continuously differentiable on the interval $[a, b]$ then $f(x)$ can be expressed in terms of a series expansion at point $x_0 \in [a, b]$:

$$f(x) = \sum_{k=0}^{n} \frac{f^{(k)}(x_0)}{k!}(x - x_0)^k + \frac{f^{(n+1)}[\zeta(x)]}{(n+1)!}(x - x_0)^{n+1}, \quad \forall x \in [a, b] . \tag{2.7}$$

Here, $\zeta(x)$ takes on a value between x and x_0.[1] The last term on the right hand side of Eq. (2.7) is commonly referred to as *truncation error*. (A more general definition of this error was given in Sect. 1.1.)

[1] Note that for $x_0 = 0$ the series expansion (2.7) is referred to as MCLAURIN series.

We introduce now the finite difference operators

$$\Delta_+ f_i = f_{i+1} - f_i \,, \tag{2.8a}$$

as the *forward difference*,

$$\Delta_- f_i = f_i - f_{i-1} \,, \tag{2.8b}$$

as the *backward difference*, and

$$\Delta_c f_i = f_{i+1} - f_{i-1} \,, \tag{2.8c}$$

as the *central difference*.[2] The derivative of $f(x)$ can be approximated with the help of TAYLOR's theorem (2.7). In a first step we consider (restricting to third order in h)

$$\begin{aligned} f_{i+1} &= f(x_i) + hf'(x_i) + \frac{h^2}{2} f''(x_i) + \frac{h^3}{6} f'''[\zeta(x_i + h)] \\ &= f_i + hf'_i + \frac{h^2}{2} f''_i + \frac{h^3}{6} f'''_{i+\epsilon_\zeta} \,, \end{aligned} \tag{2.9a}$$

with $f_{i+1} \equiv f(x_i + h)$. Here ϵ_ζ is the fractional part ϵ which has to be determined according to $\zeta(x_i + h)$. In analogue we find for f_{i-1}

$$f_{i-1} = f_i - hf'_i + \frac{h^2}{2} f''_i - \frac{h^3}{6} f'''_{i+\epsilon_\zeta} \,. \tag{2.9b}$$

Solving Eqs. (2.9) for the derivative f'_i leads directly to the definition of finite difference derivatives.

2.3 Finite Difference Derivatives

We define the *finite difference derivative* or difference approximations

$$D_+ f_i = \frac{\Delta_+ f_i}{h} = \frac{f_{i+1} - f_i}{h} \,, \tag{2.10a}$$

as the *forward difference derivative*,

[2] Please note that the symbols Δ_+, Δ_-, and Δ_c in Eqs. (2.8) are linear operators acting on f_i. For a basic introduction to the theory of linear operators see for instance [4, 5].

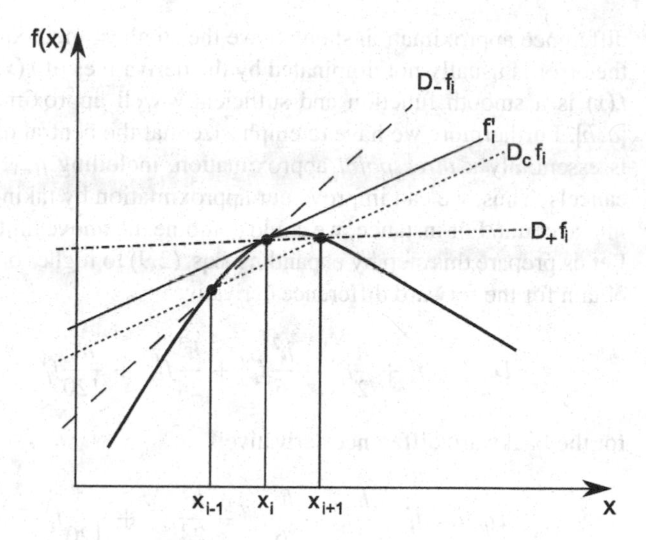

Fig. 2.2 Graphical illustration of different finite difference derivatives. The *solid line* labeled f_i' represents the real derivative for comparison

$$D_- f_i = \frac{\Delta_- f_i}{h} = \frac{f_i - f_{i-1}}{h}\,, \tag{2.10b}$$

as the *backward difference derivative*, and

$$D_c f_i = \frac{\Delta_c f_i}{2h} = \frac{f_{i+1} - f_{i-1}}{2h}\,, \tag{2.10c}$$

as the *central difference derivative*.[3] A graphical interpretation of these expressions is straight forward and is presented in Fig. 2.2.

Using the above definitions (2.10) together with the expansions (2.9) we obtain

$$\begin{aligned} f_i' &= D_+ f_i - \frac{h}{2} f_i'' - \frac{h^2}{6} f_{i+\epsilon_\zeta}''' \\ &= D_- f_i + \frac{h}{2} f_i'' - \frac{h^2}{6} f_{i+\epsilon_\zeta}''' \\ &= D_c f_i - \frac{h^2}{6} f_{i+\epsilon_\zeta}'''\,. \end{aligned} \tag{2.11}$$

We observe that in the central difference approximation of f_i' the truncation error scales like h^2 while it scales like h in the other two approximations; thus the central

---

[3]The central difference derivative is related to the forward and backward difference derivatives via:

$$D_c = \frac{1}{2}(D_+ + D_-).$$

difference approximation should have the smallest methodological error. Note that the error is usually not dominated by the derivatives of $f(x)$ since we assumed that $f(x)$ is a smooth function and sufficiently well approximated on the grid within $[a, b]$. Furthermore we have to emphasize that the central difference approximation is essentially a *three point* approximation, including f_{i-1}, f_i and f_{i+1}, although f_i cancels. Thus, we can improve our approximation by taking even more grid-points into account. For instance, we could combine the above finite difference derivatives. Let us prepare this step by expanding Eqs. (2.9) to higher order derivatives. We then obtain for the forward difference derivative

$$D_+ f_i = f_i' + \frac{h}{2} f_i'' + \frac{h^2}{6} f_i''' + \frac{h^3}{24} f_i^{IV} + \frac{h^4}{120} f_i^V + \dots , \tag{2.12}$$

for the backward difference derivative

$$D_- f_i = f_i' - \frac{h}{2} f_i'' + \frac{h^2}{6} f_i''' - \frac{h^3}{24} f_i^{IV} + \frac{h^4}{120} f_i^V \mp \dots , \tag{2.13}$$

and, finally, for the central difference derivative

$$D_c f_i = f_i' + \frac{h^2}{6} f_i''' + \frac{h^4}{120} f_i^V + \dots . \tag{2.14}$$

In order to improve the method we have to combine $D_+ f_i$, $D_- f_i$ and $D_c f_i$ from different grid-points in such a way that at least the terms proportional to h^2 cancel. This can be achieved by observing that[4]

$$8D_c f_i - D_c f_{i-1} - D_c f_{i+1} = 6 f_i' - \frac{h^4}{5} f_{i+\epsilon_\xi}^V , \tag{2.15}$$

which gives

$$\begin{aligned} f_i' &= \frac{1}{6} \left(8D_c f_i - D_c f_{i+1} - D_c f_{i-1} \right) + \frac{h^4}{30} f_i^V \\ &= \frac{1}{12h} \left(f_{i-2} - 8 f_{i-1} + 8 f_{i+1} - f_{i+2} \right) + \frac{h^4}{30} f_i^V . \end{aligned} \tag{2.16}$$

Note that this simple combination yields an improvement of two orders in h ! One can even improve the approximation in a similar fashion by simply calculating the derivative from even more points, for instance $f_{i\pm 3}$.

[4] Please note that the TAYLOR expansion of $(D_c f_{i-1} + D_c f_{i+1})/2 = (f_{i+2} - f_{i-2})/(4h)$ is equivalent to the expansion (2.14) of $D_c f_i$ with h replaced by $2h$.

2.4 A Systematic Approach: The Operator Technique

We would like to obtain a general expression which will allow to calculate the finite difference derivatives of arbitrary order up to arbitrary order of h in the truncation error. We achieve this goal by introducing the shift operator T and its inverse operator T^{-1} as[5]

$$Tf_i = f_{i+1}\ , \tag{2.18}$$

and

$$T^{-1}f_i = f_{i-1}\ , \tag{2.19}$$

where $TT^{-1} = \mathbb{1}$ is the unity operator. We can write these operators in terms of the forward and backward difference operators Δ_+ and Δ_- of Eqs. (2.8), in particular

$$T = \mathbb{1} + \Delta_+\ , \tag{2.20}$$

and

$$T^{-1} = \mathbb{1} - \Delta_-\ . \tag{2.21}$$

Moreover, if $D \equiv \mathrm{d}/\mathrm{d}x$ denotes the derivative operator and if the n-th power of this operator D is understood as the n-th successive application of it, we can rewrite the TAYLOR expansions (2.9) as

$$\begin{aligned} f_{i+1} &= \left[\mathbb{1} + hD + \frac{1}{2}h^2D^2 + \frac{1}{3!}h^3D^3 + \ldots\right] f_i \\ &\equiv \exp(hD) f_i\ , \end{aligned} \tag{2.22}$$

[5] We note in passing that the shift operators form the discrete translational group, a very important group in theoretical physics. Let $T(n) = T^n$ denote the shift by $n \in \mathbb{N}$ grid-points. We then have

$$T(n)T(m) = T(n+m)\ , \tag{2.17a}$$

$$T(0) = \mathbb{1}\ , \tag{2.17b}$$

and

$$T(n)^{-1} = T(-n)\ , \tag{2.17c}$$

which are the properties required to form a group. Here $\mathbb{1}$ denotes unity. Moreover, we have

$$T(n)T(m) = T(m)T(n)\ , \tag{2.17d}$$

i.e. it is an Abelian group. The group of discrete translations is usually denoted by $\mathbb{T}^d$ [6].

and

$$\begin{aligned} f_{i-1} &= \left[\mathbb{1} - hD + \frac{1}{2}h^2D^2 - \frac{1}{3!}h^3D^3 \pm \ldots\right] f_i \\ &\equiv \exp(-hD) f_i \ , \end{aligned} \tag{2.23}$$

Hence, we find that [7][6]

$$T = \mathbb{1} + \Delta_+ \equiv \exp(hD) \ , \tag{2.24}$$

and, accordingly, that

$$T^{-1} = \mathbb{1} - \Delta_- \equiv \exp(-hD) \ . \tag{2.25}$$

Finally, we obtain the central difference operator:

$$\Delta_c = T - T^{-1} = \exp(hD) - \exp(-hD) \equiv 2\sinh(hD) \ . \tag{2.26}$$

Equations (2.24), (2.25) and (2.26) can be inverted for hD:

$$hD = \begin{cases} \ln(\mathbb{1} + \Delta_+) = \Delta_+ - \frac{1}{2}\Delta_+^2 + \frac{1}{3}\Delta_+^3 \mp \ldots \ , \\ -\ln(\mathbb{1} - \Delta_-) = \Delta_- + \frac{1}{2}\Delta_-^2 + \frac{1}{3}\Delta_-^3 + \ldots \ , \\ \sinh^{-1}\left(\frac{\Delta_c}{2}\right) = \frac{\Delta_c}{2} - \frac{1}{3!}\left(\frac{\Delta_c}{2}\right)^3 + \frac{3^2}{5!}\left(\frac{\Delta_c}{2}\right)^5 \mp \ldots \ . \end{cases} \tag{2.27}$$

Again, the n-th power of an operator K (with $K = \Delta_+, \Delta_-, \Delta_c$) $K^n f_i$ is understood as the n-th successive action of the operator K on f_i, i.e. $K^{n-1}(Kf_i)$. Expression (2.27) allows to approximate the derivatives up to arbitrary order using finite differences. Furthermore, we can take the k-th power of Eq. (2.27) in order to get an approximate k-th derivative, $(hD)^k$ [7].

However, it turns out that the expansion (2.27) in terms of the central difference Δ_c does not optimally use the grid because it contains only odd powers of Δ_c. For instance, the third power $\Delta_c^3 f_i$ includes the function values $f_{i\pm3}$ and $f_{i\pm1}$ at 'odd' grid-points but ignores the function values f_i and $f_{i\pm2}$ at 'even' grid-points. Since this is true for all odd powers of Δ_c we observe that the expansion (2.27) uses only half of the grid. On the other hand, if one computes the square $(hD)^2$ of (2.27) only 'even' grid-points are used, while the 'odd' grid-points are ignored. This reduces the accuracy of the method and an improvement is required. The easiest remedy

[6]This representation of the shift operator T explains why the derivative operator D is frequently referred to as the *infinitesimal generator of translations* [6].

is to formally introduce function values $T^{\pm\frac{1}{2}}f_i \overset{!}{=} f_{i\pm1/2}$ at *intermediate* grid-points[7] $x_{i\pm1/2} = x_i \pm h/2$. This definition allows to introduce the central difference operator δ_c of intermediate grid-points,

$$\delta_c = T^{\frac{1}{2}} - T^{-\frac{1}{2}} = 2\sinh\left(\frac{hD}{2}\right), \tag{2.28}$$

and the average operator:

$$\mu = \frac{1}{2}\left(T^{\frac{1}{2}} + T^{-\frac{1}{2}}\right) = \cosh\left(\frac{hD}{2}\right). \tag{2.29}$$

The central difference operator Δ_c on the grid is connected to the central difference operator δ_c of intermediate grid-points by:

$$\Delta_c = 2\mu\delta_c. \tag{2.30}$$

To avoid the problem of Eq. (2.27) that only odd or even grid-points are accounted for we replace all shift operators $\Delta_c/2$ by δ_c and then multiply the right hand side of Eq. (2.27) by μ. This ensures that function values at intermediate grid-points will not appear in the final expression. Hence, we obtain for the first order derivative operator:

$$D = \frac{1}{h}\begin{cases} \Delta_+ - \frac{1}{2}\Delta_+^2 + \frac{1}{3}\Delta_+^3 \mp \dots, \\ \Delta_- + \frac{1}{2}\Delta_-^2 + \frac{1}{3}\Delta_-^3 + \dots, \\ \mu\delta_c - \frac{1}{3!}\mu\delta_c^3 + \frac{3^2}{5!}\mu\delta_c^5 \mp \dots \end{cases} \tag{2.31}$$

When higher order derivatives are calculated, we replace, again, $\Delta_c/2$ by δ_c and multiply odd powers of δ_c by μ. This procedure results, for instance, in the second order derivative operator:

$$D^2 = \frac{1}{h^2}\begin{cases} \Delta_+^2 - \Delta_+^3 + \frac{11}{12}\Delta_+^4 \mp \dots, \\ \Delta_-^2 + \Delta_-^3 + \frac{11}{12}\Delta_-^4 + \dots, \\ \delta_c^2 - \frac{1}{3}\delta_c^4 + \frac{8}{45}\delta_c^6 \mp \dots. \end{cases} \tag{2.32}$$

[7] These intermediate grid-points are virtual, auxiliary grid-points which will be eliminated in due course.

In particular, we obtain for the central difference derivative

$$f_i' = \frac{f_{i+1} - f_{i-1}}{2h} + \mathcal{O}(h^2) , \tag{2.33}$$

and

$$f_i'' = \frac{f_{i+1} - 2f_i + f_{i-1}}{h^2} + \mathcal{O}(h^2) . \tag{2.34}$$

Here, $\mathcal{O}(h^2)$ indicates that this term is of the order of h^2 and we get the important result that the truncation error is of the order $\mathcal{O}(h^2)$.[8]

2.5 Concluding Discussion

First of all, although Eq. (2.27) allows to approximate a derivative of any order k arbitrarily close, it is still an infinite series which leaves us with the decision at which order to truncate. This choice will highly depend on the choice of h which in turn depends on the function we would like to differentiate. Consider, for instance, the periodic function

$$f(x) = \exp(i\omega x) , \tag{2.35}$$

where $\omega, x \in \mathbb{R}$ and i is the imaginary unit with $i^2 = -1$. Its first derivative is

$$f'(x) = i\omega \exp(i\omega x) . \tag{2.36}$$

We now introduce grid-points by

$$x_k = x_0 + kh , \tag{2.37}$$

where h is the grid-spacing and x_0 is some finite starting point on the real axis. Accordingly,

$$f_k = \exp[i\omega(x_0 + kh)] , \tag{2.38}$$

[8]The leading order of the truncation error can be determined by inserting the dominant contribution of Eqs. (2.28) and (2.29) into the remainder of Eqs. (2.31) and (2.32), respectively. For instance, it follows from Eq. (2.29) that $\mu \sim \mathcal{O}(1)$ and from Eq. (2.28) that $\delta_c \sim \mathcal{O}(h)$ and, hence, we find with the help of Eq. (2.31) that $\mu\delta_c^3/h \sim \mathcal{O}(h^2)$. In analogue, we obtain from Eq. (2.32) that $\delta_c^4/h^2 \sim \mathcal{O}(h^2)$.

and the exact value of the first derivative is

$$f'_k = i\omega \exp\left[i\omega(x_0 + kh)\right] = i\omega f_k \; . \tag{2.39}$$

We calculate the forward, backward, and central difference derivatives according to Eqs. (2.10) and obtain

$$D_+ f_k = i\omega f_k \exp\left(\frac{ih\omega}{2}\right) \operatorname{sinc}\left(\frac{h\omega}{2}\right) , \tag{2.40a}$$

with $\operatorname{sinc}(x) = \sin(x)/x$ and

$$D_- f_k = i\omega f_k \exp\left(-\frac{ih\omega}{2}\right) \operatorname{sinc}\left(\frac{h\omega}{2}\right) , \tag{2.40b}$$

and

$$D_c f_k = i\omega f_k \operatorname{sinc}(h\omega) \; . \tag{2.40c}$$

We divide the approximate derivatives by the true value (2.39) and take the modulus. We get

$$\left|\frac{D_+ f_k}{f'_k}\right| = \left|\frac{D_- f_k}{f'_k}\right| = \operatorname{sinc}\left(\frac{h\omega}{2}\right) , \tag{2.41}$$

and

$$\left|\frac{D_c f_k}{f'_k}\right| = \operatorname{sinc}(h\omega) \; . \tag{2.42}$$

Since $|\sin(x)| \leq |x|$, $\forall x \in \mathbb{R}$ we obtain that in all three cases this ratio is less than one independent of h, unless $\omega = 0$. (Please keep in mind that $\operatorname{sinc}(x) \to 1$ as $x \to 0$.) Hence, the first order finite difference approximations underestimate the true value of the derivative. The reason is easily found: $f(x)$ oscillates with frequency ω while the finite difference derivatives applied here approximate the derivative linearly. Higher order corrections will, of course, improve the approximation significantly. Furthermore, we observe that the one-sided finite difference derivatives (2.40a) and (2.40b) are exactly zero if $h\omega = 2n\pi$, $n \in \mathbb{N}$, i.e. if the grid-spacing h matches a multiple of the frequency $2\pi\omega$ of the function $f(x)$. The same occurs when central derivatives (2.40c) are used, but now for $h\omega = \pi n$. This is not really a problem in our example because we choose the grid-spacing $h \ll 2\pi/\omega$ in order to approximate the function $f(x)$ sufficiently well. However, in many cases the analytic form of the function is unknown and we only have its representation on the grid. In this case one has to check carefully by changing h whether the function is periodic or not.

We discuss, finally, how to approximate partial derivatives of functions which depend on more than one variable. Basically this can be achieved by independently discretisizing the function of interest in each particular variable and then by defining the corresponding finite difference derivatives. We will briefly discuss the case of two variables and the extension to even more variables is straight forward. We regard a function $g(x, y)$ where $(x, y) \in [a, b] \times [c, d]$. We denote the grid-spacing in x-direction by h_x and in y-direction by h_y. The evaluation of derivatives of the form $\frac{\partial^n}{\partial x^n} g(x, y)$ or $\frac{\partial^n}{\partial y^n} g(x, y)$ for arbitrary n are approximated with the help of the schemes discussed above, only the respective grid-spacing has to be accounted for. We will now briefly discuss mixed partial derivatives, in particular the derivative $\frac{\partial^2}{\partial x \partial y} g(x, y)$. Higher orders can be easily obtained in the same fashion. Here, we will restrict to the case of the central difference derivative. Again, the extension to the other two forms of derivatives is straight forward. We would like to approximate the derivative at the point $(a + ih_x, c + jh_y)$, which will be abbreviated by (i, j). Hence, we compute

$$
\begin{aligned}
\frac{\partial}{\partial y} \frac{\partial}{\partial x} g(x, y)\bigg|_{(i,j)} &= \frac{1}{2h_x} \left[\frac{\partial}{\partial y} g(x, y)\bigg|_{(i+1,j)} - \frac{\partial}{\partial y} g(x, y)\bigg|_{(i-1,j)} \right] + \mathcal{O}(h_x^2) \\
&= \frac{1}{2h_x} \left[\frac{g_{i+1,j+1} - g_{i+1,j-1}}{2h_y} + \mathcal{O}(h_y^2)\bigg|_{(i+1,j)} \right. \\
&\qquad \left. - \frac{g_{i-1,j+1} - g_{i-1,j-1}}{2h_y} - \mathcal{O}(h_y^2)\bigg|_{(i-1,j)} \right] + \mathcal{O}(h_x^2) , \qquad (2.43)
\end{aligned}
$$

where we made use of the notation $g_{i,j} \equiv g(x_i, y_j)$. Neglecting higher order contributions yields

$$
\frac{\partial}{\partial y} \frac{\partial}{\partial x} g(x, y)\bigg|_{(i,j)} \approx \frac{1}{2h_x} \frac{g_{i+1,j+1} - g_{i+1,j-1} - g_{i-1,j+1} + g_{i-1,j-1}}{2h_y} . \qquad (2.44)
$$

This simple approximation is easily improved with the help of methods developed in the previous sections.

It should be noted that there are also other methods to approximate derivatives. One of the most powerful methods, is the method of *finite elements* [8]. The conceptual difference to the method of finite differences is that one divides the domain in finite sub-domains (elements) rather than by replacing these by sets of discrete grid-points. The function of interest, say $g(x, y)$, is then replaced within each element by an interpolating polynomial. However, this method is quite complex and definitely beyond the scope of this book. Another interesting method, which is particularly useful for the solution of hyperbolic differential equations, is the method of *finite volumes*. The interested reader is referred to the book by R. J. LEVEQUE [9].

Summary

In a first step the notion of finite differences was introduced: All functions are approximated only by their functional values at discrete grid-points and by interpolation schemes between these points. This served as a basis for the definition of finite difference derivatives. Three different types were discussed: the forward, the backward, and the central difference derivative. A more systematic approach to finite difference derivatives was then offered by the operator technique. It provided ready to use equations which allowed to approximate a particular derivative of arbitrary order to arbitrary order of grid-spacing. The two methodological errors introduced by this method, namely the subtractive cancellation error due to too dense a grid and the truncation error due to too coarse a grid were discussed in detail.

Problems

1. Derive Eq. (2.32).
2. Calculate numerically the derivative of the function

$$f(x) = \cos(\omega_1 x) + \exp(-x^2/2)\sin(\omega_2 x),$$

with $\omega_2 = 0.5$ and $\omega_2 = 10\omega_1$. Use a non-uniform grid. Calculate locally the relative error of your approximation.
3. Extend your code of the previous example to arbitrary $\omega_1 \leq 10\omega_2$ and $\omega_2 = 0.5$ by implementing an adaptive grid-spacing. In particular, write a routine which recursively finds a suitable grid-spacing.
4. Consider the finite interval $I = [-5, 5]$ on the real axis. Define N equally spaced grid-points $x_i = x_1 + (i-1)h$, $i = 1, \ldots, N$. Investigate the functions

$$g(x) = \exp\left(-x^2\right) \quad \text{and} \quad h(x) = \sin(x).$$

   a. Plot these functions within the interval I by defining these functions on the grid-points x_i.
   b. Plot the first derivative of these functions by analytical differentiation.
   c. Calculate and plot the first derivatives of these functions by employing the first order backward, forward, and central difference derivatives. For the central difference derivative use an algorithm which is based on the grid-points x_{i-1} and x_{i+1} rather than the method based on intermediate grid-points $x_{i\pm\frac{1}{2}}$.
   d. Calculate and plot the first central difference derivatives of these functions by employing second order corrections. These corrections can be obtained by applying the sum representation of the derivative operator defined in Sect. 2.4, last line of Eq. (2.31), i.e. take the term proportional to δ_c^3 into account!

e. Calculate the absolute and the relative error of the above methods. Note that the exact values are known analytically.
f. Repeat the above steps for the second derivative of the function $h(x)$. For the second order correction of the central difference derivative take the term proportional to δ_c^4 in Eq. (2.32) into account.
g. Try different values of N.

5. Consider the function:

$$f(x, y) = \cos(x)\exp(-y^2).$$

a. Calculate numerically its gradient $\nabla f(x, y)$ and compare with the analytical result.
b. Demonstrate numerically that gradient fields are curl-free, i.e. $\nabla \times \nabla f(x, y) = 0$ for all x and y.

References

1. Jordan, C.: Calculus of Finite Differences, 3rd edn. AMS Chelsea Publishing, Providence (1965)
2. Süli, E., Mayers, D.: An Introduction to Numerical Analysis. Cambridge University Press, Cambridge (2003)
3. Gautschi, W.: Numerical Analysis. Springer, Berlin/Heidelberg (2012)
4. Weidmann, J.: Lineare Operatoren in Hilberträumen, vol. I: Grundlagen. Springer, Berlin/Heidelberg (2000)
5. Davies, E.B.: Linear Operators and Their Spectra. Cambridge Studies in Advanced Mathematics. Cambridge University Press, Cambridge (2007)
6. Tung, W.K.: Group Theory in Physics. World Scientific, Hackensack (2003)
7. Lapidus, L., Pinder, G.F.: Numerical Solution of Partial Differential Equations. Wiley, New York (1982)
8. Gockenbach, M.S.: Understanding and Implementing the Finite Element Method. Cambridge University Press, Cambridge (2006)
9. LeVeque, R.J.: Finite Volume Methods for Hyperbolic Problems. Cambridge Texts in Applied Mathematics. Cambridge University Press, Cambridge (2002)

Chapter 3
Numerical Integration

3.1 Introduction

Numerical integration is certainly one of the most important concepts in computational analysis since it plays a major role in the numerical treatment of differential equations. Given a function $f(x)$ which is continuous on the interval $[a, b]$, one wishes to approximate the integral by a discrete sum of the form

$$\int_a^b \mathrm{d}x f(x) \approx \sum_{i=1}^{N} \omega_i f(x_i), \tag{3.1}$$

where the ω_i are referred to as weights and x_i are the grid-points at which the function needs to be evaluated. Such methods are commonly referred to as *quadrature* [1, 2].

We will mainly discuss two different approaches to the numerical integration of arbitrary functions. We start with a rather simple approach, the *rectangular* rule. The search of an improvement of this method will lead us first to the *trapezoidal* rule, then to the SIMPSON rule and, finally, to a general formulation of the method, the NEWTON-COTES quadrature. This will be followed by a more advanced technique, the GAUSS-LEGENDRE quadrature. At the end of the chapter we will discuss an elucidating example and briefly sketch extensions of all methods to more general problems, such as integration of non-differentiable functions or the evaluation of multiple integrals.

Another very important approach, which is based on random sampling methods, is the so called Monte-Carlo integration. This method will be presented in Sect. 14.2.

B.A. Stickler, E. Schachinger, *Basic Concepts in Computational Physics*,
DOI 10.1007/978-3-319-27265-8_3

3.2 Rectangular Rule

The straight forward approach to numerical integration is to employ the concept of finite differences developed in Sect. 2.2. We regard a smooth function $f(x)$ within the interval $[a, b]$, i.e. $f(x) \in \mathscr{C}^{\infty}[a, b]$. The RIEMANN definition of the proper integral of $f(x)$ from a to b states that:

$$\int_a^b \mathrm{d}x f(x) = \lim_{N\to\infty} \frac{b-a}{N} \sum_{i=0}^{N} f\left(a + i\frac{b-a}{N}\right). \tag{3.2}$$

We approximate the right hand side of this relation using equally spaced grid-points $x_i \in [a, b]$ according to Eq. (2.1) and find

$$\int_a^b \mathrm{d}x f(x) \approx h \sum_{i=1}^{N-1} f_i. \tag{3.3}$$

It is clear that the quality of this approach strongly depends on the discretization chosen, i.e. on the values of x_i as illustrated schematically in Fig. 3.1. Again, a non-uniform grid may be of advantage. We can estimate the error of this approximation by expanding $f(x)$ into a TAYLOR series.

We note that

$$\int_a^b \mathrm{d}x f(x) = \sum_{i=1}^{N-1} \int_{x_i}^{x_{i+1}} \mathrm{d}x f(x), \tag{3.4}$$

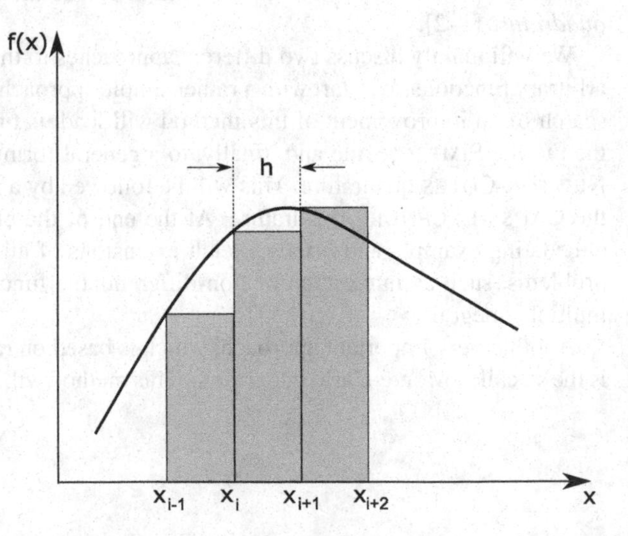

Fig. 3.1 Illustration of the numerical approximation of a proper integral according to Eq. (3.3)

hence, the approximation (3.3) is equivalent to an estimate of the area in the unit interval, *the elemental area*:

$$\int_{x_i}^{x_{i+1}} \mathrm{d}x f(x) \approx h f_i. \tag{3.5}$$

Furthermore, we find following Eq. (2.9a):

$$\begin{aligned}\int_{x_i}^{x_{i+1}} \mathrm{d}x f(x) &= \int_{x_i}^{x_{i+1}} \mathrm{d}x \left[f_i + (x - x_i) f'_{i+\varepsilon_\zeta} \right] \\ &= f_i h + \mathcal{O}(h^2).\end{aligned} \tag{3.6}$$

In this last step we applied the first mean value theorem for integration which states that if $f(x)$ is continuous in $[a, b]$, then there exists a $\zeta \in [a, b]$ such that

$$\int_a^b \mathrm{d}x f(x) = (b - a) f(\zeta). \tag{3.7}$$

(We shall come back to the mean value theorem in the course of our discussion of Monte-Carlo integration in Chap. 14.) Consequently, the error we make with approximation (3.3) can be seen from Eq. (3.6) to be of the order $\mathcal{O}(h^2)$.

This procedure corresponds to a forward difference approach and, equivalently, backward differences can be used. This results in:

$$\int_a^b \mathrm{d}x f(x) = h \sum_{i=2}^{N} f_i + \mathcal{O}(h^2). \tag{3.8}$$

Let us now define the forward and backward rectangular rule by

$${}_iI^+_{i+1} = h f_i \,, \tag{3.9}$$

and

$${}_iI^-_{i+1} = h f_{i+1}, \tag{3.10}$$

respectively. Thus, we obtain from TAYLOR's expansion that:

$$\begin{aligned}\int_{x_i}^{x_{i+1}} \mathrm{d}x f(x) &= {}_iI^+_{i+1} + \frac{h^2}{2} f'_i + \frac{h^3}{3!} f''_i + \ldots \\ &= {}_iI^-_{i+1} - \frac{h^2}{2} f'_{i+1} + \frac{h^3}{3!} f''_{i+1} \mp \ldots .\end{aligned} \tag{3.11}$$

However, the use of central differences gives more accurate results as has already been observed in Chap. 2 in which the differential operator was approximated. We make use of the concept of intermediate grid-points (see Sect. 2.4) and consider the integral

$$\int_{x_i}^{x_{i+1}} \mathrm{d}x f(x), \tag{3.12}$$

expand $f(x)$ in a TAYLOR series around the midpoint $x_{i+\frac{1}{2}}$, and obtain:

$$\begin{aligned}\int_{x_i}^{x_{i+1}} \mathrm{d}x f(x) &= \int_{x_i}^{x_{i+1}} \mathrm{d}x \left\{ f_{i+\frac{1}{2}} + \left(x - x_{i+\frac{1}{2}}\right) f'_{i+\frac{1}{2}} \right. \\ &\quad \left. + \frac{\left(x - x_{i+\frac{1}{2}}\right)^2}{2} f''_{i+\frac{1}{2}} + \mathcal{O}\left[\left(x - x_{i+\frac{1}{2}}\right)^3\right] \right\} \\ &= h f_{i+\frac{1}{2}} + \frac{h^3}{24} f''_{i+\epsilon_\xi} \\ &= {}_i I_{i+1} + \frac{h^3}{24} f''_{i+\epsilon_\xi}. \end{aligned} \tag{3.13}$$

Thus, the error generated by this method, the central rectangular rule, scales as $\mathcal{O}(h^3)$ which is a significant improvement in comparison to Eqs. (3.3) and (3.8).[1] We obtain

$$\int_a^b \mathrm{d}x f(x) = h \sum_{i=1}^{N-1} f_{i+\frac{1}{2}} + \mathcal{O}(h^3). \tag{3.14}$$

This approximation is known as the *rectangular rule*. It is illustrated in Fig. 3.2. Note that the boundary points $x_1 = a$ and $x_N = b$ do not enter Eq. (3.14). Such a procedure is commonly referred to as an *open integration rule*. On the other hand, if the end-points are taken into account by the method it is referred to as a *closed integration rule*.

[1] In this context the intermediate position $x_{i+1/2}$ is understood as a true grid-point. If, on the other hand, the function value $f_{i+1/2}$ is approximated by $\mu f_{i+1/2}$, Eq. (2.29), the method is referred to as the *trapezoidal rule*.

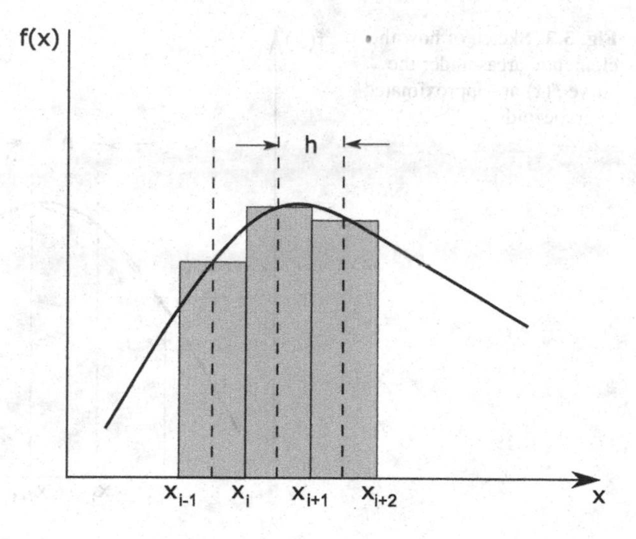

Fig. 3.2 Scheme of the rectangular integration rule according to Eq. (3.14). Note that boundary points do not enter the evaluation of the elemental areas

3.3 Trapezoidal Rule

An elegant alternative to the rectangular rule is found when the area between two grid-points is approximated by a trapezoid as is shown schematically in Fig. 3.3. The trapezoidal rule is obtained when the function values $f_{i+1/2}$ at intermediate grid-points on the right hand side of the central rectangular rule (3.13) are approximated with the help of $\mu f_{i+1/2}$, Eq. (2.29). Thus, the elemental area is calculated from

$$\int_{x_i}^{x_{i+1}} \mathrm{d}x f(x) \approx \frac{h}{2}(f_i + f_{i+1}) \,. \tag{3.15}$$

and we obtain:

$$\begin{aligned}\int_a^b \mathrm{d}x f(x) &\approx \frac{h}{2}\sum_{i=1}^{N-1}(f_i + f_{i+1}) \\ &= h\left(\frac{f_1}{2} + f_2 + \ldots + f_{N-1} + \frac{f_N}{2}\right) \\ &= \frac{h}{2}(f_1 + f_N) + h\sum_{i=2}^{N-1} f_i \\ &= {}_1I_N^T \,. \end{aligned} \tag{3.16}$$

Note that this integration rule is closed, although the boundary points f_1 and f_N enter the summation (3.16) only with half the weight in comparison to all other function values f_i. This stems from the fact that the function values f_1 and f_N contribute only to one elemental area, the first and the last one. Another noticeable

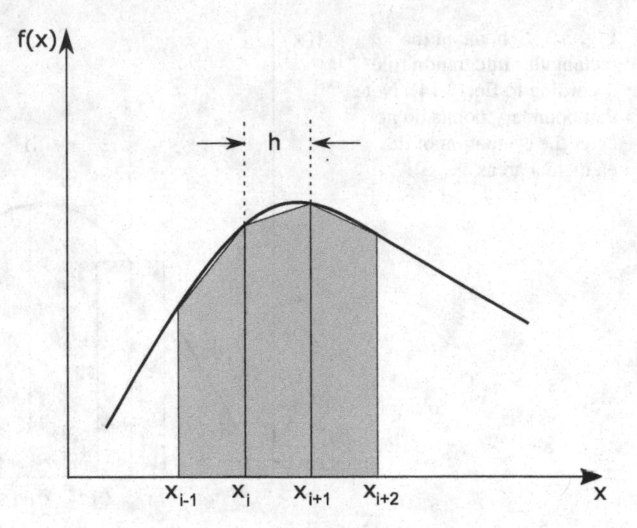

Fig. 3.3 Sketch of how the elemental areas under the curve $f(x)$ are approximated by trapezoids

feature of the trapezoidal rule is that, in contrast to the rectangular rule (3.14), only function values at grid-points enter the summation, which can be desirable in some cases.

The error of this method can be estimated by inserting expansion (2.9a) into Eq. (3.16). One obtains for an elemental area:

$$
\begin{aligned}
{}_iI_{i+1}^T &= \frac{h}{2}\left(f_i + f_{i+1}\right) \\
&= hf_i + \frac{h^2}{2}f_i' + \frac{h^3}{4}f_i'' + \dots \quad (3.17)
\end{aligned}
$$

On the other hand, we know from Eq. (3.6) that

$$
hf_i = \int_{x_i}^{x_{i+1}} \mathrm{d}x\, f(x) - \frac{h^2}{2}f_i' - \frac{h^3}{3!}f_i'' - \dots, \quad (3.18)
$$

which, when inserted into (3.17), yields

$$
{}_iI_{i+1}^T = \int_{x_i}^{x_{i+1}} \mathrm{d}x\, f(x) + \frac{h^3}{12}f_i'' + \dots \quad (3.19)
$$

Hence, we observe that the error induced by the trapezoidal rule is comparable to the error of the rectangular rule, namely $\mathcal{O}(h^3)$. However, since we do not have to compute function values at intermediate grid-points, this rule may be advantageous in many cases.

We remember from Chap. 2 that a more accurate estimate of a derivative was achieved by increasing the number of grid-points involved which in the case of integration leads us to the SIMPSON rule.

3.4 The SIMPSON Rule

The basic idea of the SIMPSON rule is to include higher order derivatives into the expansion of the integrand. These higher order derivatives, which are primarily unknown, are then approximated by expressions we obtained within the context of finite difference derivatives. Let us discuss this procedure in greater detail. To this purpose we will study the integral of $f(x)$ within the interval $[x_{i-1}, x_{i+1}]$ and expand the integrand around the midpoint x_i:

$$\begin{aligned}\int_{x_{i-1}}^{x_{i+1}} \mathrm{d}x f(x) &= \int_{x_{i-1}}^{x_{i+1}} \mathrm{d}x \left[f_i + (x - x_i) f_i' + \frac{(x - x_i)^2}{2!} f_i'' \right. \\ &\quad \left. + \frac{(x - x_i)^3}{3!} f_i''' + \ldots \right] \\ &= 2hf_i + \frac{h^3}{3} f_i'' + \mathcal{O}(h^5). \end{aligned} \tag{3.20}$$

Inserting Eq. (2.34) for f_i'' yields

$$\begin{aligned}\int_{x_{i-1}}^{x_{i+1}} \mathrm{d}x f(x) &= 2hf_i + \frac{h}{3}(f_{i+1} - 2f_i + f_{i-1}) + \mathcal{O}(h^5) \\ &= h\left(\frac{1}{3}f_{i-1} + \frac{4}{3}f_i + \frac{1}{3}f_{i+1}\right) + \mathcal{O}(h^5). \end{aligned} \tag{3.21}$$

Note that in contrast to the trapezoidal rule, the procedure described here is a three point method since the function values at three different points enter the expression. We can immediately write down the resulting integral from a to b. Since,

$$\int_a^b \mathrm{d}x f(x) = \int_{x_0}^{x_2} \mathrm{d}x f(x) + \int_{x_2}^{x_4} \mathrm{d}x f(x) + \ldots + \int_{x_{N-2}}^{x_N} \mathrm{d}x f(x), \tag{3.22}$$

where we assumed that N is even and employed the discretization $x_i = x_0 + ih$ with $x_0 = a$ and $x_N = b$. We obtain:

$$\int_a^b \mathrm{d}x f(x) = \frac{h}{3}(f_0 + 4f_1 + 2f_2 + 4f_3 + \ldots + 2f_{N-2} + 4f_{N-1} + f_N) + \mathcal{O}(h^5). \tag{3.23}$$

This expression is exact for polynomials of degree $n \leq 3$ since the first term in the error expansion involves the fourth derivative. Hence, whenever the integrand is satisfactorily reproduceable by a polynomial of degree three or less, the SIMPSON rule might give almost exact estimates, independent of the discretization h.

The arguments applied above allow for a straightforward extension to four- or even more-point rules. We find, for instance,

$$\int_{x_i}^{x_{i+3}} \mathrm{d}x\, f(x) = \frac{3h}{8}\left(f_i + 3f_{i+1} + 3f_{i+2} + f_{i+3}\right) + \mathcal{O}(h^5), \tag{3.24}$$

which is usually called SIMPSON's three-eight rule.

It is important to note that all the methods discussed so far are special cases of a more general formulation, the NEWTON-COTES rules [2] which will be discussed in the next section.

3.5 General Formulation: The NEWTON-COTES Rules

We define the LAGRANGE interpolating polynomial $p_{n-1}(x)$ of degree $n-1$ [3–5] to a function $f(x)$ as[2]

$$p_{n-1}(x) = \sum_{j=1}^{n} f_j L_j^{(n-1)}(x), \tag{3.25}$$

where

$$L_j^{(n-1)}(x) = \prod_{\substack{k=1\\k\neq j}}^{n} \frac{x - x_k}{x_j - x_k}. \tag{3.26}$$

An arbitrary smooth function $f(x)$ can then be expressed with the help of a LAGRANGE polynomial of degree n by

$$f(x) = p_{n-1}(x) + \frac{f^{(n)}[\zeta(x)]}{n!}(x - x_1)(x - x_2)\ldots(x - x_n). \tag{3.27}$$

If we neglect the second term on the right hand side of this equation and integrate the LAGRANGE polynomial of degree $n-1$ over the n grid-points from x_1 to x_n we obtain the closed n-point NEWTON-COTES formulas. For instance, if we set $n = 2$,

[2]The LAGRANGE polynomial $p_{n-1}(x)$ to the function $f(x)$ is the polynomial of degree $n-1$ that satisfies the n equations $p_{n-1}(x_j) = f(x_j)$ for $j = 1, \ldots, n$, where x_j denotes arbitrary but distinct grid-points.

then

$$
\begin{aligned}
p_1(x) &= f_1 L_1^{(1)}(x) + f_2 L_2^{(1)}(x) \\
&= f_1 \frac{x - x_2}{x_1 - x_2} + f_2 \frac{x - x_1}{x_2 - x_1} \\
&= \frac{1}{h}\left[x(f_2 - f_1) - x_1 f_2 + x_2 f_1\right],
\end{aligned}
\tag{3.28}
$$

with $f_1 \equiv f(x_1)$ and $f_2 \equiv f(x_2)$. Integration over the respective interval yields

$$
\begin{aligned}
\int_{x_1}^{x_2} \mathrm{d}x\, p_1(x) &= \frac{1}{h}\left[\frac{x^2}{2}(f_2 - f_1) + x(x_2 f_1 - x_1 f_2)\right]\Bigg|_{x_1}^{x_2} \\
&= \frac{h}{2}\left[f_2 + f_1\right],
\end{aligned}
\tag{3.29}
$$

which is exactly the trapezoidal rule. By setting $n = 3$ one obtains SIMPSON's rule and setting $n = 4$ gives the SIMPSON's three-eight rule.

The *open* NEWTON-COTES rule can be obtained by integrating the polynomial $p_{n-1}(x)$ of degree $n - 1$ which includes the grid-points $x_1, \ldots, x_n$ from x_0 to x_{n+1}. The fact that these relations are open means that the function values at the boundary points $x_0 = x_1 - h$ and $x_{n+1} = x_n + h$ do not enter the final expressions. The simplest open NEWTON-COTES formula is the central integral approximation which we encountered as the rectangular rule (3.14). A second order approximation is easily found with help of the two-point LAGRANGE polynomial (3.28)

$$
\begin{aligned}
\int_{x_0}^{x_3} \mathrm{d}x\, p_1(x) &= \frac{1}{h}\left[\frac{x^2}{2}(f_2 - f_1) + x(x_2 f_1 - x_1 f_2)\right]\Bigg|_{x_0}^{x_3} \\
&= \frac{3h}{2}\left[f_2 + f_1\right].
\end{aligned}
\tag{3.30}
$$

Higher order approximations can be obtained in a similar fashion. To conclude this section let us briefly discuss an idea which is referred to as ROMBERG's method [6].

So far, we approximated all integrals by expressions of the form

$$
I = \mathscr{I}^N + \mathscr{O}(h^m), \tag{3.31}
$$

where I is the exact, unknown, value of the integral, $\mathscr{I}^N$ is the estimate obtained from an integration scheme using N grid-points, and m is the leading order of the error. Let us review the error of the trapezoidal approximation: we learned that the error for the integral over the interval $[x_i, x_{i+1}]$ scales like h^3. Since we have N such intervals, we conclude that the total error behaves like $(b-a)h^2$. Similarly, the error of the three-point SIMPSON rule is for each sub-interval proportional to h^5 and this gives in total $(b-a)h^4$. We assume that this trend can be generalized and conclude that the error of an n-point method with the estimate $\mathcal{I}_n$ behaves like h^{2n-2}. Since, $h \propto N^{-1}$ we have

$$I = \mathcal{I}_n^N + \frac{C_N}{N^{2n-2}}, \tag{3.32}$$

where C_N depends on the number of grid-points N. Let us double the amount of grid-points and we obtain:

$$I = \mathcal{I}_n^{2N} + \frac{C_{2N}}{(2N)^{2n-2}}. \tag{3.33}$$

Obviously, Eqs. (3.32) and (3.33) can be regarded as a linear system of equations in I and C if $C_N \approx C_{2N} \approx C$. Solving Eqs. (3.32) and (3.33) for I yields

$$I \approx \frac{1}{4^{n-1}-1}\left(4^{n-1}\mathcal{I}_n^{2N} - \mathcal{I}_n^N\right). \tag{3.34}$$

It has to be emphasized that in the above expression I is no longer the exact value because of the approximation $C_N \approx C$. However, it is an improvement of the solution and it is possible to demonstrate that this new estimate is exactly the value one would have obtained with an integral approximation of order $n+1$ and $2N$ grid-points! Thus

$$\mathcal{I}_{n+1}^{2N} = \frac{1}{4^{n-1}-1}\left(4^{n-1}\mathcal{I}_n^{2N} - \mathcal{I}_n^N\right). \tag{3.35}$$

This suggests a very elegant and rapid procedure: We simply calculate the integrals using two point rules and add the results according to Eq. (3.35) to obtain more-point results. For instance, calculate $\mathcal{I}_2^2$ and $\mathcal{I}_2^4$, add these according to Eq. (3.35) and get $\mathcal{I}_3^4$. Now calculate $\mathcal{I}_2^8$, add $\mathcal{I}_2^4$, get $\mathcal{I}_3^8$, add $\mathcal{I}_3^4$ and get $\mathcal{I}_4^8$. This pyramid-like procedure can be continued until convergence is achieved, that is $|\mathcal{I}_m^N - \mathcal{I}_{m+1}^N| < \epsilon$ where $\epsilon > 0$ can be chosen arbitrarily. An illustration of this method is given in Fig. 3.4.

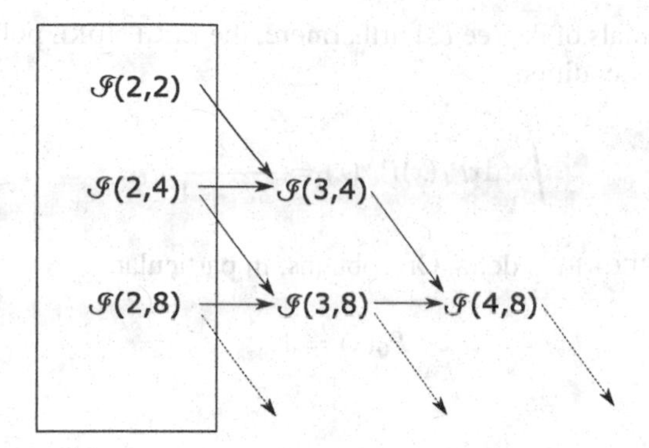

Fig. 3.4 Illustration of the ROMBERG method. Here, the $\mathscr{I}(m,n)$ are synonyms for integrals $\mathscr{I}_m^n$ where the first index m refers to the order of the quadrature while the second index n refers to the number of grid-points used. Note that we only have to use a second order integration scheme (*left row inside the box*), all other values are determined via Eq. (3.35) as indicated by the *arrows*

3.6 GAUSS-LEGENDRE Quadrature

In preparation for the GAUSS-LEGENDRE quadrature we introduce a set of orthogonal LEGENDRE polynomials $P_\ell(x)$ [3, 4, 7, 8] which are solutions of the LEGENDRE differential equation

$$\left(1-x^2\right) P_\ell''(x) - 2xP_\ell'(x) + \ell(\ell+1)P_\ell(x) = 0. \tag{3.36}$$

This equation occurs, for instance, when the LAPLACE equation $\Delta f(x) = 0$ is transformed to spherical coordinates. Here, we will introduce the most important properties of LEGENDRE polynomials which will be required for an understanding of the GAUSS-LEGENDRE quadrature.

LEGENDRE polynomials are given by

$$P_\ell(x) = \sum_{k=0}^{\infty} a_{k,\ell} x^k, \tag{3.37}$$

where the coefficients $a_{k,\ell}$ can be determined recursively:

$$a_{k+2,\ell} = \frac{k(k+1) - \ell(\ell+1)}{(k+1)(k+2)} a_{k,\ell}. \tag{3.38}$$

Hence, for even values of ℓ the LEGENDRE polynomial involves only even powers of x and for odd ℓ only odd powers of x. Note also that according to Eq. (3.38) for $k \geq \ell$ the coefficients are equal to zero and, thus, it follows from Eq. (3.37) that the

$P_\ell(x)$ are polynomials of degree ℓ. Furthermore, the LEGENDRE polynomials fulfill the orthonomality condition

$$\int_{-1}^{1} \mathrm{d}x P_\ell(x) P_{\ell'}(x) = \frac{2}{2\ell' + 1} \delta_{\ell\ell'}, \tag{3.39}$$

where δ_{ij} is KRONECKER's delta. One obtains, in particular,

$$P_0(x) = 1, \tag{3.40}$$

and

$$P_1(x) = x\,. \tag{3.41}$$

Another convenient way to calculate LEGENDRE polynomials is based on RODRIGUES' formula

$$P_\ell(x) = \frac{1}{2^\ell \ell!} \frac{\mathrm{d}^\ell}{\mathrm{d}x^\ell} \left(x^2 - 1\right)^\ell. \tag{3.42}$$

We focus now on the core of the GAUSS-LEGENDRE quadrature and introduce the function $F(x)$ as a transform of the function $f(x)$

$$F(x) = \frac{b-a}{2} f\left(\frac{b-a}{2}x + \frac{b+a}{2}\right), \tag{3.43}$$

in such a way that we can rewrite the integral of interest as:

$$\int_a^b \mathrm{d}x f(x) = \int_{-1}^{1} \mathrm{d}x F(x). \tag{3.44}$$

If the function $F(x)$ can be well approximated by some polynomial of degree $2n-1$, like

$$F(x) \approx p_{2n-1}(x)\,, \tag{3.45}$$

then this means that according to TAYLOR's theorem (2.7) the error introduced by this approximation is proportional to $F^{(2n)}(x)$. If the polynomial $p_{2n-1}(x)$ is explicitly given then we can apply the methods discussed in the previous sections to approximate the integral (3.44). However, even if the polynomial is not explicitly given we write the integral (3.44) as

$$\int_{-1}^{1} \mathrm{d}x F(x) = \sum_{i=1}^{n} \omega_i F(x_i)\,, \tag{3.46}$$

with weights ω_i and grid-points x_i, $i = 1,\ldots,n$ which are yet undetermined! Therefore, we will determine the weights ω_i and grid-points x_i in such a way, that the integral is well approximated *even* if the polynomial p_{2n-1} in Eq. (3.45) is unknown. For this purpose we decompose $p_{2n-1}(x)$ into

$$p_{2n-1}(x) = p_{n-1}(x)P_n(x) + q_{n-1}(x) , \tag{3.47}$$

where $P_n(x)$ is the LEGENDRE polynomial of degree n and $p_{n-1}(x)$ and $q_{n-1}(x)$ are polynomials of degree $n-1$. Since $p_{n-1}(x)$ itself is a polynomial of degree $n-1$, it can also be expanded in LEGENDRE polynomials of degrees up to $n-1$ by

$$p_{n-1}(x) = \sum_{i=0}^{n-1} a_i P_i(x) . \tag{3.48}$$

Using Eq. (3.48) in (3.47) we obtain together with normalization relation (3.39)

$$\int_{-1}^{1} \mathrm{d}x\, p_{2n-1}(x) = \sum_{i=0}^{n-1} a_i \int_{-1}^{1} \mathrm{d}x\, P_i(x)P_n(x) + \int_{-1}^{1} \mathrm{d}x\, q_{n-1}(x) = \int_{-1}^{1} \mathrm{d}x\, q_{n-1}(x) . \tag{3.49}$$

Moreover, since $P_n(x)$ is a LEGENDRE polynomial of degree n it has n-zeros in the interval $[-1, 1]$ and Eq. (3.47) results in

$$p_{2n-1}(x_i) = q_{n-1}(x_i) , \tag{3.50}$$

where $x_1, x_2, \ldots, x_n$ denote the zeros of $P_n(x)$ and these zeros determine the grid-points of our integration routine. It is interesting to note, that these zeros are independent of the function $F(x)$ we want to integrate. We also expand $q_{n-1}(x)$ in terms of LEGENDRE polynomials

$$q_{n-1}(x) = \sum_{i=0}^{n-1} b_i P_i(x) , \tag{3.51}$$

and use it in Eq. (3.50) to obtain

$$p_{2n-1}(x_i) = \sum_{k=0}^{n-1} b_k P_k(x_i) , \quad i = 1,\ldots,n , \tag{3.52}$$

which can be written in a more compact form by defining $p_i \equiv p_{2n-1}(x_i)$ and $P_{ki} \equiv P_k(x_i)$:

$$p_i = \sum_{k=0}^{n-1} b_k P_{ki} , \quad i = 1, \dots, n . \tag{3.53}$$

It has to be emphasized again that the grid-points x_i are independent of the polynomial $p_{2n-1}(x)$ and, therefore, independent of $F(x)$. Furthermore, we can replace $p_i \approx F(x_i) \equiv F_i$ according to Eq. (3.45). We recognize that Eq. (3.53) corresponds to a system of linear equations which can be solved for the weights b_k. We obtain

$$b_k = \sum_{i=1}^{n} F_i \left[\mathbf{P}^{-1}\right]_{ik} , \tag{3.54}$$

where $\mathbf{P}$ is the matrix $\mathbf{P} = \{P_{ij}\}$, which is known to be non-singular. We can now rewrite the integral (3.44) with the help of Eqs. (3.45), (3.49), and (3.51) together with the properties of the zeros of LEGENDRE polynomials [7, 8] as

$$\int_{-1}^{1} \mathrm{d}x\, F(x) \approx \int_{-1}^{1} \mathrm{d}x\, p_{2n-1}(x) = \sum_{k=0}^{n-1} b_k \int_{-1}^{1} \mathrm{d}x\, P_k(x) . \tag{3.55}$$

Since $P_0(x) = 1$ according to Eq. (3.40), we deduce from Eq. (3.39)

$$\int_{-1}^{1} \mathrm{d}x\, P_k(x) = \int_{-1}^{1} \mathrm{d}x\, P_k(x) P_0(x) = \frac{2}{2k+1} \delta_{k0} = 2\delta_{k0} . \tag{3.56}$$

Hence, Eq. (3.55) reads

$$\int_{-1}^{1} \mathrm{d}x\, F(x) \approx 2b_0 = 2 \sum_{i=1}^{n} F_i \left[\mathbf{P}^{-1}\right]_{i0} . \tag{3.57}$$

By defining

$$\omega_i = 2 \left[\mathbf{P}^{-1}\right]_{i0} , \tag{3.58}$$

we arrive at the desired expansion

$$\int_{-1}^{1} \mathrm{d}x F(x) \approx \sum_{i=1}^{n} \omega_i F_i . \tag{3.59}$$

Moreover, since we approximated $F(x)$ by a polynomial of degree $2n-1$, the GAUSS-LEGENDRE quadrature is exact for polynomials of degree $2n-1$, i.e. the error is proportional to a derivative of $F(x)$ of order $2n$. Furthermore, expression (3.58) can be put in a more convenient form. One can show that

$$\omega_i = \frac{2}{(1-x_i^2)\left[P_n'(x_i)\right]^2}\,, \tag{3.60}$$

where

$$P_n'(x_i) = \left.\frac{\mathrm{d}}{\mathrm{d}x}P_n(x)\right|_{x=x_i}. \tag{3.61}$$

Let us make some concluding remarks. The grid-points x_i as well as the weights ω_i are independent of the actual function $F(x)$ we want to integrate. This means, that one can table these values once and for all [7, 8] and use them for different types of problems. The grid-points x_i are symmetrically distributed around the point $x=0$, i.e. for every x_j there is a $-x_j$. Furthermore, these two grid-points have the same weight ω_j. The density of grid-points increases approaching the boundary, however, the boundary points themselves are not included, which means that the GAUSS-LEGENDRE quadrature is an *open method*. Furthermore, it has to be emphasized that low order GAUSS-LEGENDRE parameters can easily be calculated by employing relation (3.42). This makes the GAUSS-LEGENDRE quadrature the predominant integration method. In comparison to the trapezoidal rule or even the ROMBERG method, it needs in many cases a smaller number of grid-points, is simpler to implement, converges faster and yields more accurate results. One drawback of this method is that one has to compute the function $F(x)$ at the zeros of the LEGENDRE polynomial x_i. This can be a problem if the integrand at hand is not known analytically.

It is important to note at this point that comparable procedures exist which use other types of orthogonal polynomials, such as HERMITE polynomials. This procedure is known as the GAUSS-HERMITE quadrature.

Table 3.1 lists the methods, discussed in the previous sections, which allow to calculate numerically an estimate of integrals of the form:

$$\int_a^b \mathrm{d}x f(x)\,. \tag{3.62}$$

Equal grid-spacing h is assumed, with the GAUSS-LEGENDRE method as the only exception. The particular value of h depends on the order of the method employed and is given in Table 3.1.

Table 3.1 Summary of the quadrature methods discussed in this chapter applied to the integral $\int_a^b \mathrm{d}x f(x)$. For a detailed description consult the corresponding sections. Equal grid-spacing is assumed for all methods except for the GAUSS-LEGENDRE quadrature. The explicit values of h depend on the order of the method and are listed in the table. Furthermore, we use $x_i = a + ih$ and denote $f(x_i) = f_i$. The function $P^{(m)}(x)$ which appears in the description of the NEWTON-COTES rules denotes the m-th order LAGRANGE interpolating polynomial and $P_m(x)$ is the m-th degree LEGENDRE polynomial

n	h	$\mathscr{I}$	Method	Comment
1	$\frac{b-a}{2}$	hf_1	Rectangular	Open
2	$b-a$	$\frac{h}{2}(f_0 + f_1)$	Trapezoidal	Closed
3	$\frac{b-a}{2}$	$\frac{h}{3}(f_0 + 4f_1 + f_2)$	SIMPSON	Closed
4	$\frac{b-a}{3}$	$\frac{3h}{8}(f_0 + 3f_1 + 3f_2 + f_3)$	SIMPSON $\frac{3}{8}$	Closed
m	$\frac{b-a}{m-1}$	$\int_{x_0}^{x_{m-1}} \mathrm{d}x P^{(m)}(x)$	NEWTON-COTES	Closed
m	$\frac{b-a}{m+1}$	$\int_{x_0}^{x_{m+1}} \mathrm{d}x P^{(m)}(x)$	NEWTON-COTES	Open
m	$P_m(x_j) = 0$	$\frac{b-a}{2}\sum_{j=1}^{m} \omega_j f(z_j)$ $z_j = \frac{a+b}{2} + \frac{a-b}{2}x_j$ $\omega_j = \frac{2}{(1-x_j)^2[P_m'(x_j)]^2}$	GAUSS-LEGENDRE	Open

3.7 An Example

Let us discuss as an example the following proper integral:

$$I = \int_{-1}^{1} \frac{\mathrm{d}x}{x+2} = \ln(3) - \ln(1) \approx 1.09861\,. \tag{3.63}$$

We will now apply the various methods of Table 3.1 to approximate Eq. (3.63). Note that these methods could give better results if a finer grid had been chosen. However, since this is only an illustrative example, we wanted to keep it as simple as possible. The rectangular rule gives

$$\mathscr{I}_R = 1 \cdot \frac{1}{2} = 0.5\,, \tag{3.64}$$

the trapezoidal rule

$$\mathscr{I}_T = \frac{2}{2}\left(\frac{1}{1} + \frac{1}{3}\right) = \frac{4}{3} = 1.333\ldots\,, \tag{3.65}$$

and an application of the SIMPSON rule yields

$$\mathscr{I}_S = \frac{1}{3}\left(\frac{1}{1} + \frac{4}{2} + \frac{1}{3}\right) = \frac{10}{9} = 1.111\ldots\,. \tag{3.66}$$

Finally, we apply the GAUSS-LEGENDRE quadrature in a second order approximation. We could look up the parameters in [7, 8], however, for illustrative reasons we will calculate those in this simple case. For a second order approximation we need the LEGENDRE polynomial of second degree. It can be obtained from RODRIGUES' formula (3.42):

$$\begin{aligned} P_2(x) &= \frac{1}{2^2 2!} \frac{d^2}{dx^2} \left(x^2 - 1\right)^2 \\ &= \frac{1}{2} \left(3x^2 - 1\right) . \end{aligned} \tag{3.67}$$

In a next step the zeros x_1 and x_2 of $P_2(x)$ are determined from Eq. (3.67) which results immediately in:

$$x_{1,2} = \pm \frac{1}{\sqrt{3}} \approx \pm 0.57735 . \tag{3.68}$$

The weights ω_1 and ω_2 can now be evaluated according to Eq. (3.60):

$$\omega_i = \frac{2}{\left(1 - x_i^2\right) \left[P_2'(x_i)\right]^2} . \tag{3.69}$$

It follows from Eq. (3.67) that

$$P_2'(x) = 3x , \tag{3.70}$$

and, thus,

$$P_2'(x_1) = -\sqrt{3} \qquad \text{and} \qquad P_2'(x_2) = \sqrt{3} . \tag{3.71}$$

This is used to calculate the weights from Eq. (3.69):

$$\omega_1 = \omega_2 = 1 . \tag{3.72}$$

We combine the results (3.68) and (3.72) to arrive at the GAUSS-LEGENDRE estimate of the integral (3.63):

$$\mathscr{I}_{GL} = \frac{1}{-\frac{1}{\sqrt{3}} + 2} + \frac{1}{\frac{1}{\sqrt{3}} + 2} = 1.090909\ldots . \tag{3.73}$$

Obviously, a second order GAUSS-LEGENDRE approximation results already in a much better estimate of the integral (3.63) than the trapezoidal rule which is also of second order. It is also better than the estimate by the SIMPSON rule which is of third order.

3.8 Concluding Discussion

Let us briefly discuss some further aspects of numerical integration. In many cases one is confronted with improper integrals of the form

$$\int_a^\infty \mathrm{d}x f(x), \qquad \int_{-\infty}^a \mathrm{d}x f(x), \quad \text{or} \quad \int_{-\infty}^\infty \mathrm{d}x f(x) . \tag{3.74}$$

The question arises whether or not we can treat such an integral with the methods discussed so far. The answer is yes, it is possible as we will demonstrate using the integral

$$I = \int_a^\infty \mathrm{d}x f(x) . \tag{3.75}$$

as an example; other integrals can be treated in a similar fashion. We rewrite Eq. (3.75) as

$$I = \lim_{b\to\infty} \int_a^b \mathrm{d}x f(x) = \lim_{b\to\infty} I(b) . \tag{3.76}$$

One now calculates $I(b_1)$ for some $b_1 > a$ and $I(b_2)$ for some $b_2 > b_1$. If $|I(b_2) - I(b_1)| < \epsilon$, where $\epsilon > 0$ is the required accuracy, the resulting value $I(b_2)$ can be regarded as the appropriate estimate to I.[3] However, in many cases it is easier to perform an integral transform in order to map the infinite interval onto a finite interval. For instance, consider [9]

$$I = \int_0^\infty \mathrm{d}x \frac{1}{(1+x^2)^{\frac{4}{3}}} . \tag{3.77}$$

The transformation

$$t = \frac{1}{1+x} \tag{3.78}$$

gives

$$I = \int_0^1 \mathrm{d}t \frac{t^{\frac{2}{3}}}{[t^2 + (1-t)^2]^{\frac{4}{3}}} . \tag{3.79}$$

[3] Particular care is required when dealing with periodic functions!

Thus, we mapped the interval $[0, \infty) \rightarrow [0, 1]$. Integral (3.79) can now be approximated with help of the methods discussed in the previous sections. These can also be applied to approximate convergent integrals whose integrand shows singular behavior within $[a, b]$.

If the integrand $f(x)$ is not smooth within the interval $I : x \in [a, b]$ we can split the total integral into a sum over sub-intervals. For instance, if we consider the function

$$f(x) = \begin{cases} x \cos(x), & x < 0, \\ x \sin(x), & x \geq 0, \end{cases}$$

we can calculate the integral over the interval $I : x \in [-10, 10]$ as

$$\int_{-10}^{10} \mathrm{d}x f(x) = \int_{-10}^{0} \mathrm{d}x\, x \cos(x) + \int_{0}^{10} \mathrm{d}x\, x \sin(x) .$$

We generalize this result and write

$$\int_I \mathrm{d}x f(x) = \sum_k \int_{I_k} \mathrm{d}x f(x) , \tag{3.80}$$

with sub-intervals $I_k \in I, \forall k$ and the integrand $f(x)$ is assumed to be smooth within each sub-interval I_k but not necessarily within the interval I. We can then apply one of the methods discussed in this chapter to calculate an estimate of the integral over any of the sub-intervals I_k.

Similar to the discussion in Sect. 2.5 about the approximation of partial derivatives on the basis of finite differences, one can apply the rules of quadrature developed here for different dimensions to obtain an estimate of multi-dimensional integrals. However, the complexity of the problem is significantly increased if the integration boundaries are functions of the variables rather than constants. For instance,

$$\int_a^b \mathrm{d}x \int_{\varphi_1(x)}^{\varphi_2(x)} \mathrm{d}y f(x, y) . \tag{3.81}$$

Such cases are rather difficult to handle and the method to choose depends highly on the form of the functions $\varphi_1(x)$, $\varphi_2(x)$ and $f(x, y)$. We will not deal with integrals of this kind because this is beyond the scope of this book. The interested reader is referred to books by DAHLQUIST and BJÖRK [10] and by PRESS et al. [11].

In a final remark we would like to point out that it can be of advantage to utilize the properties of FOURIER transforms when integrals of the convolution type are to be approximated numerically (see Appendix D).

Summary

The starting point was the concept of finite differences (Sect. 2.2). Based on this concept proper integrals over smooth functions $f(x)$ were approximated by a sum over elemental areas with the elemental area defined as the area under $f(x)$ between two consecutive grid-points. The simplest method, the *rectangular rule*, was based on forward/backward differences. It was a *closed* method, i.e. the functional values at the boundaries were included. On the other hand, a rectangular rule based on central differences was an *open* method, i.e. the functional values at the boundaries were not included. Application of the TAYLOR expansion (2.7) revealed that the methodological error of the rectangular rule was of order $\mathcal{O}(h^2)$. With the elemental area approximated by a trapezoid we arrived at the *trapezoidal rule*. It was a closed method and the methodological error was of order $\mathcal{O}(h^3)$. The inclusion of higher order derivatives of $f(x)$ allowed the derivation of the SIMPSON rules of quadrature. They resulted a remarkable reduction of the methodological error. A more general formulation of all these methods was based on the interpolation of the function $f(x)$ using LAGRANGE interpolating polynomials of degree n and resulted in the class of NEWTON-COTES rules. For various orders of n of the interpolating polynomial all the above rules were derived. Within this context a particularly useful method, the ROMBERG method, was discussed. By adding diligently only two-point rules the error of the numerical estimate of the integral has been made arbitrarily small. An even more general approach was offered by the GAUSS-LEGENDRE quadrature which used LEGENDRE polynomials of degree ℓ to approximate the function $f(x)$. The grid-points were defined by the zeros of the ℓ-th degree polynomial and the weights ω_i in Eq. (3.1) were proportional to the square of the inverse first derivative of the polynomial. This method had the enormous advantage that the grid-points and weights were independent of the function $f(x)$ and, thus, could be determined once and for all for any polynomial degree ℓ. Error analysis proved that this method had the smallest methodological error.

Problems

We consider the interval $I = [-5, 5]$ together with the functions $g(x)$ and $h(x)$:

$$g(x) = \exp\left(-x^2\right) \quad \text{and} \quad h(x) = \sin(x) \,.$$

We discretize the interval I by introducing N equally spaced grid-points. The corresponding $N-1$ sub-intervals are denoted by $I_j, j = 1, \ldots N-1$. In the following we wish to calculate estimates of the integrals

$$\mathcal{I}_1 = \int_I \mathrm{d}x\, g(x) \qquad \text{and} \qquad \mathcal{I}_2 = \int_I \mathrm{d}x\, h(x) \,.$$

Furthermore, we add a third integral of the form

$$\mathscr{I}_3 = \int_I \mathrm{d}x\, h^2(x) = \int_I \mathrm{d}x\, \sin^2(x)\,,$$

to our discussion.

1. Evaluate $\mathscr{I}_1$ with the help of the *error function* erf(x), which should be supplied by the environment you use as an intrinsic function. Note that the error function is defined as

$$\mathrm{erf}(x) = \frac{2}{\sqrt{\pi}} \int_0^x \mathrm{d}z \exp(-z^2). \tag{3.82}$$

   Hence you should be able to express $\mathscr{I}_1$ in terms of erf(x).
2. Calculate $\mathscr{I}_2$ and $\mathscr{I}_3$ analytically.
3. In order to approximate $\mathscr{I}_1$, $\mathscr{I}_2$ and $\mathscr{I}_3$ with the help of the two second order methods we discussed in this chapter, employ the following strategy: First the integrals are rewritten as

$$\int_I \mathrm{d}x \cdot = \sum_i \int_{I_i} \mathrm{d}x \cdot\,,$$

   where $\cdot$ is a placeholder for $g(x)$, $h(x)$ and $h^2(x)$ and I_i, $i = 1, \ldots, N$ are suitable intervals. In a second step the integrals are approximated with (i) the central rectangular rule and (ii) the trapezoidal rule.
4. In addition, we approximate the integrals $\mathscr{I}_1$, $\mathscr{I}_2$ and $\mathscr{I}_3$ by employing SIMPSON's rule for odd N. Here

$$\int_I \mathrm{d}x \cdot = \int_{I_1 \cup I_2} \mathrm{d}x \;\cdot + \int_{I_3 \cup I_4} \mathrm{d}x \;\cdot + \ldots + \int_{I_{N-2} \cup I_{N-1}} \mathrm{d}x \cdot\,,$$

   is used as it was discussed in Sect. 3.4.
5. Compare the results obtained with different algorithms and different numbers of grid-points, N. Plot the absolute and the relative error as a function of N.
6. Approximate numerically the integral

$$I = \int_{-\infty}^{\infty} \mathrm{d}x\, g(x).$$

7. Calculate the integral over the function

$$f(x, y) = \exp\left(-\frac{y^2}{2}\right) \cos(x)$$

within the intervals $x \in [-10\pi, 10\pi]$ and $y \in [-10, 10]$ with the help of an approximation of your choice.

8. Demonstrate numerically that the line integral over closed loops $\mathscr{C}$ of the function $f(x, y)$ of the previous problem vanishes:

$$\oint_{\mathscr{C}} \mathrm{d}s \cdot \nabla f(x, y) = 0.$$

References

1. Jordan, C.: Calculus of Finite Differences, 3rd edn. AMS Chelsea, Providence (1965)
2. Ueberhuber, C.W.: Numerical Computation 2: Methods, Software and Analysis. Springer, Berlin/Heidelberg (1997)
3. Mathai, A.M., Haubold, H.J.: Special Functions for Applied Scientists. Springer, Berlin/Heidelberg (2008)
4. Beals, R., Wong, R.: Special Functions. Cambridge Studies in Advanced Mathematics. Cambridge University Press, Cambridge (2010)
5. Fornberg, B.: A Practical Guide to Pseudospectral Methods. Cambridge Monographs on Applied and Computational Mathematics. Cambridge University Press, Cambridge (1999)
6. Stoer, J., Bulirsch, R.: Introduction to Numerical Analysis, 2nd edn. Springer, Berlin/Heidelberg (1993)
7. Abramovitz, M., Stegun, I.A. (eds.): Handbook of Mathemathical Functions. Dover, New York (1965)
8. Olver, F.W.J., Lozier, D.W., Boisvert, R.F., Clark, C.W.: NIST Handbook of Mathematical Functions. Cambridge University Press, Cambridge (2010)
9. Sormann, H.: Numerische Methoden in der Physik. Lecture Notes. Institute of Theoretical and Computational Physics, Graz University of Technology (2011)
10. Dahlquist, G., Björk, Å.: Numerical Methods in Scientific Computing. Cambridge University Press, Cambridge (2008)
11. Press, W.H., Teukolsky, S.A., Vetterling, W.T., Flannery, B.P.: Numerical Recipes in C++, 2nd edn. Cambridge University Press, Cambridge (2002)

Chapter 4
The KEPLER Problem

4.1 Introduction

The KEPLER problem [1–6] is certainly one of the most important problems in the history of physics and natural sciences in general. We will study this problem for several reasons: (i) it is a nice demonstration of the applicability of the methods introduced in the previous chapters, (ii) important concepts of the numerical treatment of ordinary differential equations can be introduced quite naturally, and (iii) it allows to revisit some of the most important aspects of classical mechanics.

The KEPLER problem is a special case of the two-body problem which is discussed in Appendix A. Let us summarize the main results. We consider two point particles interacting via the rotationally symmetric two body potential U which is solely a function of the distance between the particles. The symmetries of this problem allow several simplifications: (i) The problem can be reduced to the two dimensional motion of a point particle with reduced mass m in the central potential U. (ii) By construction, the total energy E is conserved. (iii) The length ℓ of the angular momentum vector is also conserved because of the symmetry of the potential U. Due to this rotational symmetry it is a natural choice to describe the particle's motion in polar coordinates (ρ, φ).

The final differential equations which have to be solved are of the form

$$\dot{\varphi} = \frac{\ell}{m\rho^2} , \tag{4.1}$$

and

$$\dot{\rho} = \pm\sqrt{\frac{2}{m}\left[E - U(\rho) - \frac{\ell^2}{2m\rho^2}\right]} . \tag{4.2}$$

B.A. Stickler, E. Schachinger, *Basic Concepts in Computational Physics*,
DOI 10.1007/978-3-319-27265-8_4

Here, one usually defines the effective potential

$$U_{\text{eff}}(\rho) = U(\rho) + \frac{\ell^2}{2m\rho^2}, \tag{4.3}$$

as the sum of the interaction potential and the centrifugal barrier $U_{\text{mom}}(\rho) = \ell^2/2m\rho^2$. Equation (4.2) can be transformed into an implicit equation for ρ

$$t = t_0 \pm \int_{\rho_0}^{\rho} d\rho' \left\{ \frac{2}{m} \left[E - U_{\text{eff}}(\rho') \right] \right\}^{-\frac{1}{2}}, \tag{4.4}$$

with $\rho_0 \equiv \rho(t_0)$ the initial condition at time t_0. Furthermore, the angle φ is related to the radius ρ by

$$\varphi = \varphi_0 \pm \int_{\rho_0}^{\rho} d\rho' \frac{\ell}{m\rho'^2} \left\{ \frac{2}{m} \left[E - U_{\text{eff}}(\rho') \right] \right\}^{-\frac{1}{2}}, \tag{4.5}$$

with the initial condition $\varphi_0 \equiv \varphi(t_0)$.

The KEPLER problem is defined by the gravitational interaction potential

$$U(\rho) = -\frac{\alpha}{\rho}, \quad \alpha > 0. \tag{4.6}$$

For this case, we show in Fig. 4.1 schematically the effective potential (4.3) (solid black line), together with the gravitational potential $U(\rho)$ (dashed-dotted line) and the centrifugal barrier U_{mom} (dashed line). The gravitational potential (4.6) is now

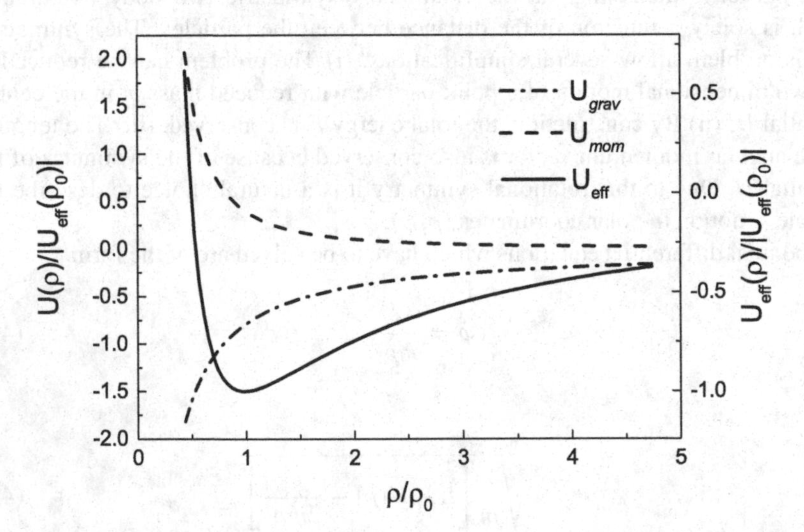

Fig. 4.1 Schematic illustration of the effective potential $U_{\text{eff}}(\rho)/U_{\text{eff}}(\rho_0)$ vs ρ/ρ_0 (*solid line, right hand scale*). Here, ρ_0 is the distance of the minimum in $U_{\text{eff}}(\rho)$. $U_{\text{grav}}(\rho)$ (*dashed-dotted line*) denotes the gravitational contribution while $U_{\text{mom}}(\rho)$ (*dashed line*) denotes the centrifugal barrier. Both potentials are normalized to $U_{\text{eff}}(\rho_0)$ (*Left hand scale applies*)

inserted into Eq. (4.5):

$$\varphi = \varphi_0 \pm \int_{\rho_0}^{\rho} d\rho' \frac{\ell}{m\rho'^2} \left[\frac{2}{m} \left(E + \frac{\alpha}{\rho'} - \frac{\ell^2}{2m\rho'^2} \right) \right]^{-\frac{1}{2}} . \tag{4.7}$$

The substitution $u = 1/\rho$ simplifies Eq. (4.7) to

$$\varphi = \varphi_0 \mp \int_{u_1}^{u_2} du \left[\frac{2mE}{\ell^2} + \frac{2m\alpha}{\ell^2} u - u^2 \right]^{-\frac{1}{2}} , \tag{4.8}$$

where the integration boundaries u_1 and u_2 are $1/\rho_0$ and $1/\rho$, respectively. The integral can now be evaluated with the help of a simple substitution[1] and we obtain the angle φ as a function of ρ:

$$\varphi = \varphi_0 \pm \cos^{-1} \left(\frac{\frac{\ell}{\rho} - \frac{m\alpha}{\ell}}{\sqrt{2mE + \frac{m^2\alpha^2}{\ell^2}}} \right) + \text{const} . \tag{4.9}$$

This solution can conveniently be characterized by the introduction of two parameters, namely

$$a = \frac{\ell^2}{m\alpha} \tag{4.10}$$

and the *eccentricity* e

$$e = \sqrt{1 + \frac{2E\ell^2}{m\alpha^2}} . \tag{4.11}$$

Hence, by neglecting the integration constant and setting $\varphi_0 = 0$ we arrive at

$$\frac{a}{\rho} = 1 + e \cos(\varphi) \tag{4.12}$$

as the final form of Eq. (4.9). It describes for $e > 1$ a hyperbola, for $e = 1$ a parabola, and for $e < 1$ an ellipse. The case $e = 0$ is a special case of the ellipse and describes a circle with radius $\rho = a$. A more detailed discussion of this result, in particular the

---

[1] In particular, we substitute

$$w = \left(u - \frac{m\alpha}{\ell^2} \right) \left(\frac{2mE}{\ell^2} + \frac{m^2\alpha^2}{\ell^4} \right)^{-\frac{1}{2}} .$$

derivation of KEPLER's laws can be found in any textbook on classical mechanics [1–6]. We discuss now some numerical aspects.

4.2 Numerical Treatment

In the previous section we solved the KEPLER problem by evaluating the integrand (4.7) expressing the angle φ as a function of the radius ρ. However, in this section we aim at solving the integral equation (4.4) numerically with the help of the methods discussed in the previous chapter. Remember that Eq. (4.4) expresses the time t as a function of the radius ρ. This equation has to be inverted, in order to obtain $\rho(t)$, which, in turn, is then inserted into Eq. (4.1) in order to determine the angle $\varphi(t)$ as a function of time. This discussion will lead us in a natural way to the most common techniques applied to solve ordinary differential equations, which is of no surprise since Eq. (4.4) is the integral representation of Eq. (4.2).

We give a short outline of what we plan to do: We discretize the time axis in equally spaced time steps Δt, i.e. $t_n = t_0 + n\Delta t$. Accordingly, we define the radius ρ at time $t = t_n$ as $\rho(t_n) \equiv \rho_n$. We can use the methods introduced in Chap. 3 to approximate the integral (4.4) from some ρ_n to ρ_{n+1}. According to this chapter the absolute error introduced will behave like $\delta = |\rho_n - \rho_{n+1}|^K$ where the explicit value of K depends on the method used. However, since the radius ρ changes continuously with time t we know that for sufficiently small values of Δt the error δ will also become arbitrarily small. If we start from some initial values t_0 and ρ_0, we can successively calculate the values $\rho_1, \rho_2, \ldots,$ by applying a small time step Δt.

Let us start by rewriting Eq. (4.4) as:

$$t - t_0 = \int_{\rho_0}^{\rho} d\rho' f(\rho') . \tag{4.13}$$

As we discretized the time axis in equally spaced increments and defined $\rho_n \equiv \rho(t_n)$, we can rewrite (4.13) as

$$\Delta t = t_n - t_{n-1} = \int_{\rho_n}^{\rho_{n+1}} d\rho' f(\rho') . \tag{4.14}$$

The forward rectangular rule, (3.9), results in the approximation

$$\Delta t = (\rho_{n+1} - \rho_n) f(\rho_n) . \tag{4.15}$$

We solve this equation for ρ_{n+1} and obtain

$$\rho_{n+1} = h(\rho_n)\Delta t + \rho_n \,, \tag{4.16}$$

where we defined

$$h(\rho) = \frac{1}{f(\rho)} = \sqrt{\frac{2}{m}\left[E - U_{\text{eff}}(\rho)\right]} \,, \tag{4.17}$$

following Eqs. (4.2) and (4.3). As Eq. (4.4) is the integral representation of the ordinary differential equation (4.2), approximation (4.16) corresponds to the approximation

$$D_+\rho_n = h(\rho_n) \,, \tag{4.18}$$

where $D_+\rho_n$ is the forward difference derivative (2.10a). Since the left hand side of the discretized differential equation (4.18) is independent of ρ_{n+1}, this method is referred to as an *explicit* method. In particular, consider an ordinary differential equation of the form

$$\dot{y} = F(y) \,. \tag{4.19}$$

Then the approximation method is referred to as an *explicit* EULER *method* if it is of the form

$$y_{n+1} = y_n + F(y_n)\Delta t \,. \tag{4.20}$$

Note that y might be a vector.

Let us use the backward rectangular rule (3.10) to solve Eq. (4.14). We obtain

$$t_{n+1} - t_n = (\rho_{n+1} - \rho_n) f(\rho_{n+1}) \,, \tag{4.21}$$

or equivalently

$$\rho_{n+1} = \rho_n + h(\rho_{n+1})\Delta t \,. \tag{4.22}$$

Again, this corresponds to an approximation of the differential equation (4.2) by

$$D_-\rho_{n+1} = h(\rho_{n+1}) \,, \tag{4.23}$$

where $D_-(\rho_{n+1})$ is the backward difference derivative (2.10b). In this case the quantity of interest ρ_{n+1} still appears in the argument of the function $h(\rho)$ and Eq. (4.22) is an *implicit* equation for ρ_{n+1} which has to be solved. In general, if the problem (4.19) is approximated by an algorithm of the form

$$y_{n+1} = y_n + F(y_{n+1})\Delta t \,, \tag{4.24}$$

it is referred to as an *implicit* EULER *method*. Note that the implicit equation (4.24) might be analytically unsolvable. Hence, one has to employ a numerical method to solve (4.24) which will also imply a numerical error. However, in the particular case of Eq. (4.22) we can solve it analytically since it is a fourth order polynomial in ρ_{n+1} of the form

$$\rho_{n+1}^4 - 2\rho_n\rho_{n+1}^3 + \left(\rho_n^2 - \frac{2E\Delta t^2}{m}\right)\rho_{n+1}^2 - \frac{2\alpha\Delta t^2}{m}\rho_{n+1} + \frac{\ell^2\Delta t^2}{m^2} = 0\,. \tag{4.25}$$

The solution of this equation is quite tedious and will not be discussed here, however, the method one employs is referred to as FERRARI's method [7].

A natural way to proceed is to regard the central rectangular rule (3.13) in a next step. Within this approximation we obtain for Eq. (4.13)

$$\Delta t = (\rho_{n+1} - \rho_n)f\left(\frac{\rho_{n+1} + \rho_n}{2}\right)\,, \tag{4.26}$$

which is equivalent to the implicit equation

$$\rho_{n+1} = \rho_n + h\left(\frac{\rho_{n+1} + \rho_n}{2}\right)\Delta t\,. \tag{4.27}$$

It can be written as an approximation to Eq. (4.2) with help of the central difference derivative $D_c\rho_{n+\frac{1}{2}}$:

$$D_c\rho_{n+\frac{1}{2}} = h\left(\frac{\rho_{n+1} + \rho_n}{2}\right)\,. \tag{4.28}$$

In general, for a problem of the form (4.19) a method of the form

$$y_{n+1} = y_n + F\left(\frac{y_{n+1} + y_n}{2}\right)\Delta t\,, \tag{4.29}$$

is referred to as the *implicit midpoint rule*. We note that this method might be more accurate since the error of the rectangular rule scales like $\mathcal{O}(\Delta t^2)$ while the error of the forward and backward rectangular rules scale like $\mathcal{O}(\Delta t)$. Nevertheless, in case of the KEPLER problem, one can solve the implicit equation (4.27) analytically for ρ_{n+1} which is certainly of advantage.

In this chapter the KEPLER problem was instrumental in introducing three common methods which can be employed to solve numerically ordinary differential equations of the form (4.19). More general and advanced methods to solve ordinary differential equations and a more systematic description of these methods will be offered in the next chapter.

However, let us discuss another point before proceeding to the chapter on the numerics of ordinary differential equations. As demonstrated in Sect. 1.3 the approximation of the integral (4.4) involves a numerical error. What will be the consequence of this error? Since we demonstrated that the approximations we discussed result in a differential equation in finite difference form, i.e. Eqs. (4.18), (4.23), and (4.27), we know that the derivative $\dot{p}$ will exhibit an error. Consequently, energy conservation [see Appendix A, Eq. (A.27)] will be violated with the implication that deviations from the trajectory (4.12) can be expected. This is definitely not desirable.

A solution is provided by a special class of methods, known as *symplectic integrators*, which were specifically designed for such cases. They are based on a formulation of the problem using HAMILTON's equations of motion. (See, for instance, Refs. [1–5].) In the particular case of the KEPLER problem the HAMILTON function is equivalent to the total energy of the system and reads (in some scaled units):

$$H(p,q) = \frac{1}{2}\left(p_1^2 + p_2^2\right) - \frac{1}{\sqrt{q_1^2 + q_2^2}} . \tag{4.30}$$

Here $p = (p_1, p_2)$ are the generalized momentum coordinates of the point particle in the two-dimensional plane and (q_1, q_2) are the generalized position coordinates. From this HAMILTON's equations of motion

$$\begin{pmatrix} \dot{q} \\ \dot{p} \end{pmatrix} = \begin{pmatrix} \nabla_p H(p,q) \\ -\nabla_q H(p,q) \end{pmatrix} = \begin{pmatrix} a(q,p) \\ b(q,p) \end{pmatrix} , \tag{4.31}$$

follow, where the functions $a(q,p)$ and $b(q,p)$ have been introduced for a more convenient notation. Note that these functions are two dimensional vectors in the case of KEPLER's problem. The so called *symplectic* EULER *method* is given by

$$\begin{aligned} q_{n+1} &= q_n + a(q_n, p_{n+1})\Delta t , \\ p_{n+1} &= p_n + b(q_n, p_{n+1})\Delta t . \end{aligned} \tag{4.32}$$

Obviously, the first equation is explicit while the second is implicit. An alternative formulation reads

$$\begin{aligned} q_{n+1} &= q_n + a(q_{n+1}, p_n)\Delta t , \\ p_{n+1} &= p_n + b(q_{n+1}, p_n)\Delta t , \end{aligned} \tag{4.33}$$

where the first equation is implicit and the second equation is explicit. Of course, Eq. (4.31) may be solved with the help of the explicit EULER method (4.20), the implicit EULER method (4.24) or the implicit midpoint rule (4.29). The solution should be equivalent to solving Eq. (4.4) with the respective method and then calculating (4.1) successively. Again, a more systematic discussion of symplectic integrators can be found in the following chapters.

Let us conclude this chapter with a final remark. We decided to solve Eqs. (4.4) and (4.1) because we wanted to reproduce the dynamics of the system, i.e. we wanted to obtain $\rho(t)$ and $\varphi(t)$. This directed us to the numerical solution of two integrals. If we wanted to employ symplectic methods, which provide several advantages, we would have to solve four differential equations (4.31) instead of two integrals. Moreover, if we are not interested in the time evolution of the system but in the form of the trajectory in general, we could simply evaluate the integral (4.5) analytically or, if an analytical solution is not feasible for the potential $U(\rho)$ one is interested in, numerically. Methods to approximate such an integral were extensively discussed in Chap. 3.

Summary

KEPLER's two-body problem was used as an incentive to introduce intuitively numerical methods to solve ordinary first order differential equations. To serve this purpose the basic differential equations were transformed into integral form. These integrals were then solved with the help of the rules discussed in Sect. 3.2. Three basic methods have been identified, namely the *explicit* EULER method (based on the forward difference derivative), the *implicit* EULER method (based on the backward difference derivative), and the *explicit midpoint rule* (based on the central rectangular rule). Shortcomings of such methods have been discussed briefly as were remedies to overcome these.

Problems

1. Planetary-Orbits: Apply the methods of numerical integration to the integral (4.4) and compare it to the analytical result. Identify the three different cases of elliptic, parabolic or hyperbolic orbits by varying the initial conditions.
2. LENNARD-JONES Scattering: Consider the scattering of two point particles which interact via the LENNARD-JONES potential $U(r) = 4\sigma[(\varepsilon/r)^{12} - (\varepsilon/r)^6]$ with $\sigma, \varepsilon > 0$ (see Chapter 7). Calculate the orbit $\varphi(\rho)$.
3. Harmonic-Motion: Consider the motion of a point particle in the radial harmonic oscillator $U(\rho) = m\omega^2\rho^2/2$. According to BERTRAND's theorem (see Refs. [1–3]) the particle's trajectories should be closed orbits. Demonstrate this numerically as well as analytically.

References

1. Arnol'd, V.I.: Mathematical Methods of Classical Mechanics, 2nd edn. Graduate Texts in Mathematics, vol. 60. Springer, Berlin/Heidelberg (1989)
2. Fetter, A.L., Walecka, J.D.: Theoretical Mechanics of Particles and Continua. Dover, New York (2004)
3. Scheck, F.: Mechanics, 5th edn. Springer, Berlin/Heidelberg (2010)
4. Goldstein, H., Poole, C., Safko, J.: Classical Mechanics, 3rd edn. Addison-Wesley, Menlo Park (2013)
5. Fließbach, T.: Mechanik, 7th edn. Lehrbuch zur Theoretischen Physik I. Springer, Berlin/Heidelberg (2015)
6. Ó'Mathúna, D.: Integrable Systems in Celestial Mechanics. Progress in Mathematical Physics, vol. 51. Birkhäuser Basel, Basel (2008)
7. Clark, A.: Elements of Abstract Algebra. Dover, New York (1971)

Chapter 5
Ordinary Differential Equations: Initial Value Problems

5.1 Introduction

This chapter introduces common numeric methods designed to solve *initial value problems*. The discussion of the KEPLER problem in the previous chapter allowed the introduction of three concepts, namely the implicit EULER method, the explicit EULER method, and the implicit midpoint rule. Furthermore, we mentioned the symplectic EULER method. In this chapter we plan to put these methods into a more general context and to discuss more advanced techniques.

Let us define the problem: We consider initial value problems of the form

$$\begin{cases} \dot{y}(t) = f(y, t) , \\ y(0) = y_0 , \end{cases} \tag{5.1}$$

where $y(t) \equiv y$ is an n-dimensional vector and y_0 is referred to as the *initial value* of y. Some remarks about the form of Eq. (5.1) are required:

(i) We note that by posing Eq. (5.1), we assume that the differential equation is *explicit* in $\dot{y}$, i.e. initial value problems of the form

$$\begin{cases} G(\dot{y}) = f(y, t) , \\ y(0) = y_0 , \end{cases} \tag{5.2}$$

are only considered if $G(\dot{y})$ is analytically invertible. For instance, we will not deal with differential equations of the form

$$\dot{y} + \log(\dot{y}) = 1 . \tag{5.3}$$

B.A. Stickler, E. Schachinger, *Basic Concepts in Computational Physics*,
DOI 10.1007/978-3-319-27265-8_5

(ii) We note that Eq. (5.1) is a *first order* differential equation in y. However, this is in fact not a restriction since we can transform every explicit differential equation of order n into a coupled set of explicit first order differential equations. Let us demonstrate this. We regard an explicit differential equation of the form

$$y^{(n)} = f(t; y, \dot{y}, \ddot{y}, \dots, y^{(n-1)}) , \tag{5.4}$$

where we defined $y^{(k)} \equiv \frac{d^k}{dt^k} y$. This equation is equivalent to the set

$$\begin{aligned} \dot{y}_1 &= y_2 , \\ \dot{y}_2 &= y_3 , \\ \vdots &\quad \vdots \\ \dot{y}_{n-1} &= y_n , \\ \dot{y}_n &= f(t, y_1, y_2, \dots, y_n) , \end{aligned} \tag{5.5}$$

which can be written as Eq. (5.1). Hence, we can attenuate the criterion discussed in point (i), that the differential equation has to be explicit in $\dot{y}$, to the criterion that the differential equation of order n has to be explicit in the n-th derivative of y, namely $y^{(n)}$.

There is another point required to be discussed before moving on. The numerical treatment of initial value problems is of eminent importance in physics because many differential equations, which appear unspectacular at first glance, cannot be solved analytically. For instance, consider a first order differential equation of the type:

$$\dot{y} = t^2 + y^2 . \tag{5.6}$$

Although this equation appears to be simple, one has to rely on numerical methods to obtain a solution. However, Eq. (5.6) is not *well posed* since the solution is ambiguous as long as no initial values are given. A numerical solution is only possible if the problem is completely defined. In many cases, one uses numerical methods although the problem is solvable with the help of analytic methods, simply because the solution would be too complicated. A numerical approach might be justified, however, one should always remember that, quote [1]:

Numerical methods are no excuse for poor analysis.

This chapter will be augmented by a chapter on the double pendulum, which will serve as a demonstration of the applicability of RUNGE-KUTTA methods and by a chapter on molecular dynamics which will demonstrate the applicability of the leap-frog algorithm.

5.2 Simple Integrators

We start by reintroducing the methods already discussed in the previous chapter. Again, we discretize the time coordinate t via the relation $t_n = t_0 + n\Delta t$ and define $f_n \equiv f(t_n)$ accordingly. In the following we will refrain from noting the initial condition explicitly for a more compact notation. We investigate Eq. (5.1) at some particular time t_n:

$$\dot{y}_n = f(y_n, t_n) . \tag{5.7}$$

Integrating both sides of (5.7) over the interval $[t_n, t_{n+1}]$ gives

$$y_{n+1} = y_n + \int_{t_n}^{t_{n+1}} \mathrm{d}t' f[y(t'), t'] . \tag{5.8}$$

Note that Eq. (5.8) is exact and it will be our starting point in the discussion of several paths to a numeric solution of initial value problems. These solutions will be based on an approximation of the integral on the right hand side of Eq. (5.8) with the help of the methods already discussed in Chap. 3.

In the following we list four of the best known simple integration methods for initial value problems:

(1) Applying the forward rectangular rule (3.9) to Eq. (5.8) yields

$$y_{n+1} = y_n + f(y_n, t_n)\Delta t + \mathcal{O}(\Delta t^2) , \tag{5.9}$$

which is the explicit EULER method we encountered already in Sect. 4.2. This method is also referred to as the *forward* EULER *method*. In accordance to the forward rectangular rule, the leading term of the error of this method is proportional to Δt^2 as was pointed out in Sect. 3.2.

(2) We use the backward rectangular rule (3.10) in Eq. (5.8) and obtain

$$y_{n+1} = y_n + f(y_{n+1}, t_{n+1})\Delta t + \mathcal{O}(\Delta t^2) , \tag{5.10}$$

which is the implicit EULER method, also referred to as *backward* EULER *method*. As already highlighted in Sect. 4.2, it may be necessary to solve Eq. (5.10) numerically for y_{n+1}. (Some notes on the numeric solution of non-linear equations can be found in Appendix B.)

(3) The central rectangular rule (3.13) approximates Eq. (5.8) by

$$y_{n+1} = y_n + f(y_{n+\frac{1}{2}}, t_{n+\frac{1}{2}})\Delta t + \mathcal{O}(\Delta t^3) , \tag{5.11}$$

and we rewrite this equation in the form:

$$y_{n+1} = y_{n-1} + 2f(y_n, t_n)\Delta t + \mathcal{O}(\Delta t^3) . \tag{5.12}$$

This method is sometimes referred to as the *leap-frog* routine or STÖRMER-VERLET method. We will come back to this point in Chap. 7. Note that the approximation

$$y_{n+\frac{1}{2}} \approx \frac{y_n + y_{n+1}}{2}, \tag{5.13}$$

in Eq. (5.11) gives the implicit midpoint rule as it was introduced in Sect. 4.2.

(4) Employing the trapezoidal rule (3.15) in an approximation to Eq. (5.8) yields

$$y_{n+1} = y_n + \frac{\Delta t}{2}\left[f(y_n, t_n) + f(y_{n+1}, t_{n+1})\right] + \mathcal{O}(\Delta t^3) . \tag{5.14}$$

This is an implicit method which has to be solved for y_{n+1}. It is generally known as the CRANK-NICOLSON *method* [2] or simply as *trapezoidal method*.

Methods (1), (2), and (4) are also known as one-step methods, since only function values at times t_n and t_{n+1} are used to propagate in time. In contrast, the leap-frog method is already a *multi-step* method since three different times appear in the expression. Basically, there are three different strategies to improve these rather simple methods:

- TAYLOR series methods: Use more terms in the TAYLOR expansion of y_{n+1}.
- Linear Multi-Step methods: Use data from previous time steps y_k, $k < n$ in order to cancel terms in the truncation error.
- RUNGE-KUTTA method: Use intermediate points within one time step.

We will briefly discuss the first two alternatives and then turn our attention to the RUNGE-KUTTA methods in the next section.

TAYLOR *Series Methods*

From Chap. 2 we are already familiar with the TAYLOR expansion (2.7) of the function y_{n+1} around the point y_n,

$$y_{n+1} = y_n + \Delta t\, \dot{y}_n + \frac{\Delta t^2}{2}\ddot{y}_n + \mathcal{O}(\Delta t^3) . \tag{5.15}$$

We insert Eq. (5.7) into Eq. (5.15) and obtain

$$y_{n+1} = y_n + \Delta t f(y_n, t_n) + \frac{\Delta t^2}{2}\ddot{y}_n + \mathcal{O}(\Delta t^3) . \tag{5.16}$$

So far nothing has been gained since the truncation error is still proportional to Δt^2. However, calculating $\ddot{y}_n$ with the help of Eq. (5.7) gives

$$\ddot{y}_n = \frac{\mathrm{d}}{\mathrm{d}t} f(y_n, t_n) = \dot{f}(y_n, t_n) + f'(y_n, t_n)\dot{y}_n = \dot{f}(y_n, t_n) + f'(y_n, t_n) f(y_n, t_n) , \qquad (5.17)$$

and this results together with Eq. (5.16) in:

$$y_{n+1} = y_n + \Delta t f(y_n, t_n) + \frac{\Delta t^2}{2} \left[\dot{f}(y_n, t_n) + f'(y_n, t_n) f(y_n, t_n) \right] + \mathcal{O}(\Delta t^3) . \qquad (5.18)$$

This manipulation reduced the local truncation error to orders of Δt^3. The derivatives of $f(y_n, t_n)$, $f'(y_n, t_n)$ and $\dot{f}(y_n, t_n)$ can be approximated with the help of the methods discussed in Chap. 2, if an analytic differentiation is not feasible. The above procedure can be repeated up to arbitrary order in the TAYLOR expansion (5.15).

Linear Multi-step Methods

A k-th order linear multi-step method is defined by the approximation

$$y_{n+1} = \sum_{j=0}^{k} a_j y_{n-j} + \Delta t \sum_{j=0}^{k+1} b_j f(y_{n+1-j}, t_{n+1-j}) , \qquad (5.19)$$

of Eq. (5.8). The coefficients a_j and b_j have to be determined in such a way that the local truncation error is reduced. Two of the best known techniques are the so called second order ADAMS-BASHFORD method

$$y_{n+1} = y_n + \frac{\Delta t}{2} \left[3 f(y_n, t_n) - f(y_{n-1}, t_{n-1}) \right] , \qquad (5.20)$$

and the second order rule (*backward differentiation formula*)

$$y_{n+1} = \frac{1}{3} \left[4 y_n - y_{n-1} + \frac{\Delta t}{2} f(y_{n+1}, t_{n+1}) \right] . \qquad (5.21)$$

(For details please consult Refs. [3–5].)

We note in passing that the backward differentiation formula of arbitrary order can easily be obtained with the help of the operator technique introduced in Sect. 2.4, Eq. (2.27). One simply inserts the backward difference series (2.27) to arbitrary order into the right hand side of the differential equation (5.7).

In many cases, multi-step methods are based on the interpolation of previously computed values y_k by LAGRANGE polynomials. This interpolation is then inserted into Eq. (5.8) and integrated. However, a detailed discussion of such procedures is beyond the scope of this book. The interested reader is referred to Refs. [6, 7].

Nevertheless, let us make one last point. We note that Eq. (5.19) is explicit for $b_0 = 0$ and implicit for $b_0 \neq 0$. In many numerical realizations one combines implicit and explicit multi-step methods in such a way that the explicit result [solve Eq. (5.19) with $b_0 = 0$] is used as a guess to solve the implicit equation [solve Eq. (5.19) with $b_0 \neq 0$]. Hence, the explicit method *predicts* the value y_{n+1} and the implicit method *corrects* it. Such methods yield very good results and are commonly referred to as *predictor–corrector* methods [8].

5.3 RUNGE-KUTTA Methods

In contrast to linear multi-step methods, the idea in RUNGE-KUTTA methods (see, for instance, Ref. [6]) is to improve the accuracy by calculating intermediate grid-points within the interval $[t_n, t_{n+1}]$. We note that the approximation (5.11) resulting from the central rectangular rule is already such a method since the function value $y_{n+1/2}$ at the grid-point $t_{n+1/2} = t_n + \Delta t/2$ is taken into account. We investigate this in more detail and rewrite Eq. (5.11):

$$y_{n+1} = y_n + f(y_{n+\frac{1}{2}}, t_{n+\frac{1}{2}})\Delta t + \mathcal{O}(\Delta t^3) . \tag{5.22}$$

We now have to find appropriate approximations to $y_{n+1/2}$ which will increase the accuracy of Eq. (5.11). Our first choice is to replace $y_{n+1/2}$ with the help of the explicit EULER method, Eq. (5.9),

$$y_{n+\frac{1}{2}} = y_n + \frac{\Delta t}{2}\dot{y}_n = y_n + \frac{\Delta t}{2}f(y_n, t_n) , \tag{5.23}$$

which, inserted into Eq. (5.22) yields

$$y_{n+1} = y_n + f\left[y_n + \frac{\Delta t}{2}f(y_n, t_n), t_n + \frac{\Delta t}{2}\right]\Delta t + \mathcal{O}(\Delta t^2) . \tag{5.24}$$

We note that Eq. (5.24) is referred to as the *explicit midpoint rule*. In analogy we could have approximated $y_{n+1/2}$ with the help of the averaged function value $\mu y_{n+1/2}$ which results in

$$y_{n+1} = y_n + f\left(\frac{y_n + y_{n+1}}{2}, t_n + \frac{\Delta t}{2}\right)\Delta t + \mathcal{O}(\Delta t^2) . \tag{5.25}$$

This equation is referred to as the *implicit midpoint rule*. Let us explain how we obtain an estimate for the error in Eqs. (5.24) and (5.25). In case of Eq. (5.24) we investigate the term

$$y_{n+1} - y_n - f\left[y_n + \frac{\Delta t}{2} f(y_n, t_n), t_n + \frac{\Delta t}{2}\right] \Delta t \,.$$

The TAYLOR expansion of y_{n+1} and $f(\cdot)$ around the point $\Delta t = 0$ yields

$$\Delta t \left[\dot{y}_n - f(y_n, t_n)\right] + \frac{\Delta t^2}{2} \left[\ddot{y} - \dot{f}(y_n, t_n) - f'(y_n, t_n)\dot{y}_n\right] + \ldots \,. \tag{5.26}$$

We observe that the first term cancels because of Eq. (5.7). Consequently, the error is of order Δt^2. A similar argument holds for Eq. (5.25).

Let us introduce a more convenient notation for the above examples before we concentrate on a more general topic. It is presented in algorithmic form, i.e. it defines the sequence in which one should calculate the various terms. This is convenient for two reasons, first of all it increases the readability of complex methods such as Eq. (5.25) and, secondly, it is easy to identify which part of the method involves an implicit step and which part has to be solved separately for the corresponding variable. For this purpose let us introduce variables Y_i of some index $i \geq 1$ and we use a simple example to illustrate this notation. Consider the explicit EULER method (5.9). It can be written as

$$\begin{aligned} Y_1 &= y_n \,, \\ y_{n+1} &= y_n + f(Y_1, t_n)\Delta t \,. \end{aligned} \tag{5.27}$$

In a similar fashion we write the implicit EULER method (5.10) as

$$\begin{aligned} Y_1 &= y_n + f(Y_1, t_{n+1})\Delta t \,, \\ y_{n+1} &= y_n + f(Y_1, t_{n+1})\Delta t \,. \end{aligned} \tag{5.28}$$

It is understood that the first equation of (5.28) has to be solved for Y_1 first and this result is then plugged into the second equation in order to obtain y_{n+1}. One further example: the CRANK-NICOLSON (5.14) method can be rewritten as

$$\begin{aligned} Y_1 &= y_n \,, \\ Y_2 &= y_n + \frac{\Delta t}{2} \left[f(Y_1, t_n) + f(Y_2, t_{n+1})\right] \,, \\ y_{n+1} &= y_n + \frac{\Delta t}{2} \left[f(Y_1, t_n) + f(Y_2, t_{n+1})\right] \,, \end{aligned} \tag{5.29}$$

where the second equation is to be solved for Y_2 in the second step.

In analogy, the algorithmic form of the explicit midpoint rule (5.24) is defined as

$$\begin{aligned} Y_1 &= y_n , \\ Y_2 &= y_n + \frac{\Delta t}{2} f\left(Y_1, t_n + \frac{\Delta t}{2}\right) , \\ y_{n+1} &= y_n + \frac{\Delta t}{2} f\left(Y_2, t_n + \frac{\Delta t}{2}\right) , \end{aligned} \tag{5.30}$$

and we find for the implicit midpoint rule (5.25):

$$\begin{aligned} Y_1 &= y_n + \frac{\Delta t}{2} f\left(Y_1, t_n + \frac{\Delta t}{2}\right) , \\ y_{n+1} &= y_n + \Delta t f\left(Y_1, t_n + \frac{\Delta t}{2}\right) . \end{aligned} \tag{5.31}$$

The above algorithms are all examples of the so called RUNGE-KUTTA methods. We introduce the general representation of a d-stage RUNGE-KUTTA method:

$$\begin{aligned} Y_i &= y_n + \Delta t \sum_{j=1}^{d} a_{ij} f\left(Y_j, t_n + c_j \Delta t\right) , \qquad i = 1, \ldots, d , \\ y_{n+1} &= y_n + \Delta t \sum_{j=1}^{d} b_j f\left(Y_j, t_n + c_j \Delta t\right) . \end{aligned} \tag{5.32}$$

We note that Eq. (5.32) it is completely determined by the coefficients a_{ij}, b_j and c_j. In particular $a = \{a_{ij}\}$ is a $d \times d$ matrix, while $b = \{b_j\}$ and $c = \{c_j\}$ are d dimensional vectors.

BUTCHER tableaus are a very useful tool to characterize such methods. They provide a structured representation of the coefficient matrix a and the coefficient vectors b and c:

$$\begin{array}{c|cccc} c_1 & a_{11} & a_{12} & \ldots & a_{1d} \\ c_2 & a_{21} & a_{22} & \ldots & a_{2d} \\ \vdots & \vdots & \vdots & \ddots & \vdots \\ c_d & a_{d1} & a_{d2} & \ldots & a_{dd} \\ \hline & b_1 & b_2 & \ldots & b_d \end{array} \tag{5.33}$$

We note that the RUNGE-KUTTA method (5.32) or (5.33) is explicit if the matrix a is zero on and above the diagonal, i.e. $a_{ij} = 0$ for $j \geq i$. Let us rewrite all the methods described here in the form of BUTCHER tableaus:

***Explicit* EULER:**

$$\begin{array}{c|c} 0 & 0 \\ \hline & 1 \end{array} \tag{5.34}$$

***Implicit* EULER:**

$$\begin{array}{c|c} 1 & 1 \\ \hline & 1 \end{array} \tag{5.35}$$

CRANK-NICOLSON:

$$\begin{array}{c|cc} 0 & 0 & 0 \\ 1 & \frac{1}{2} & \frac{1}{2} \\ \hline & \frac{1}{2} & \frac{1}{2} \end{array} \tag{5.36}$$

Explicit Midpoint:

$$\begin{array}{c|cc} 0 & 0 & 0 \\ \frac{1}{2} & \frac{1}{2} & 0 \\ \hline & \frac{1}{2} & \frac{1}{2} \end{array} \tag{5.37}$$

Implicit Midpoint:

$$\begin{array}{c|c} \frac{1}{2} & \frac{1}{2} \\ \hline & 1 \end{array} \tag{5.38}$$

With the help of RUNGE-KUTTA methods of the general form (5.32) one can develop methods of arbitrary accuracy. One of the most popular methods is the explicit four stage method (we will call it *e-RK-4*) which is defined by the algorithm:

$$\begin{aligned}
Y_1 &= y_n, \\
Y_2 &= y_n + \frac{\Delta t}{2} f(Y_1, t_n), \\
Y_3 &= y_n + \frac{\Delta t}{2} f\left(Y_2, t_n + \frac{\Delta t}{2}\right), \\
Y_4 &= y_n + \Delta t f\left(Y_3, t_n + \frac{\Delta t}{2}\right), \\
y_{n+1} &= y_n + \frac{\Delta t}{6}\left[f(Y_1, t_n) + 2f\left(Y_2, t_n + \frac{\Delta t}{2}\right)\right. \\
&\quad \left. +2f\left(Y_3, t_n + \frac{\Delta t}{2}\right) + f(Y_4, t_n)\right] . \qquad (5.39)
\end{aligned}$$

This method is an analogue to the SIMPSON rule of numerical integration as discussed in Sect. 3.4. However, a detailed compilation of the coefficient array a and coefficient vectors b, and c is quite complicated. A closer inspection reveals that the methodological error of this method behaves as Δt^5. The algorithm *e-RK-4*, Eq. (5.39), is represented by a BUTCHER tableau of the form

$$\begin{array}{c|cccc}
0 & 0 & 0 & 0 & 0 \\
\frac{1}{2} & \frac{1}{2} & 0 & 0 & 0 \\
\frac{1}{2} & 0 & \frac{1}{2} & 0 & 0 \\
1 & 0 & 0 & 1 & 0 \\
\hline
 & \frac{1}{6} & \frac{1}{3} & \frac{1}{3} & \frac{1}{6}
\end{array} \qquad (5.40)$$

Another quite popular method is given by the BUTCHER tableau

$$\begin{array}{c|cc}
\frac{1}{2} - \frac{\sqrt{3}}{6} & \frac{1}{4} & \frac{1}{4} - \frac{\sqrt{3}}{6} \\
\frac{1}{2} + \frac{\sqrt{3}}{6} & \frac{1}{4} + \frac{\sqrt{3}}{6} & \frac{1}{4} \\
\hline
 & \frac{1}{2} & \frac{1}{2}
\end{array} \qquad (5.41)$$

We note that this method is implicit and mention that it corresponds to the two point GAUSS-LEGENDRE quadrature of Sect. 3.6.

A further improvement of implicit RUNGE-KUTTA methods can be achieved by *choosing* the Y_i in such a way that they correspond to solutions of the differential equation (5.7) at intermediate time steps. The intermediate time steps at which one wants to reproduce the function are referred to as *collocation points*. At these points the functions are approximated by interpolation on the basis of LAGRANGE polynomials, which can easily be integrated analytically. However, the discussion of such collocation methods [8] is far beyond the scope of this book.

In general RUNGE-KUTTA methods are very useful. However one always has to keep in mind that there could be better methods for the problem at hand. Let us close this section with a quote from the book by PRESS et al. [9]:

> For many scientific users, fourth-order Runge-Kutta is not just the first word on ODE integrators, but the last word as well. In fact, you can get pretty far on this old workhorse, especially if you combine it with an adaptive step-size algorithm. Keep in mind, however, that the old workhorse's last trip may well take you to the poorhouse: Bulirsch-Stoer or predictor-corrector methods can be very much more efficient for problems where high accuracy is a requirement. Those methods are the high-strung racehorses. Runge-Kutta is for ploughing the fields.

5.4 Hamiltonian Systems: Symplectic Integrators

Let us define a symplectic integrator as a numerical integration in which the mapping

$$\Phi_{\Delta t} : y_n \mapsto y_{n+1} , \tag{5.42}$$

is symplectic. Here $\Phi_{\Delta t}$ is referred to as the *numerical flow* of the method. If we regard the initial value problem (5.1) we can define in an analogous way the *flow of the system* φ_t as

$$\varphi_t(y_0) = y(t) . \tag{5.43}$$

For instance, if we consider the initial value problem

$$\begin{cases} \dot{y} = Ay , \\ y(0) = y_0 , \end{cases} \tag{5.44}$$

where $y \in \mathbb{R}^n$ and $A \in \mathbb{R}^{n \times n}$, then the flow φ_t of the system is given by:

$$\varphi_t(y_0) = \exp(At)y_0 . \tag{5.45}$$

On the other hand, if we regard two vectors $v, w \in \mathbb{R}^2$, we can express the area ω of the parallelogram spanned by these vectors as

$$\omega(v, w) = \det(vw) = v \begin{pmatrix} 0 & 1 \\ -1 & 0 \end{pmatrix} w = ad - bc , \tag{5.46}$$

where we put $v = (a, b)^T$ and $w = (c, d)^T$. More generally, if $v, w \in \mathbb{R}^{2d}$, we have

$$\omega(v, w) = v \begin{pmatrix} 0 & I \\ -I & 0 \end{pmatrix} w \equiv vJw \,, \tag{5.47}$$

where I is the $d \times d$ dimensional unity matrix. Hence (5.47) represents the sum of the projected areas of the form

$$\det \begin{pmatrix} v_i & w_i \\ v_{i+d} & w_{i+d} \end{pmatrix} . \tag{5.48}$$

If we regard a mapping $M : \mathbb{R}^{2d} \mapsto \mathbb{R}^{2d}$ and require that

$$\omega(Mv, Mw) = \omega(v, w) \,, \tag{5.49}$$

i.e. the area is preserved, we obtain the condition that

$$M^T JM = J \,, \tag{5.50}$$

which is equivalent to $\det(M) = 1$. Finally, a differentiable mapping $f : \mathbb{R}^{2d} \mapsto \mathbb{R}^{2d}$ is referred to as *symplectic* if the linear mapping $f'(x)$ (JACOBI matrix) conserves ω for all $x \in \mathbb{R}^{2d}$. One can easily prove that the flow of Hamiltonian systems is symplectic, i.e. area preserving in phase space. Every Hamiltonian system is characterized by its HAMILTON function $H(p, q)$ and the corresponding HAMILTON equations of motion [10–14]:

$$\dot{p} = -\nabla_q H(p, q) \quad \text{and} \quad \dot{q} = \nabla_p H(p, q) \,. \tag{5.51}$$

We define the flow of the system via

$$\varphi_t(x_0) = x(t) \,, \tag{5.52}$$

where

$$x_0 = \begin{pmatrix} p_0 \\ q_0 \end{pmatrix} \quad \text{and} \quad x(t) = \begin{pmatrix} p(t) \\ q(t) \end{pmatrix} . \tag{5.53}$$

Hence we rewrite (5.51) as

$$\dot{x} = J^{-1} \nabla_x H(x) \,, \tag{5.54}$$

and note that $x \equiv x(t, x_0)$ is a function of time and initial conditions. In a next step we define the Jacobian of the flow via

$$P_t(x_0) = \nabla_{x_0} \varphi_t(x_0) \,, \tag{5.55}$$

and calculate

$$\begin{aligned}
\dot{P}_t(x_0) &= \nabla_{x_0}\dot{x} \\
&= J^{-1}\nabla_{x_0}\nabla_x H(x) \\
&= J^{-1}\Delta_x H(x)\nabla_{x_0}x \\
&= J^{-1}\Delta_x H(x) P_t(x_0) \\
&= \begin{pmatrix} -\nabla_{qp}H(p,q) & -\nabla_{qq}H(p,q) \\ \nabla_{pp}H(p,q) & \nabla_{pq}H(p,q) \end{pmatrix} P_t(x_0) \, .
\end{aligned} \tag{5.56}$$

Hence, P_t is given by the solution of the equation

$$\dot{P}_t = J^{-1}\Delta_x H(x) P_t \, . \tag{5.57}$$

Symplecticity ensures that the area

$$P_t^T J P_t = \text{const} \, , \tag{5.58}$$

which can be verified by calculating $\frac{\mathrm{d}}{\mathrm{d}t}\left(P_t^T J P_t\right)$ where we keep in mind that $J^T = -J$. Hence,

$$\begin{aligned}
\frac{\mathrm{d}}{\mathrm{d}t} P_t^T J P_t &= \dot{P}_t^T J P_t + P_t^T J \dot{P}_t \\
&= P_t^T \Delta_x H(x) (J^{-1})^T J P_t + P_t^T J J^{-1} \Delta_x H(x) P_t \\
&= 0 \, ,
\end{aligned} \tag{5.59}$$

even if the HAMILTON function is not conserved. This means that the flow of a Hamiltonian system is symplectic, i.e. area preserving in phase space [10, 11, 13].

Since this conservation law is violated by methods like *e-RK-4* or explicit EULER, one introduces so called *symplectic integrators*, which have been particularly designed as a remedy to this shortcoming. A detailed investigation of these techniques is far too engaged for this book. The interested reader is referred to Refs. [12, 15–17].

However, we provide a list of the most important integrators.

Symplectic EULER

$$q_{n+1} = q_n + a(q_n, p_{n+1})\Delta t \, , \tag{5.60a}$$

$$p_{n+1} = p_n + b(q_n, p_{n+1})\Delta t \, . \tag{5.60b}$$

Here $a(p,q) = \nabla_p H(p,q)$ and $b(p,q) = -\nabla_q H(p,q)$ have already been defined in Sect. 4.2.

Symplectic RUNGE-KUTTA

It can be demonstrated that a RUNGE-KUTTA method is symplectic if the coefficients fulfill

$$b_i a_{ij} + b_j a_{ji} = b_i b_j , \tag{5.61}$$

for all i, j [16, 18]. This is a property of the collocation methods based on GAUSS points c_i.

5.5 An Example: The KEPLER Problem, Revisited

It has already been discussed in Sect. 4.2 that the HAMILTON function of this system takes on the form [19]

$$H(p,q) = \frac{1}{2}\left(p_1^2 + p_2^2\right) - \frac{1}{\sqrt{q_1^2 + q_2^2}} , \tag{5.62}$$

and HAMILTON's equations of motion read

$$\dot{p}_1 = -\nabla_{q_1} H(p,q) = \frac{q_1}{(q_1^2 + q_2^2)^{\frac{3}{2}}} , \tag{5.63a}$$

$$\dot{p}_2 = -\nabla_{q_2} H(p,q) = -\frac{q_2}{(q_1^2 + q_2^2)^{\frac{3}{2}}} , \tag{5.63b}$$

$$\dot{q}_1 = \nabla_{p_1} H(p,q) = p_1 , \tag{5.63c}$$

$$\dot{q}_2 = \nabla_{p_2} H(p,q) = p_2 . \tag{5.63d}$$

We now introduce the time instances $t_n = t_0 + n\Delta t$ and define $q_i^n \equiv q_i(t_n)$ and $p_i^n \equiv p_i(t_n)$ for $i = 1, 2$. In the following we give the discretized recursion relation for three different methods, namely explicit EULER, implicit EULER, and symplectic EULER.

Explicit EULER

In case of the explicit EULER method we have simple recursion relations

$$p_1^{n+1} = p_1^n - \frac{q_1^n \Delta t}{[(q_1^n)^2 + (q_2^n)^2]^{\frac{3}{2}}} , \tag{5.64a}$$

$$p_2^{n+1} = p_2^n - \frac{q_2^n \Delta t}{[(q_1^n)^2 + (q_2^n)^2]^{\frac{3}{2}}} , \tag{5.64b}$$

$$q_1^{n+1} = q_1^n + p_1^n \Delta t , \tag{5.64c}$$

$$q_2^{n+1} = q_2^n + p_2^n \Delta t . \tag{5.64d}$$

Implicit EULER

We obtain the implicit equations

$$p_1^{n+1} = p_1^n - \frac{q_1^{n+1} \Delta t}{[(q_1^{n+1})^2 + (q_2^{n+1})^2]^{\frac{3}{2}}} , \tag{5.65a}$$

$$p_2^{n+1} = p_2^n - \frac{q_2^{n+1} \Delta t}{[(q_1^{n+1})^2 + (q_2^{n+1})^2]^{\frac{3}{2}}} , \tag{5.65b}$$

$$q_1^{n+1} = q_1^n + p_1^{n+1} \Delta t , \tag{5.65c}$$

$$q_2^{n+1} = q_2^n + p_2^{n+1} \Delta t . \tag{5.65d}$$

These implicit equations can be solved, for instance, by the use of the NEWTON method discussed in Appendix B.

Symplectic EULER

Employing Eqs. (5.60) gives

$$p_1^{n+1} = p_1^n - \frac{q_1^n \Delta t}{[(q_1^n)^2 + (q_2^n)^2]^{\frac{3}{2}}} , \tag{5.66a}$$

$$p_2^{n+1} = p_2^n - \frac{q_2^n \Delta t}{[(q_1^n)^2 + (q_2^n)^2]^{\frac{3}{2}}} , \tag{5.66b}$$

$$q_1^{n+1} = q_1^n + p_1^{n+1}\Delta t\,, \tag{5.66c}$$

$$q_2^{n+1} = q_2^n + p_2^{n+1}\Delta t\,. \tag{5.66d}$$

These implicit equations can be solved analytically and we obtain

$$p_1^{n+1} = p_1^n - \frac{q_1^n \Delta t}{[(q_1^n)^2 + (q_2^n)^2]^{\frac{3}{2}}}\,, \tag{5.67a}$$

$$p_2^{n+1} = p_2^n - \frac{q_2^n \Delta t}{[(q_1^n)^2 + (q_2^n)^2]^{\frac{3}{2}}}\,, \tag{5.67b}$$

$$q_1^{n+1} = q_1^n + p_1^n \Delta t - \frac{q_1^n \Delta t^2}{[(q_1^n)^2 + (q_2^n)^2]^{\frac{3}{2}}}\,, \tag{5.67c}$$

$$q_2^{n+1} = q_2^n + p_2^n \Delta t - \frac{q_2^n \Delta t^2}{[(q_1^n)^2 + (q_2^n)^2]^{\frac{3}{2}}}\,. \tag{5.67d}$$

A second possibility of the symplectic EULER is given by Eq. (4.33). It reads

$$p_1^{n+1} = p_1^n - \frac{q_1^{n+1} \Delta t}{[(q_1^{n+1})^2 + (q_2^{n+1})^2]^{\frac{3}{2}}}\,, \tag{5.68a}$$

$$p_2^{n+1} = p_2^n - \frac{q_2^{n+1} \Delta t}{[(q_1^{n+1})^2 + (q_2^{n+1})^2]^{\frac{3}{2}}}\,, \tag{5.68b}$$

$$q_1^{n+1} = q_1^n + p_1^n \Delta t\,, \tag{5.68c}$$

$$q_2^{n+1} = q_2^n + p_2^n \Delta t\,. \tag{5.68d}$$

The trajectories calculated using these four methods are presented in Figs. 5.1 and 5.2, the time evolution of the total energy of the system is plotted in Fig. 5.3. The initial conditions were [16]

$$p_1(0) = 0, \qquad q_1(0) = 1 - e\,, \tag{5.69}$$

and

$$p_2(0) = \sqrt{\frac{1+e}{1-e}}, \qquad q_2(0) = 0\,, \tag{5.70}$$

with $e = 0.6$ which gives $H = -1/2$. Furthermore, we set $\Delta t = 0.01$ for the symplectic EULER methods and $\Delta t = 0.005$ for the forward and backward EULER methods in order to reduce the methodological error. The implicit equations were solved with help of the NEWTON method as discussed in Appendix B. The JACOBI

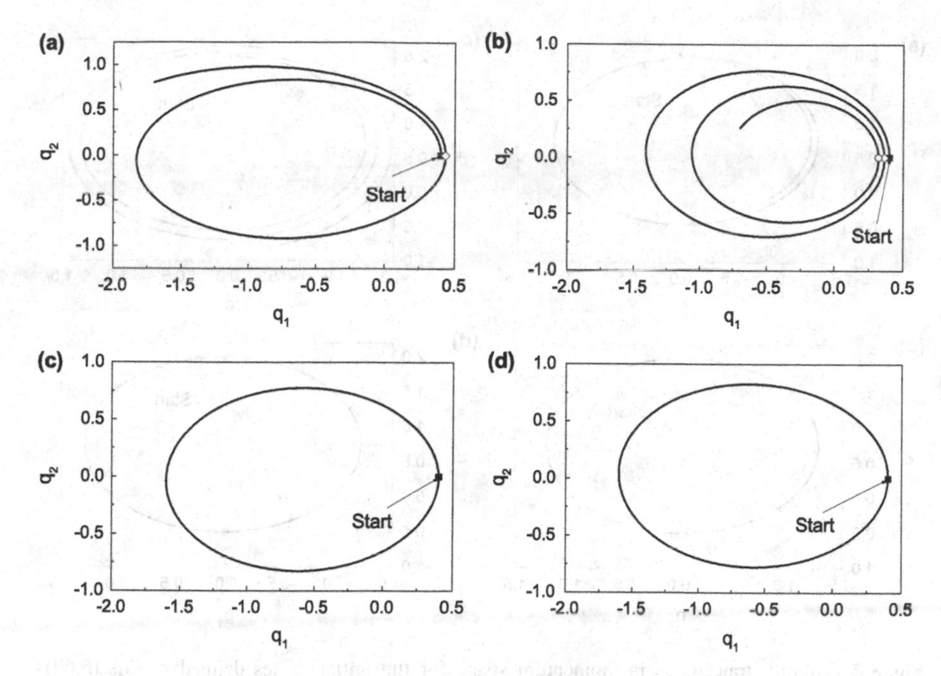

Fig. 5.1 KEPLER trajectories in position space for the initial values defined in Eqs. (5.69) and (5.70). They are indicated by a *solid square*. Solutions have been generated (**a**) by the explicit EULER method (5.64), (**b**) by the implicit EULER method (5.65), (**c**) by the symplectic EULER method (5.67), and (**d**) by the symplectic EULER method (5.68)

matrix was calculated analytically, hence no methodological error enters because approximations of derivatives were unnecessary.

According to theory [19] the q-space and p-space projections of the phase space trajectory are ellipses. Furthermore, energy and angular momentum are conserved. Thus, the numerical solutions of HAMILTON's equations of motion (5.63) should reflect these properties. Figures 5.1a, b and 5.2a, b present the results of the explicit EULER method, Eqs. (5.64), and the implicit EULER method, Eqs. (5.65), respectively. Obviously, the result does not agree with the theoretical expectation and the trajectories are open instead of closed. The reason for this behavior is the methodological error of the method which is accumulative and, thus, causes a violation of energy conservation. This violation becomes apparent in Fig. 5.3 where the total energy $H(t)$ is plotted vs time t. Neither the explicit EULER method (dashed line) nor the implicit EULER method (short dashed line) conform to the requirement of energy conservation. We also see step-like structures of $H(t)$. At the center of these steps an open diamond symbol and in the case of the implicit EULER method an additional open circle indicate the position in time of the perihelion of the point-mass (point of closest approach to the center of attraction). It is

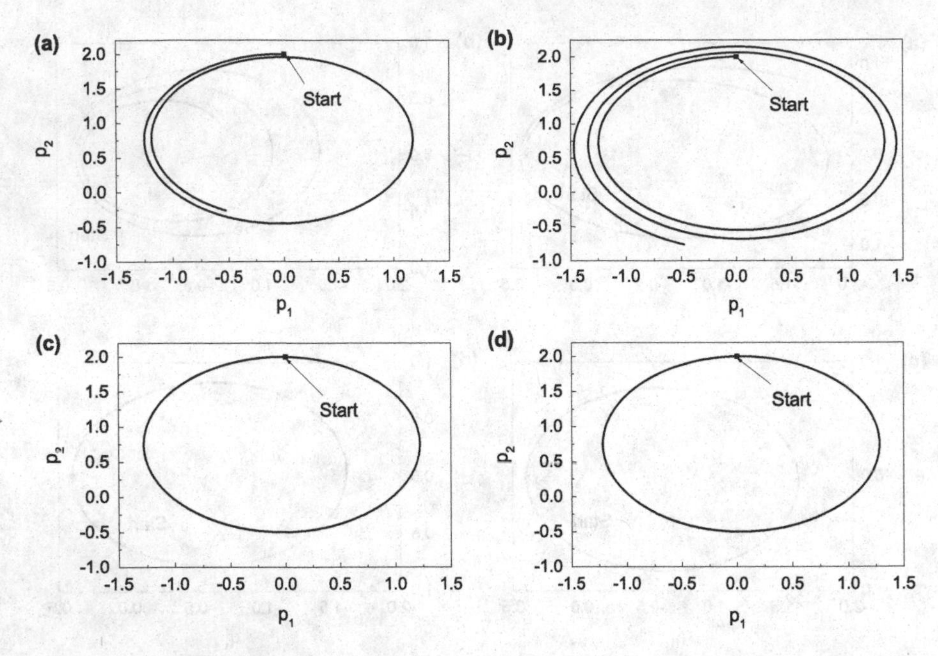

Fig. 5.2 KEPLER trajectories in momentum space for the initial values defined in Eqs. (5.69) and (5.70). They are indicated by a *solid square*. Solutions have been generated **(a)** by the explicit EULER method (5.64), **(b)** by the implicit EULER method (5.65), **(c)** by the symplectic EULER method (5.67), and **(d)** by the symplectic EULER method (5.68)

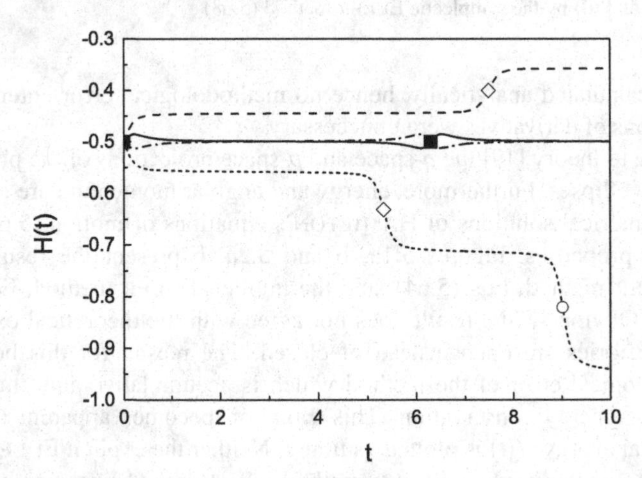

Fig. 5.3 Time evolution of the total energy H calculated with the help of the four methods discussed in the text. The initial values are given by Eqs. (5.69) and (5.70). Solutions have been generated (i) by the explicit EULER method (5.64) (*dashed line*), (ii) by the implicit EULER method (5.65) (*dotted line*), (iii) by the symplectic EULER method (5.67) (*solid line*), and (iv) by the symplectic EULER method (5.68) (*dashed-dotted line*)

indicated by the same symbols in Fig. 5.1a, b. At this point the point-mass reaches its maximum velocity, the pericenter velocity, and it covers the biggest distances along its trajectory per time interval Δt. Consequently, the methodological error is biggest in this part of the trajectory which manifests itself in those steps in $H(t)$. As the point-mass moves 'faster' when the implicit EULER method is applied, again, the distances covered per time interval are greater than those covered by the point-mass in the explicit EULER method. Thus, it is not surprising that the error of the implicit EULER method is bigger as well when $H(t)$ is determined.

These results are in strong contrast to the numerical solutions of Eqs. (5.63) obtained with the help of symplectic EULER methods which are presented in Figs. 5.1c, d and 5.2c, d. The trajectories are almost perfect ellipses for both symplectic methods which follow Eqs. (5.67) and (5.68). Moreover, the total energy $H(t)$ (solid and dashed-dotted lines in Fig. 5.3) varies very little as a function of t. Deviations from the mean value can only be observed around the perihelion which is indicated by a solid square. Moreover, these deviations compensate because of the symplectic nature of the method. This demonstrates that symplectic integrators are the appropriate technique to solve the equations of motion of Hamiltonian systems.

Summary

We concentrated on numerical methods to solve the initial value problem of ordinary differential equations. The methods discussed here rely heavily on the various methods developed for numerical integration because we can always find an integral representation of this kind of equations. The simple integrators known from Chap. 4 were augmented by the more general CRANK-NICHOLSON method which was based on the trapezoidal rule introduced in Sect. 3.3. The simple single-step methods were improved in their methodological error by TAYLOR series methods, linear multi-step methods, and by the RUNGE-KUTTA method. The latter took intermediate points within the time interval $[t_n, t_{n+1}]$ into account. In principle, it is possible to achieve almost arbitrary accuracy with such a method. Nevertheless, all those methods had the disadvantage that because of their methodological error energy conservation was violated when applied to Hamiltonian systems. As this problem can be remedied by symplectic integrators a short introduction into this topic was provided and the most important symplectic integrators were presented. The final discussion of KEPLER's two-body problem elucidated the various points discussed throughout this chapter.

Problems

1. Write a program to solve numerically the KEPLER problem. The HAMILTON function of the problem is defined as

$$H(p,q) = \frac{1}{2}\left(p_1^2 + p_2^2\right) - \frac{1}{\sqrt{q_1^2 + q_2^2}},$$

and the initial conditions are given by

$$p_1(0) = 0, \quad q_1(0) = 1 - e, \quad p_2(0) = \sqrt{\frac{1+e}{1-e}}, \quad q_2(0) = 0,$$

where $e = 0.6$. Derive HAMILTON's equations of motion and implement an algorithm which solves these equations based on the following methods

   (a) Explicit EULER,
   (b) Symplectic EULER.

2. Plot the trajectories and the total energy as a function of time. You can use the results presented in Figs. 5.1 and 5.2 to check your code. Modify the initial conditions and discuss the results! Try to confirm KEPLER's laws of planetary motion with the help of your algorithm.
3. Use a symplectic integrator to study LENNARD-JONES scattering; see Problems of Chap. 4
4. Solve the differential equation (5.6) numerically with different methods. Use also the TAYLOR series method (5.18).

References

1. Dorn, W.S., McCracken, D.D.: Numerical Methods with Fortran IV Case Studies. Wiley, New York (1972)
2. Crank, J., Nicolson, P.: A practical method for numerical evaluation of solutions of partial differential equations of the heat-conduction type. Proc. Camb. Philos. Soc. **43**, 50–67 (1947). doi:10.1017/S0305004100023197
3. Hairer, E., Wanner, G.: Solving Ordinary Differential Equations II. Springer Series in Computational Mathematics, vol. 14. Springer, Berlin/Heidelberg (1991)
4. Hairer, E., Nørsett, S.P., Wanner, G.: Solving Ordinary Differential Equations I, 2nd edn. Springer Series in Computational Mathematics, vol. 8. Springer, Berlin/Heidelberg (1993)
5. Süli, E., Mayers, D.: An Introduction to Numerical Analysis. Cambridge University Press, Cambridge (2003)
6. Collatz, L.: The Numerical Treatment of Differential Equations. Springer, Berlin/Heidelberg (1960)
7. van Winckel, G.: Numerical methods for differential equations. Lecture Notes, Karl-Franzens Universität Graz (2012)

8. Ascher, U.M., Petzold, L.R.: Computer Methods for Ordinary Differential Equations and Differential-Algebraic Equations. Society for Industrial and Applied Mathematics, Philadelphia (1998)
9. Press, W.H., Teukolsky, S.A., Vetterling, W.T., Flannery, B.P.: Numerical Recipes in C++, 2nd edn. Cambridge University Press, Cambridge (2002)
10. Arnol'd, V.I.: Mathematical Methods of Classical Mechanics, 2nd edn. Graduate Texts in Mathematics, vol. 60. Springer, Berlin/Heidelberg (1989)
11. Fetter, A.L., Walecka, J.D.: Theoretical Mechanics of Particles and Continua. Dover, New York (2004)
12. Scheck, F.: Mechanics, 5th edn. Springer, Berlin/Heidelberg (2010)
13. Goldstein, H., Poole, C., Safko, J.: Classical Mechanics, 3rd edn. Addison-Wesley, Menlo Park (2013)
14. Fließbach, T.: Mechanik, 7th edn. Lehrbuch zur Theoretischen Physik I. Springer, Berlin/Heidelberg (2015)
15. Guillemin, V., Sternberg, S.: Symplectic Techniques in Physics. Cambridge University Press, Cambridge (1990)
16. Hairer, E.: Geometrical Integration – Symplectic Integrators. Lecture Notes, TU München (2010)
17. Levi, D., Oliver, P., Thomova, Z., Winteritz, P. (eds.): Symmetries and Integrability of Difference Equations. London Mathematical Society Lecture Note Series. Cambridge University Press, Cambridge (2011)
18. Feng, K., Qin, M.: Symplectic Runge-Kutta methods. In: Feng, K., Qin, M. (eds.) Symplectic Geometric Algorithms for Hamiltonian Systems, pp. 277–364. Springer, Berlin/Heidelberg (2010)
19. Ó'Mathúna, D.: Integrable Systems in Celestial Mechanics. Progress in Mathematical Physics, vol. 51. Birkhäuser Basel, Basel (2008)

Chapter 6
The Double Pendulum

6.1 HAMILTON's Equations

We investigate the dynamics of the double pendulum in two spacial dimensions as illustrated schematically in Fig. 6.1. It is the aim of this section to derive HAMILTON's equations of motion for this system. In a first step we introduce generalized coordinates and determine the LAGRANGE function of the system from its kinetic and potential energy [1–5]. We then introduce generalized momenta and, finally, derive the HAMILTON function from which HAMILTON's equations of motion follow. They will serve as a starting point for the formulation of a numerical method.

From Fig. 6.1 we find the coordinates of the two point masses m:

$$x_1 = \ell \sin(\varphi_1) , \qquad z_1 = 2\ell - \ell \cos(\varphi_1) , \tag{6.1}$$

and

$$x_2 = \ell \left[\sin(\varphi_1) + \sin(\varphi_2)\right] , \qquad z_2 = 2\ell - \ell \left[\cos(\varphi_1) + \cos(\varphi_2)\right] . \tag{6.2}$$

Here, 2ℓ is the pendulum's total length. The angles φ_i, $i = 1, 2$ are defined in Fig. 6.1.

We note that $\ell = \text{const}$ and obtain the time derivatives of the coordinates (6.1) and (6.2):

$$\dot{x}_1 = \ell \dot{\varphi}_1 \cos(\varphi_1) , \tag{6.3}$$

$$\dot{z}_1 = \ell \dot{\varphi}_1 \sin(\varphi_1) , \tag{6.4}$$

$$\dot{x}_2 = \ell \left[\dot{\varphi}_1 \cos(\varphi_1) + \dot{\varphi}_2 \cos(\varphi_2)\right] , \tag{6.5}$$

$$\dot{z}_2 = \ell \left[\dot{\varphi}_1 \sin(\varphi_1) + \dot{\varphi}_2 \sin(\varphi_2)\right] . \tag{6.6}$$

B.A. Stickler, E. Schachinger, *Basic Concepts in Computational Physics*,
DOI 10.1007/978-3-319-27265-8_6

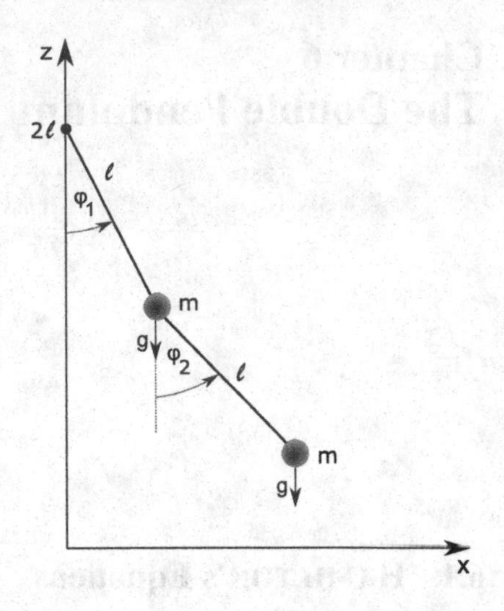

Fig. 6.1 Schematic illustration of the double pendulum. m are the point-masses, 2ℓ is the total length of the pendulum and φ_1, φ_2 are the corresponding angles

The LAGRANGE function of the system is defined by

$$L = T - U\,, \tag{6.7}$$

with the kinetic energy T and the potential U. The kinetic energy T is given by[1]

$$\begin{aligned} T &= \frac{m}{2}\left(\dot{x}_1^2 + \dot{z}_1^2 + \dot{x}_2^2 + \dot{z}_2^2\right) \\ &= \frac{m\ell^2}{2}\left[2\dot{\varphi}_1^2 + \dot{\varphi}_2^2 + 2\dot{\varphi}_1\dot{\varphi}_2\cos(\varphi_1 - \varphi_2)\right] . \end{aligned} \tag{6.8}$$

The potential energy U is determined by the gravitational force

$$\begin{aligned} U &= mgz_1 + mgz_2 \\ &= mg\ell\left[4 - 2\cos(\varphi_1) - \cos(\varphi_2)\right] , \end{aligned} \tag{6.9}$$

where g is the acceleration due to gravity. Hence, we get for the LAGRANGE function L:

$$L = \frac{m\ell^2}{2}\left[2\dot{\varphi}_1^2 + \dot{\varphi}_2^2 + 2\dot{\varphi}_1\dot{\varphi}_2\cos(\varphi_1 - \varphi_2)\right] - mg\ell\left[4 - 2\cos(\varphi_1) - \cos(\varphi_2)\right] . \tag{6.10}$$

---

[1] We make use of the relation:

$$\sin(x)\sin(y) + \cos(x)\cos(y) = \cos(x - y)\,.$$

We find a description of the motion in phase space by calculating the generalized momenta p_i, $i = 1, 2$ as

$$p_1 = \frac{\partial}{\partial \dot{\varphi}_1} L = m\ell^2 \left[2\dot{\varphi}_1 + \dot{\varphi}_2 \cos(\varphi_1 - \varphi_2)\right] , \tag{6.11}$$

and

$$p_2 = \frac{\partial}{\partial \dot{\varphi}_2} L = m\ell^2 \left[\dot{\varphi}_2 + \dot{\varphi}_1 \cos(\varphi_1 - \varphi_2)\right] . \tag{6.12}$$

The aim is now to express the kinetic energy (6.8) in terms of generalized momenta p_1 and p_2. To accomplish this we solve in a first step Eq. (6.12) for $\dot{\varphi}_2$ and obtain

$$\dot{\varphi}_2 = \frac{p_2}{m\ell^2} - \dot{\varphi}_1 \cos(\varphi_1 - \varphi_2) . \tag{6.13}$$

This is used to rewrite Eq. (6.11). Solving for $\dot{\varphi}_1$ gives:

$$\dot{\varphi}_1 = \left[2 - \cos^2(\varphi_1 - \varphi_2)\right]^{-1} \left[\frac{p_1}{m\ell^2} - \frac{p_2}{m\ell^2} \cos(\varphi_1 - \varphi_2)\right] . \tag{6.14}$$

The trigonometric identity $\cos^2(x) + \sin^2(x) = 1$ changes Eq. (6.14) into

$$\dot{\varphi}_1 = \frac{1}{m\ell^2} \frac{p_1 - p_2 \cos(\varphi_1 - \varphi_2)}{1 + \sin^2(\varphi_1 - \varphi_2)} . \tag{6.15}$$

This is then used to transform Eq. (6.13) into

$$\begin{aligned} \dot{\varphi}_2 &= \frac{1}{m\ell^2} \left[p_2 - \frac{p_1 \cos(\varphi_1 - \varphi_2) - p_2 \cos^2(\varphi_1 - \varphi_2)}{1 + \sin^2(\varphi_1 - \varphi_2)}\right] \\ &= \frac{1}{m\ell^2} \frac{2p_2 - p_1 \cos(\varphi_1 - \varphi_2)}{1 + \sin^2(\varphi_1 - \varphi_2)} . \end{aligned} \tag{6.16}$$

Hence, with help of Eqs. (6.15) and (6.16) we can reevaluate the kinetic energy (6.8) to give

$$\begin{aligned} T &= \frac{m\ell^2}{2} \left[2\dot{\varphi}_1^2 + \dot{\varphi}_2^2 + 2\dot{\varphi}_1 \dot{\varphi}_2 \cos(\varphi_1 - \varphi_2)\right] \\ &= \frac{1}{2m\ell^2} \frac{p_1^2 + 2p_2^2 - 2p_1 p_2 \cos(\varphi_1 - \varphi_2)}{1 + \sin^2(\varphi_1 - \varphi_2)} . \end{aligned} \tag{6.17}$$

The HAMILTON function $H(p_1, p_2, \varphi_1, \varphi_2)$ is the sum of the kinetic energy (6.17) and the potential energy (6.9) and we get:

$$\begin{aligned} H &= T + U \\ &= \frac{1}{2m\ell^2} \frac{p_1^2 + 2p_2^2 - 2p_1 p_2 \cos(\varphi_1 - \varphi_2)}{1 + \sin^2(\varphi_1 - \varphi_2)} \\ &\quad + mg\ell \left[4 - 2\cos(\varphi_1) - \cos(\varphi_2)\right] . \end{aligned} \tag{6.18}$$

Thus, we are now, finally, in a position to formulate HAMILTON's equations of motion from

$$\dot{\varphi}_i = \frac{\partial}{\partial p_i} H , \quad \dot{p}_i = -\frac{\partial}{\partial \varphi_i} H , \quad i = 1, 2, \tag{6.19}$$

and the dynamics of the double pendulum are determined by the solutions of the following set of differential equations:

$$\dot{\varphi}_1 = \frac{1}{m\ell^2} \frac{p_1 - p_2 \cos(\varphi_1 - \varphi_2)}{1 + \sin^2(\varphi_1 - \varphi_2)} , \tag{6.20a}$$

$$\dot{\varphi}_2 = \frac{1}{m\ell^2} \frac{2p_2 - p_1 \cos(\varphi_1 - \varphi_2)}{1 + \sin^2(\varphi_1 - \varphi_2)} , \tag{6.20b}$$

$$\begin{aligned} \dot{p}_1 = \frac{1}{m\ell^2} \frac{1}{1 + \sin^2(\varphi_1 - \varphi_2)} \Bigg[&-p_1 p_2 \sin(\varphi_1 - \varphi_2) \\ &+ \frac{p_1^2 + 2p_2^2 - 2p_1 p_2 \cos(\varphi_1 - \varphi_2)}{1 + \sin^2(\varphi_1 - \varphi_2)} \cos(\varphi_1 - \varphi_2) \sin(\varphi_1 - \varphi_2) \Bigg] \\ &- 2mg\ell \sin(\varphi_1) , \end{aligned} \tag{6.20c}$$

and

$$\begin{aligned} \dot{p}_2 = \frac{1}{m\ell^2} \frac{1}{1 + \sin^2(\varphi_1 - \varphi_2)} \Bigg[&p_1 p_2 \sin(\varphi_1 - \varphi_2) \\ &- \frac{p_1^2 + 2p_2^2 - 2p_1 p_2 \cos(\varphi_1 - \varphi_2)}{1 + \sin^2(\varphi_1 - \varphi_2)} \sin(\varphi_1 - \varphi_2) \cos(\varphi_1 - \varphi_2) \Bigg] \\ &- mg\ell \sin(\varphi_2) . \end{aligned} \tag{6.20d}$$

The following section is dedicated to the numerical solution of Eqs. (6.20) with the help of the explicit RUNGE-KUTTA algorithm *e-RK-4* introduced in Sect. 5.3.

6.2 Numerical Solution

In a first step we recognize that Eqs. (6.20) are of the form

$$\dot{y} = F(y) , \tag{6.21}$$

where $y \in \mathbb{R}^4$. Let us define

$$y = \begin{pmatrix} y_1 \\ y_2 \\ y_3 \\ y_4 \end{pmatrix} \equiv \begin{pmatrix} \varphi_1 \\ \varphi_2 \\ p_1 \\ p_2 \end{pmatrix} , \tag{6.22}$$

and consequently

$$\begin{pmatrix} \dot{\varphi}_1 \\ \dot{\varphi}_2 \\ \dot{p}_1 \\ \dot{p}_2 \end{pmatrix} = F(y) \equiv \begin{pmatrix} f_1(y) \\ f_2(y) \\ f_3(y) \\ f_4(y) \end{pmatrix} . \tag{6.23}$$

We introduce time instances $t_n = n\Delta t$, $n \in \mathbb{N}$ and use the notation $y_n \equiv y(t_n) = (y_1^n, y_2^n, y_3^n, y_4^n)^T$. Furthermore, $F(y)$ is not an explicit function of time t and we reformulate the *e-RK-4* algorithm of Eq. (5.39) as:

$$\begin{aligned} Y_1 &= y_n , \\ Y_2 &= y_n + \frac{\Delta t}{2} F(Y_1) , \\ Y_3 &= y_n + \frac{\Delta t}{2} F(Y_2) , \\ Y_4 &= y_n + \Delta t F(Y_3) , \\ y_{n+1} &= y_n + \frac{\Delta t}{6} \left[F(Y_1) + 2F(Y_2) + 2F(Y_3) + F(Y_4) \right] . \end{aligned} \tag{6.24}$$

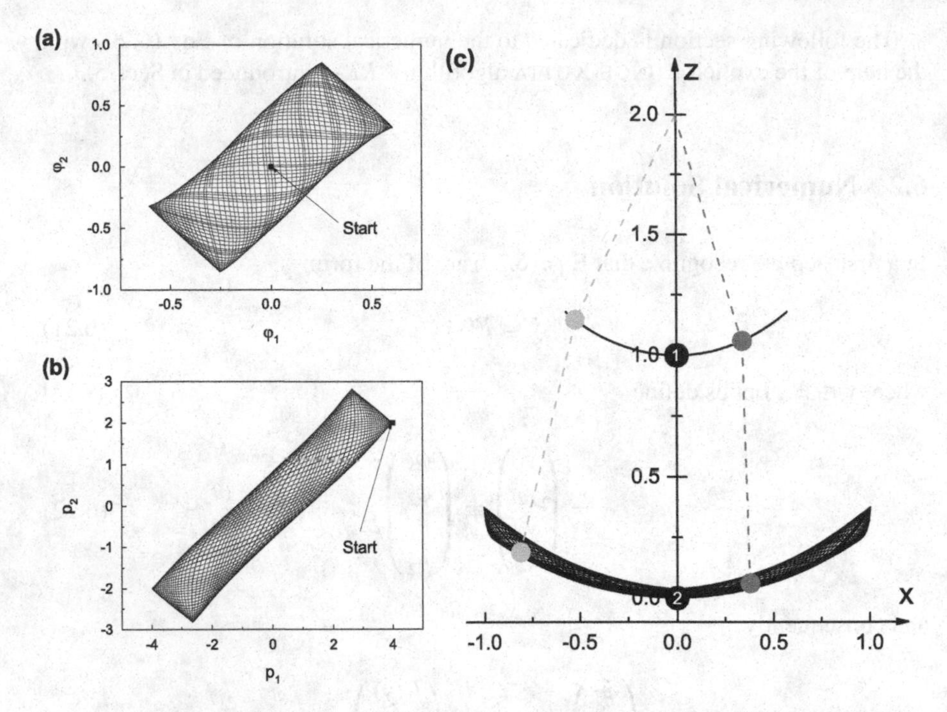

Fig. 6.2 Numerical solution of the double pendulum with initial conditions $\varphi_1(0) = \varphi_2(0) = 0.0$, $p_1(0) = 4.0$ and $p_2(0) = 2.0$. **(a)** Trajectory in φ-space, **(b)** trajectory in p-space, and **(c)** trajectory in local (x, z)-space. The *solid circles* numbered 1 and 2 represent the two masses in their initial configuration

Hence, the only remaining challenge is to correctly implement the function $F(y) = [f_1(y), f_2(y), f_3(y), f_4(y)]^T$ according to Eqs. (6.20).

The following graphs discuss the dynamics (trajectories in φ- and p-space, as well as in configuration space) of the pendulum and for this purpose we defined the parameters $m = \ell = 1$ and $g = 9.8067$. The time step was chosen to be $\Delta t = 0.001$ and we calculated $N = 60{,}000$ time steps.

We start with Fig. 6.2. The two masses numbered 1 and 2 are initially in the equilibrium position (solid circles). Both masses are pushed to the right but the push on mass 1 $[p_1(0) = 4.0]$ is much stronger than the one mass 2 experiences $[p_2(0) = 2.0]$. Thus, mass 2 is 'dragged' along in the process. This is made transparent by two 'snapshots' indicated by solid light gray circles and solid gray circles. The motion of the whole system is quite regular.

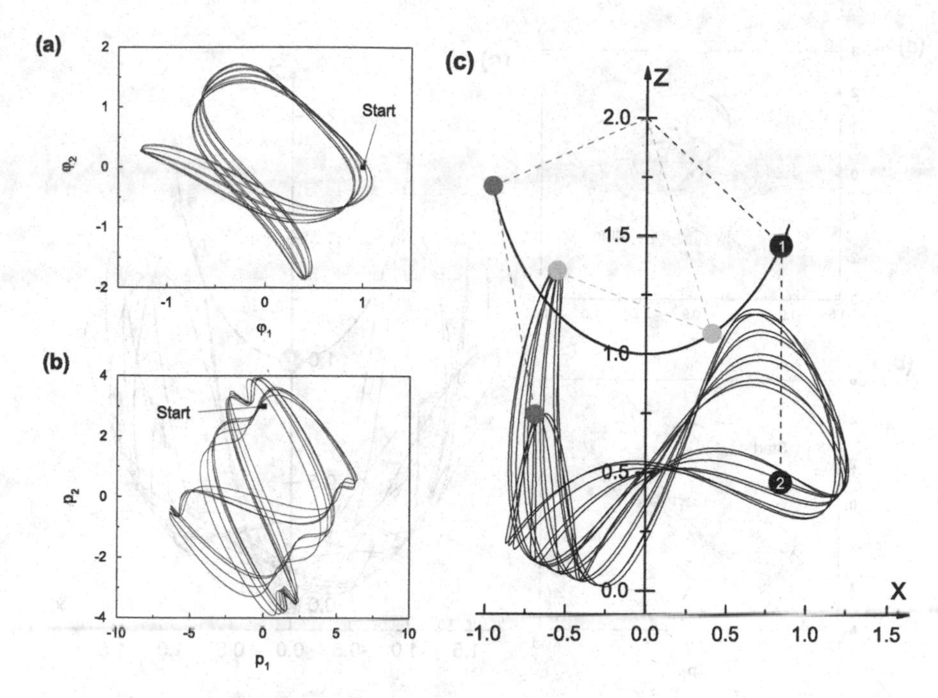

Fig. 6.3 Numerical solution of the double pendulum with initial conditions $\varphi_1(0) = 1.0, \varphi_2(0) = 0.0, p_1(0) = 0.0$ and $p_2(0) = 3.0$. **(a)** Trajectory in φ-space, **(b)** trajectory in p-space, and **(c)** trajectory in local (x, z)-space. The *solid circles* numbered 1 and 2 represent the two masses in their initial configuration

We proceed with Fig. 6.3. In this case mass 1 is displaced from its position by the initial angular displacement $\varphi_1 = 1.0$. This initial configuration is indicated by the solid circles numbered 1 and 2 representing the two point-masses. Mass 2 is then pushed to the right with $p_2(0) = 3.0$. Again, mass 1 remains on a trajectory centered around the point (0,2) in configuration space. But in contrast to the previous situation it follows now mass 2. Mass 2, on the other hand, develops a very lively trajectory, Fig. 6.3c. Two snapshots indicated by solid light gray circles and solid gray circles illustrate configurations of particular interest.

The dynamics depicted in Fig. 6.4 is quite similar to the one already discussed in Fig. 6.2. Initially both masses are in the equilibrium position and then mass 2 is pushed to the right $[p_2(0) = 4.0]$. Thus, mass 1 is trailing behind. In contrast to the previous Fig. 6.3 the trajectory of mass 2 will now be symmetric around the z-axis given enough time. Again, snapshots indicated by solid light gray circles and solid gray circles indicate interesting configurations.

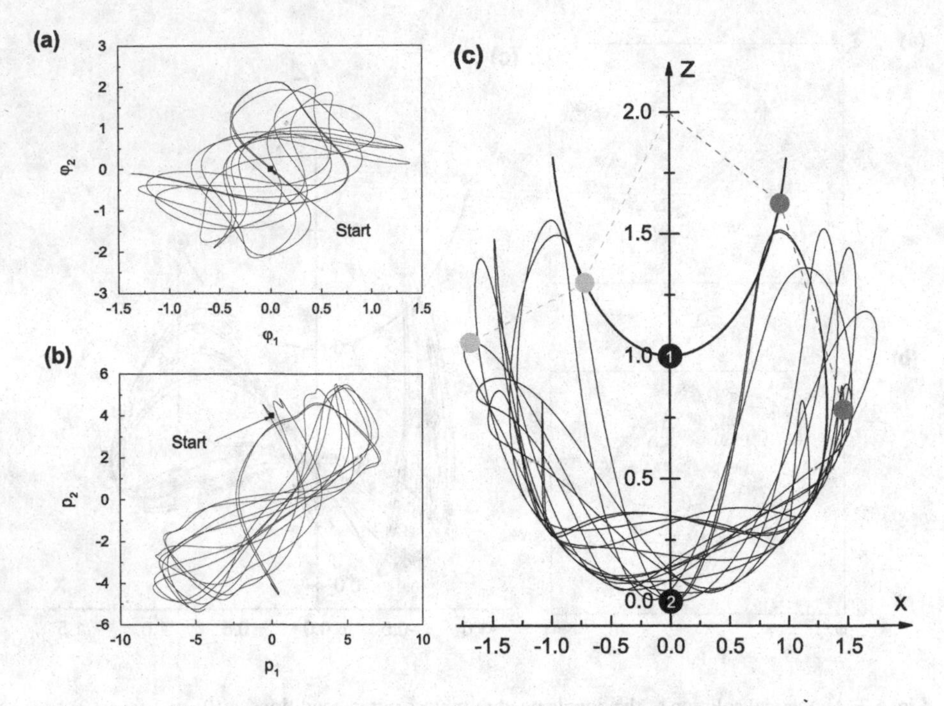

Fig. 6.4 Numerical solution of the double pendulum with initial conditions $\varphi_1(0) = \varphi_2(0) = 0.0$, $p_1(0) = 0.0$ and $p_2(0) = 4.0$. **(a)** Trajectory in φ-space, **(b)** trajectory in p-space, and **(c)** trajectory in local (x, z)-space. The *solid circles* numbered 1 and 2 represent the two masses in their initial configuration

The initial condition which resulted in the trajectory shown in Fig. 6.5 differs only for mass 2 from the initial conditions which lead to the trajectory in Fig. 6.4. Mass 2 is now pushed even more strongly to the right [$p_2(0) = 5.0$]. Of course, mass 1 is again dragging behind mass 2. In contrast to Fig. 6.4 the initial momentum of mass 2 is now sufficient to allow mass 2 to pass through the center of the inner mass' circular trajectory. Snapshots indicated by light gray solid circles and solid gray circles emphasize interesting configurations.

The situation shown in Fig. 6.6 differs from the one of Fig. 6.5 only by the initial condition for mass 2. It is now pushed even more strongly to the right [$p_2(0) = 6.5$] and this initial momentum is sufficient to cause mass 1 to rotate around the point (0,2). Nevertheless, mass 1 is permanently dragging behind mass 2. Two interesting configurations are depicted by snapshots (solid light gray circles and solid gray circles).

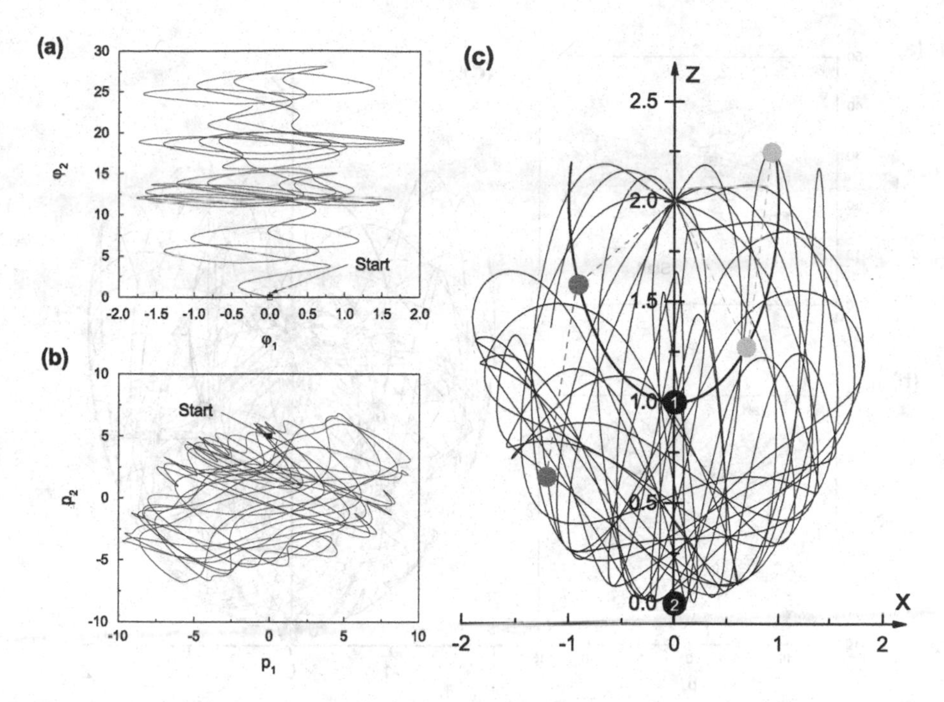

Fig. 6.5 Numerical solution of the double pendulum with initial conditions $\varphi_1(0) = \varphi_2(0) = 0.0$, $p_1(0) = 0.0$ and $p_2(0) = 5.0$. **(a)** Trajectory in φ-space, **(b)** trajectory in p-space, and **(c)** trajectory in local (x, z)-space. The *solid circles* numbered 1 and 2 represent the two masses in their initial configuration (The angles $\varphi_2 > \pi$ correspond to complete rotations of the pendulum)

A comparison between trajectories as a result of different initial conditions reveals that the physical system is highly sensitive to the choice of the initial conditions $y_0 = [\varphi_1(0), \varphi_2(0), p_1(0), p_2(0)]^T$. For instance, consider Figs. 6.4, 6.5, and 6.6. In all three cases we chose y_0 in such a way that the initial angles $\varphi_1(0) = \varphi_2(0) = 0$ and the generalized momentum coordinate $p_1(0) = 0$. The only difference is that we used different values for the initial value of the second momentum coordinate p_2. However, the resulting dynamics of φ_1 vs. φ_2 as well as p_1 vs. p_2 are entirely different and so are the local (x, z)-space trajectories. Hence, the system is *chaotic*. In the following section we will briefly discuss a method designed to characterize chaotic behavior of physical systems [6–10].

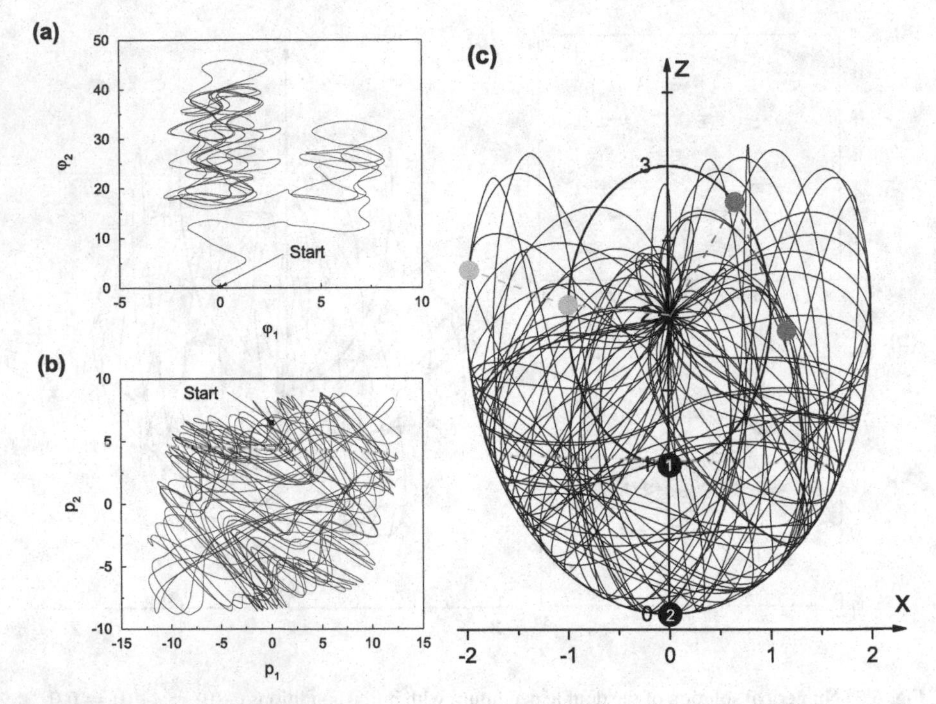

Fig. 6.6 Numerical solution of the double pendulum with initial conditions $\varphi_1(0) = \varphi_2(0) = 0.0$, $p_1(0) = 0.0$ and $p_2(0) = 6.5$. **(a)** Trajectory in φ-space, **(b)** trajectory in p-space, and **(c)** trajectory in configuration space. The *solid circles* numbered 1 and 2 represent the two masses in their initial configuration (The angles $\varphi_2 > \pi$ correspond to complete rotations of the pendulum)

6.3 Numerical Analysis of Chaos

It is the aim of this section to analyze in more detail the chaotic behavior observed in the dynamics of the double pendulum. This requires the introduction of some basic notations. We consider a physical system with f degrees of freedom where $q_1(t), \ldots, q_f(t)$ denote the generalized coordinates and $p_1(t), \ldots, p_f(t)$ denote the corresponding generalized momenta. Together, both fully characterize the state of the system at time t. Consequently, the f-dimensional vector $q(t) = [q_1(t), q_2(t), \ldots, q_f(t)]^T$ describes a point in *configuration space* of the physical system. In case of a pendulum consisting of f point-masses connected in a similar fashion as the double pendulum discussed above, which corresponds to the particular case $f = 2$, the configuration space is constrained to values $\varphi_i \in (-\pi, \pi]$, $i = 1, \ldots, f$. This resembles an f-dimensional torus.

The $2f$-dimensional vector $x(t) = [q_1(t), \ldots, q_f(t), p_1(t), \ldots, p_f(t)]^T$ describes a point in the *phase space* of the physical system at some particular time t. The time evolution of a physical system is represented by its *phase space trajectory*. Of course, the phase space trajectories $x(t)$ are differentiable with respect to t.[2]

We define an *autonomous system* as a system which is *time-invariant*, i.e. the HAMILTON function $H(x, t)$ does not depend explicitly on time t, $H(x, t) \equiv H(x)$. Hence, a physical system is referred to as autonomous if the HAMILTON function $H(x, t)$ of the system obeys

$$\frac{\partial}{\partial t} H(x, t) = 0 \, . \tag{6.25}$$

Thus, the total energy is conserved.

An autonomous system is referred to as *integrable* if it has f independent invariants $I_1, \ldots, I_f$

$$I_j(x) = I_j = \text{const}, \quad j = 1, \ldots, f \, . \tag{6.26}$$

One of these is the energy. Each particular invariant I_j reduces the dimension of the manifold on which the phase space trajectories can propagate. Hence, an integrable system propagates on an f-dimensional subspace of the $2f$-dimensional phase space. We note that a one-dimensional autonomous system is integrable since the conservation of energy delivers the required invariant.

On the other hand, non-integrable systems can show chaotic behavior. In this case the trajectories develop a strong dependence on the initial conditions which makes an analytic calculation of the dynamics extremely difficult. However, since the trajectories can be computed without problems by numeric means, we discuss now how to characterize chaotic behavior on the computer.

For this sake we investigate the dynamics of an autonomous Hamiltonian system starting with one of two initial conditions, namely x_0 and x_0'. Then the system arrives at time t at the phase space points $x(t) = \varphi_t(x_0)$ and $x'(t) = \varphi_t(x_0')$, respectively, as a solution of HAMILTON's equations of motion. Here $\varphi_t(x_0)$ denotes the flow of the system as defined in Sect. 5.4. Since the trajectories in a chaotic system strongly depend on the initial conditions x_0 and x_0' we introduce the separation between the two trajectories $\varphi_t(x_0)$ and $\varphi_t(x_0')$ at time t as $d(t) = |\varphi_t(x_0) - \varphi_t(x_0')|$ where $|\cdot|$ denotes some suitable norm. This length can now, for instance, be used to characterize the *stability* of the trajectory $\varphi_t(x_0)$ [11]. In particular, a solution $\varphi_t(x_0)$ is referred to as stable if

$$\forall \epsilon > 0 \; \exists \delta(\epsilon) > 0 : \forall x_0' : d(0) < \delta \Rightarrow d(t) < \epsilon, \quad \forall t > 0 \, . \tag{6.27}$$

In words: We speak of a stable solution if the trajectory $\varphi_t(x_0')$ which corresponds to the perturbed initial condition x_0' stays within a tube of radius ϵ around the

[2] The symplectic mapping $\varphi_t : x_0 \mapsto x(t)$ from the initial conditions x_0 to the phase space point $x(t)$ at time t is referred to as *Hamiltonian flow* of the system. This was discussed in Sect. 5.4.

unperturbed trajectory $\varphi_t(x_0)$ *for all* $t > 0$. Alternatively, a solution is referred to as asymptotically stable if the distance to adjacent trajectories tends to zero, i.e. $d(t) \to 0$ as $t \to \infty$. Such solutions tend to attract trajectories from their neighborhood and, hence, they are referred to as *attractors*. Finally, a *periodic orbit* is defined as a trajectory for which one can find a time τ such that:

$$\varphi_\tau(x) = x\,, \quad \forall x. \tag{6.28}$$

To find an easy answer to the question whether or not a particular solution of a non-integrable system is stable, the clear, topological method of POINCARÉ maps was introduced. The idea was to reduce the investigation of the complete $2f$-dimensional phase space trajectory $x(t) = \varphi_t(x_0)$ to the investigation of its *intersection* points through a plane Σ which is transverse to the flow of the system. This plane is a subspace of dimension $2f - 1$ and is commonly referred to as POINCARÉ section [3]. The transversality of the POINCARÉ section Σ means that periodic flows intersect this section and never flow parallel to or within it.

Consider a trajectory which is bound to a finite domain, i.e. it does not tend to infinity in any phase space coordinate. In this case it is possible to define the POINCARÉ section in such a way that the trajectory will intersect this section not only once but several times. Thus, a POINCARÉ map is then the mapping of one intersection point P onto the next intersection point P'.

Let us substantiate this idea: we consider the initial condition x_0 for which a trajectory Γ is periodic. We choose the initial time $t = 0$ in such a way that $x_0 \in \Sigma$, where Σ is the POINCARÉ section, Fig. 6.7. We suppose that after a time $\tau(x_0)$ the trajectory intersects this POINCARÉ section again.[3] Since we demanded that the trajectory which started in x_0 is periodic, we deduce that it intersects the POINCARÉ section again at some point $\varphi_{\tau(x_0)}(x_0) = x_0$. We consider now a slightly perturbed initial condition $x' \in U_0(x_0)$, where $U_0(x_0)$ is referred to as the *neighborhood* of x_0. In this case the trajectory will in general not be periodic, and the next intersection point $\varphi_{\tau(x')}(x') \neq x'$. The mapping from one intersection point x' onto the next intersection point $\varphi_{\tau(x')}(x')$ is called the POINCARÉ map $P(x') = \varphi_{\tau(x')}(x')$. We note that the particular point x_0 is a fixed point of this mapping, $P(x_0) = x_0$. Furthermore, we note that if $x' \in U_0(x_0)$ we will have $P(x') \in U_1(x_0)$, where $U_1(x_0)$ is the neighborhood of first return. This is indicated schematically in Fig. 6.7.

[3]Note that we denoted $\tau \equiv \tau(x_0)$ in order to emphasize that the recurrence time τ will depend on the initial condition x_0.

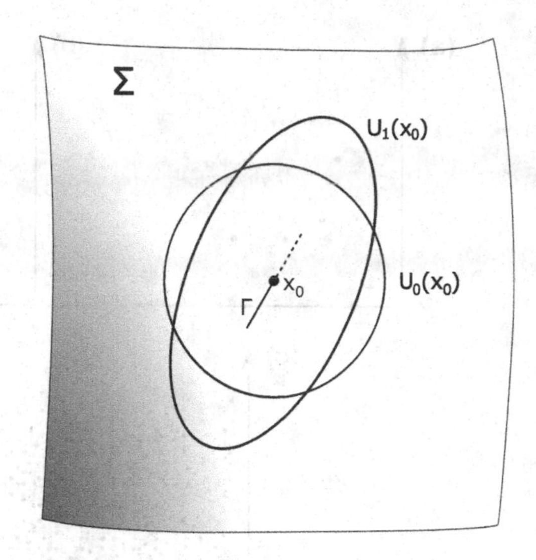

Fig. 6.7 Schematic illustration of the neighborhood $U_0(x_0)$ and the neighborhood of first return $U_1(x_0)$ of a periodic trajectory Γ. The intersection point x_0 of Γ with Σ is a fixed point of this mapping

We utilize now these concepts and analyze the dynamics of the double pendulum. We have four generalized coordinates which, with the help of conservation of energy, are constrained to a three-dimensional manifold within the four-dimensional phase space. Since the investigation of these three-dimensional trajectories is very complex we consider a two-dimensional POINCARÉ section. For instance, the coordinates $[\varphi_1(t), p_1(t)]^T$ can be 'measured' whenever $\varphi_2(t) = 0$ and $p_2 > 0$. Thus, the system's state is registered whenever mass 2 crosses the vertical plane from the left-hand side.

We discuss now some of the most typical scenarios for POINCARÉ plots. (Such a plot represents the POINCARÉ section together with all intersection points of a particular trajectory.) Note that this discussion is, of course, not restricted to the case of the double pendulum. Two different scenarios can be distinguished for integrable systems: (i) the set of intersection points $(\eta_1, \eta_2, \ldots, \eta_N)$ is finite. (ii) In the more general case, the dimension N of the set of intersection points is infinite. In both cases the intersection points form one-dimensional lines which do not have to be connected. Figure 6.8a, b discuss this schematically. However, if the system is non-integrable, a third scenario is possible: chaotic behavior. In this case the intersection points appear to be randomly distributed on the two-dimensional POINCARÉ section and one observes space-filling behavior. This is illustrated schematically in Fig. 6.8c. Whether one observes chaotic behavior or not depends on the choice of the initial conditions.

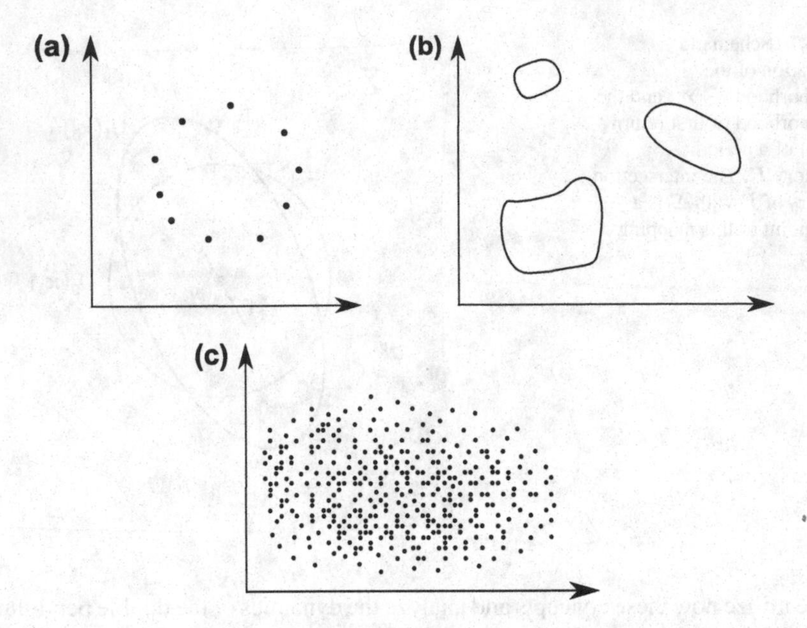

Fig. 6.8 Schematic illustration of the three types of POINCARÉ plots as discussed in the text. **(a)** Finite number of intersection points, **(b)** infinite number of intersection points which, however, form closed lines, **(c)** space-filling and, consequently, chaotic behavior

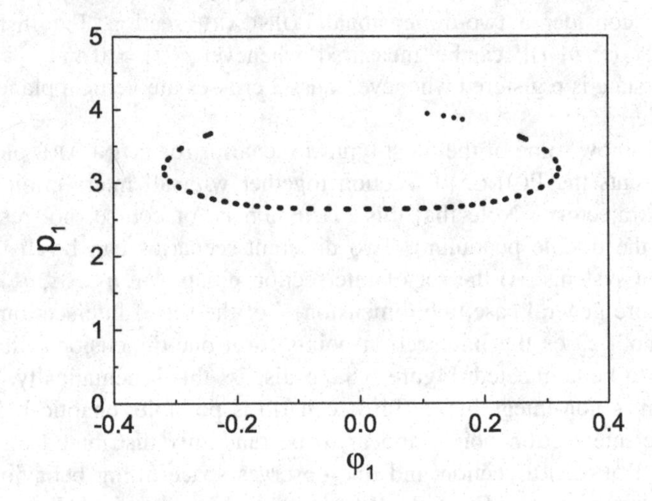

Fig. 6.9 POINCARÉ plot of the double pendulum with initial conditions $\varphi_1(0) = \varphi_2(0) = 0.0$, $p_1(0) = 4.0$ and $p_2(0) = 2.0$. It corresponds to the situation discussed in Fig. 6.2

In Figs. 6.9, 6.10, and 6.11 we present POINCARÉ plots of the double pendulum. The graphs were obtained with help of the method discussed above, i.e. $\varphi_2 = 0$ and $p_2 > 0$. Again, we set $m = \ell = 1$ and $g = 9.8067$. The time step was chosen to be $\Delta t = 0.001$ and we calculated $N = 36 \times 10^4$ time steps. In Figs. 6.9 and 6.10

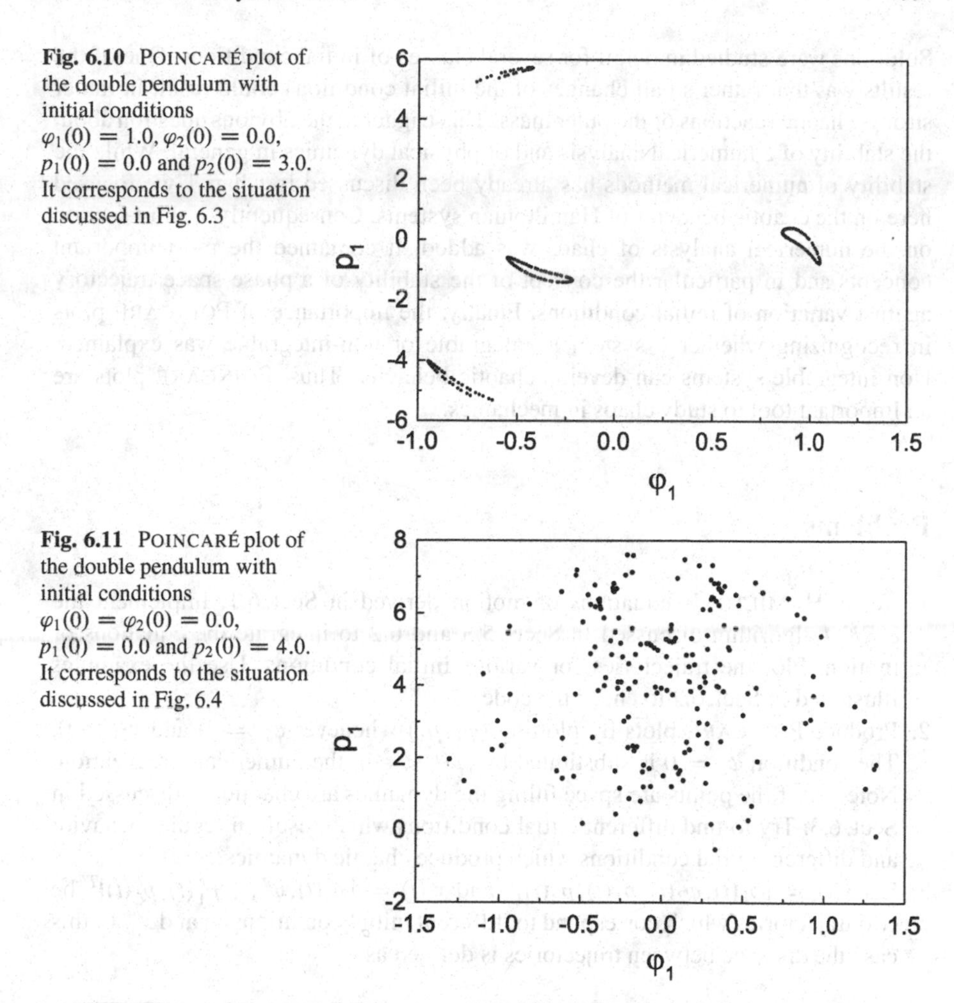

Fig. 6.10 POINCARÉ plot of the double pendulum with initial conditions $\varphi_1(0) = 1.0, \varphi_2(0) = 0.0$, $p_1(0) = 0.0$ and $p_2(0) = 3.0$. It corresponds to the situation discussed in Fig. 6.3

Fig. 6.11 POINCARÉ plot of the double pendulum with initial conditions $\varphi_1(0) = \varphi_2(0) = 0.0$, $p_1(0) = 0.0$ and $p_2(0) = 4.0$. It corresponds to the situation discussed in Fig. 6.4

we observe regular behavior as it was illustrated in Fig. 6.8b. In Fig. 6.11 the points are space filling and, consequently, chaotic behavior is observed in this particular case. Keeping in mind that this particular POINCARÉ plot refers to the initial value problem of Fig. 6.4 we conclude that all problems of this series, i.e. Figs. 6.4, 6.5, and 6.6, are non-integrable and chaotic.

Summary

The dynamics of the double pendulum is described by a system of four ordinary first order differential equations. It is a typical initial value problem and, thus, the methods introduced in Chap. 5 are all candidates to find a numerical solution. Here we concentrated on the explicit RUNGE-KUTTA algorithm *e-RK-4* of Sect. 5.3.

Solutions were studied in detail for several classes of initial conditions. One of the results was that rather small changes of the initial conditions could result in rather strong, chaotic reactions of the outer mass. This triggered the obvious question about the stability of a numerical analysis and of physical dynamics in general. While the stability of numerical methods has already been discussed in Chap. 1 we focused here on the chaotic behavior of Hamiltonian systems. Consequently, a short section on the numerical analysis of chaos was added. It contained the most important concepts and in particular the concept of the stability of a phase space trajectory against variation of initial conditions. Finally, the importance of POINCARÉ plots in recognizing whether a system is integrable or non-integrable was explained. Non-integrable systems can develop chaotic behavior. Thus, POINCARÉ plots are an important tool to study chaos in mechanics.

Problems

1. Verify HAMILTON's equations of motion derived in Sect. 6.1. Implement the *e-RK-4* algorithm discussed in Sects. 5.3 and 6.2 to integrate the equations of motion. Plot the trajectories for various initial conditions. Use the examples illustrated in Sect. 6.2 to check the code.
2. Produce POINCARÉ plots by plotting (φ_1, p_1) whenever $\varphi_2 = 0$ and $p_2 > 0$. The condition $\varphi_2 = 0$ is substituted by $|\varphi_2| < \epsilon$ in the numerically realization. Note that if the points are space filling the dynamics are chaotic, as discussed in Sect. 6.3. Try to find different initial conditions which result in regular behavior and different initial conditions which produce chaotic dynamics.
3. Let $x(t) = [\varphi_1(t), \varphi_2(t), p_1(t), p_2(t)]^T$ and $x'(t) = [\varphi_1'(t), \varphi_2'(t), p_1'(t), p_2'(t)]^T$ be two trajectories which correspond to different initial conditions x_0 and x_0'. In this case the distance between trajectories is defined as

$$d(t) = \sqrt{[\varphi_1(t) - \varphi_1'(t)]^2 + [\varphi_2(t) - \varphi_2'(t)]^2 + [p_1(t) - p_1'(t)]^2 + [p_2(t) - p_2'(t)]^2}\,.$$

   Plot the distance $d(t)$ as a function of time t for two different initial conditions.
4. Extend the code of the double pendulum of equal mass and equal length to cover the case when the lengths and masses of the individual pendula are different, $\ell_1 \neq \ell_2$ and $m_1 \neq m_2$. What happens? For instance, one can choose a certain initial condition and keep ℓ_1 and ℓ_2 fixed. What is the influence of ℓ_1 and ℓ_2 on the dynamics?
5. Show that the dynamics become integrable in the absence of a gravitational force, i.e. $g = 0$. What are the conserved quantities? How do the POINCARÉ plots look like? Again, try different initial conditions.

References

1. Arnol'd, V.I.: Mathematical Methods of Classical Mechanics, 2nd edn. Graduate Texts in Mathematics, vol. 60. Springer, Berlin/Heidelberg (1989)
2. Fetter, A.L., Walecka, J.D.: Theoretical Mechanics of Particles and Continua. Dover, New York (2004)
3. Scheck, F.: Mechanics, 5th edn. Springer, Berlin/Heidelberg (2010)
4. Goldstein, H., Poole, C., Safko, J.: Classical Mechanics, 3rd edn. Addison-Wesley, Menlo Park (2013)
5. Fließbach, T.: Mechanik, 7th edn. Lehrbuch zur Theoretischen Physik I. Springer, Berlin/Heidelberg (2015)
6. Arnol'd, V.I.: Catastrophe Theory. Springer, Berlin/Heidelberg (1992)
7. McCauley, J.L.: Chaos, Dynamics and Fractals. Cambridge University Press, Cambridge (1994)
8. Devaney, R.L.: An Introduction to Chaotic Dynamical Systems, 2nd edn. Westview, Boulder (2003)
9. Schuster, H.G., Just, W.: Deterministic Chaos, 4th edn. Wiley, New York (2006)
10. Teschl, G.: Ordinary Differential Equations and Dynamical Systems. Graduate Studies in Mathematics, vol. 140. American Mathematical Society, Providence (2012)
11. Lyapunov, A.M.: The General Problem of Stability of Motion. Taylor & Francis, London (1992)

2. [illegible]
3. [illegible]
4. [illegible]
5. [illegible]
6. [illegible]
7. [illegible]
8. [illegible]
9. [illegible] 4th edn. [illegible] New York (20[illegible])

11. [illegible] Stability of Motion [illegible]

Chapter 7
Molecular Dynamics

7.1 Introduction

It is the aim of many branches of research in physics to describe macroscopic properties of matter on the basis of microscopic dynamics. However, a description of the simultaneous motion of a large number of interacting particles is in most cases not feasible by analytic methods. Moreover, a description is particularly difficult if the interaction between the particles is strong. Within the framework of statistical mechanics one tries to remedy these difficulties by employing some simplifying assumptions and by treating the system from a statistical point of view [1–4]. However, most of these simplifying assumptions are only justified within certain limits, such as the *weak coupling limit* or the *low density limit*. Nevertheless, it is not easy to establish how the solutions acquired are influenced by these limits and how the physics beyond these limits can be perceived. This makes the necessity of numerical solutions quite apparent. There are essentially two methods to determine physical quantities over a restricted set of states, namely *molecular dynamics* [5–7] and *Monte Carlo* methods. The technique of molecular dynamics will be discussed within this chapter while an introduction into some basic features of Monte Carlo algorithms is postponed to the second part of this book.

We strictly focus on a particular sub-field of molecular dynamics, namely on *classical molecular dynamics*, i.e. the treatment of classical physical systems. Extensions to quantum mechanical systems, which are commonly referred to as quantum molecular dynamics [8], will not be discussed here.

B.A. Stickler, E. Schachinger, *Basic Concepts in Computational Physics*,
DOI 10.1007/978-3-319-27265-8_7

7.2 Classical Molecular Dynamics

The classical model system for molecular dynamics consists of N particles with positions $r_i \equiv r_i(t)$, velocities $v_i \equiv v_i(t) = \dot{r}_i(t)$ and masses m_i, where $i = 1, 2, \ldots, N$. We note that r_i and v_i are vectors of the same dimension. We can write NEWTON's equations of motion as

$$m_i \ddot{r}_i = f_i(r_1, r_2, \ldots, r_N) , \tag{7.1}$$

where we introduced the forces $f_i \equiv f_i(r_1, r_2, \ldots, r_N)$. Again, we note that the forces f_i are vectors of the same dimension as r_i and v_i. We specify the forces f_i by demanding them to be *conservative*. Thus, we write

$$f_i(r_1, r_2, \ldots, r_N) = -\nabla_i U(r_1, r_2, \ldots, r_N) , \tag{7.2}$$

where ∇_i is the gradient pertaining to the spatial components of the i-th particle and $U(r_1, r_2, \ldots, r_N)$ is some potential which we will abbreviate by dropping its arguments: $U \equiv U(r_1, r_2, \ldots r_N)$. We then specify this potential U as the sum of two-particle interactions U_{ij} and some external potential U_{ext} as, for instance, the gravitational field or a static electric potential applied to the system:

$$U = \frac{1}{2} \sum_i \sum_{j \neq i} U_{ij} + U_{\text{ext}} . \tag{7.3}$$

In our discussion of the two-body problem (Appendix A) and, in particular, of the KEPLER problem in Chap. 4 we considered a central potential, which was proportional to $-1/r$. Due to the conservation of angular momentum, it was convenient to introduce an effective potential U_{eff} as the sum of an attractive and repulsive part as it was defined in Eq. (4.3) and illustrated in Fig. 4.1. In contrast, in molecular dynamics the most prominent two-body interaction potential is known as the LENNARD-JONES potential [9]. It is of the form

$$U(|r|) = 4\sigma \left[\left(\frac{\epsilon}{|r|} \right)^{12} - \left(\frac{\epsilon}{|r|} \right)^{6} \right] , \tag{7.4}$$

where ϵ and σ are real parameters and $|r|$ is the distance between two particles. The significance of the parameters ϵ and σ as well as the form of $U(|r|)$ defined by Eq. (7.4) is illustrated in Fig. 7.1. The LENNARD-JONES potential was particularly developed to model the interaction between neutral atoms or molecules. The repulsive term, which is proportional to $|r|^{-12}$, describes the PAULI repulsion while the attractive $|r|^{-6}$ term accounts for attractive VAN DER WAALS forces.

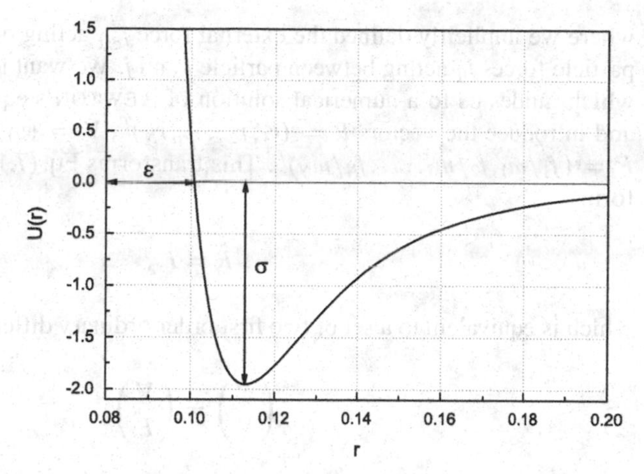

Fig. 7.1 Illustration of the LENNARD-JONES potential, Eq. (7.4). σ describes the depth of the potential well and ϵ is the position of the root of the LENNARD-JONES potential

We introduce the distance between particles i and j via

$$r_{ij} = |r_i - r_j| = |r_j - r_i| = r_{ji} \,, \tag{7.5}$$

and define the two-body potential

$$U_{ij} \equiv U(r_{ij}) \,, \tag{7.6}$$

where U is approximated by the LENNARD-JONES potential (7.4). Furthermore, we deduce from Eq. (7.4) that

$$f(|r|) = -\nabla_r U(|r|) = \frac{24\sigma}{|r|^2}\left[2\left(\frac{\epsilon}{|r|}\right)^{12} - \left(\frac{\epsilon}{|r|}\right)^6\right] r \,, \tag{7.7}$$

where we keep in mind that r is a vector. Hence, we write the forces f_i which appear in NEWTON's equations of motion (7.1) with the help of (7.3) in the form

$$\begin{aligned} f_i &= -\nabla_i U \\ &= -\nabla_i \left(\frac{1}{2} \sum_k \sum_{l \neq k} U_{kl} + U_{\text{ext}} \right) \\ &= -\sum_{j \neq i} \nabla_i U_{ij} - \nabla_i U_{\text{ext}} \\ &= \sum_{j \neq i} f(r_{ij}) + f_{\text{ext}}^i \\ &= \sum_{j \neq i} f_{ij} + f_{\text{ext}}^i \,, \end{aligned} \tag{7.8}$$

where we implicitly defined the external force f_{ext}^i acting on particle i and the two-particle forces f_{ij} acting between particle i and j. We want to make the road visible, which guides us to a numerical solution of NEWTON's equations of motion (7.1), and introduce the vectors $R = (r_1, r_2, \ldots, r_N)^T$, $V = (v_1, v_2, \ldots, v_N)^T = \dot{R}$, and $F = (f_1/m_1, f_2/m_2, \ldots, f_N/m_N)^T$. This transforms Eq. (7.1) into the very compact form

$$\ddot{R} = F\ , \tag{7.9}$$

which is equivalent to a set of two first order ordinary differential equations:

$$\begin{pmatrix} \dot{R} \\ \dot{V} \end{pmatrix} = \begin{pmatrix} V \\ F \end{pmatrix}. \tag{7.10}$$

This set is already of the standard form (5.1) of initial value problems.

We are now in a position to proceed with a discussion of some numerical methods which have been developed in Chap. 5 to solve this initial value problem. For this sake, we regard discrete time instances $t_k = k\Delta t$, where $k \in \mathbb{N}$ and function values at these discrete time instances t_k are denoted by a subscript k, as for instance $R_k \equiv R(t_k)$.

(i) In a first approximation we apply the symplectic EULER method [see Eq. (4.33)] to Eq. (7.10) and obtain

$$\begin{pmatrix} R_{k+1} \\ V_{k+1} \end{pmatrix} = \begin{pmatrix} R_k \\ V_k \end{pmatrix} + \begin{pmatrix} V_{k+1} \\ F_k \end{pmatrix} \Delta t\ . \tag{7.11}$$

Inserting the second into the first equation results in

$$R_{k+1} = R_k + V_k \Delta t + F_k \Delta t^2\ . \tag{7.12}$$

The velocity V_k at time t_k is then approximated by the backward difference derivative (2.10b) and we find the recursion relation:

$$R_{k+1} = 2R_k - R_{k-1} + F_k \Delta t^2\ . \tag{7.13}$$

We note that it is only valid for $k \geq 1$. The initialization step necessary to complete the analysis is found by expanding R_1 in a TAYLOR series up to second order:

$$R_1 = R_0 + \Delta t V_0 + \frac{1}{2} F_0 \Delta t^2\ . \tag{7.14}$$

This method is referred to as the STÖRMER-VERLET algorithm [10]. Note that Eq. (7.14) serves as the initialization of the sequence of time steps.

Furthermore, we remark that Eq. (7.13) could have also been obtained using the central difference derivative to approximate the second time derivative in Eq. (7.1):

$$\ddot{R}_k \approx \frac{R_{k+1} - 2R_k + R_{k-1}}{\Delta t^2} = F_k \,. \tag{7.15}$$

In summary, the VERLET or STÖRMER-VERLET algorithm is defined by the following set of equations:

$$\begin{aligned} R_{k+1} &= 2R_k - R_{k-1} + F_k \Delta t^2 \,, \qquad k \geq 1, \\ R_1 &= R_0 + \Delta t V_0 + \frac{1}{2} F_0 \Delta t^2 \,. \end{aligned} \tag{7.16}$$

(ii) We employ the central rectangular rule of integration (Sect. 3.2) in order to obtain approximations which are formally equivalent to Eq. (5.11). In particular, we obtain from Eq. (7.10):

$$R_{k+1} = R_k + V_{k+\frac{1}{2}} \Delta t \,. \tag{7.17}$$

We note that the value of $V_{k+1/2}$ is yet undetermined. However, it can bc determined in a similar fashion via

$$V_{k+\frac{1}{2}} = V_{k-\frac{1}{2}} + F_k \Delta t \,. \tag{7.18}$$

This method is referred to as the *leap-frog* algorithm and is initialized by the relation

$$V_{\frac{1}{2}} = V_0 + \frac{\Delta t}{2} F_0 \,. \tag{7.19}$$

This equation can also be obtained by expanding $V_{1/2}$ in a TAYLOR series up to first order around the point $t_0 = 0$ and by noting that $\dot{V}_k = F_k$. In summary we write the leap-frog algorithm as

$$\begin{aligned} R_{k+1} &= R_k + V_{k+\frac{1}{2}} \Delta t \,, \\ V_{k+\frac{1}{2}} &= V_{k-\frac{1}{2}} + F_k \Delta t \,, \\ V_{\frac{1}{2}} &= V_0 + \frac{1}{2} F_0 \Delta t \,. \end{aligned} \tag{7.20}$$

(iii) A third, very elegant alternative is the so-called *velocity* VERLET algorithm. We expand R_{k+1}:

$$R_{k+1} = R_k + V_k \Delta t + \frac{1}{2} F_k \Delta t^2 . \tag{7.21}$$

This allows to calculate the spatial coordinates at time t_{k+1} if R_k and V_k are given. Note that $F_k \equiv F(R_k)$ is completely determined by the positions R_k. Nevertheless, we need one more relation in order to determine the velocities at times t_{k+1}. Again, we expand V_{k+1} in a TAYLOR series. However, we approximate the remainder by the arithmetic mean between t_k and t_{k+1}:

$$V_{k+1} = V_k + \frac{1}{2} \left(F_k + F_{k+1} \right) \Delta t . \tag{7.22}$$

The strategy is clear: we calculate the positions R_{k+1} from Eq. (7.21) for given values of R_k and V_k. With the help of R_{k+1} we compute F_{k+1}, which is then inserted into Eq. (7.22) which determines V_{k+1}. In summary, the complete algorithm of the velocity VERLET method is defined by the steps:

$$\begin{aligned} R_{k+1} &= R_k + V_k \Delta t + \frac{1}{2} F_k \Delta t^2 , \\ V_{k+1} &= V_k + \frac{1}{2} \left(F_k + F_{k+1} \right) \Delta t . \end{aligned} \tag{7.23}$$

We note some properties of these methods. The STÖRMER-VERLET algorithm of Eq. (7.16) is time-reversal symmetric (invariant under the transformation $\Delta t \to -\Delta t$), hence reversible. This is a direct consequence of its relation to the symplectic EULER method. Moreover, the positions R_k obtained with this method are highly accurate, however, the procedure suffers under an inaccurate approximation of the velocities V_k. This shortcoming is clearly remedied by the leap-frog algorithm (7.20) or the velocity VERLET algorithm (7.23). However, these methods are not time-reversal invariant. Hence, one has to decide whether or not very accurate values for the velocities are required for the problem at hand. In many cases the velocity VERLET algorithm is the most popular choice.

7.3 Numerical Implementation

The rough structure of a molecular dynamics code consists of three crucial steps, namely

- Initialization,
- start simulation and equilibrate,
- continue simulation and store results.

In the following we discuss some of the most important subtleties associated with these three parts. In particular we will focus on the choice of appropriate boundary conditions and on the choice of the scales of characteristic quantities.

Boundary Conditions

Basically, there are two possibilities: (i) The system is of finite size and the implementation of boundary conditions might be straightforward. For instance, let us assume that we regard N particles within a finite box of reflecting boundaries, we simply propagate the particle-coordinates in time and if a particle tries to leave the box, we correct its trajectory according to a reflection law. The velocity is adjusted accordingly. This is illustrated in Fig. 7.2 for a two-dimensional case and the particular situation that the particle is reflected from the right hand boundary of the box. The corresponding equations read

$$r_{k+1} = \begin{pmatrix} x_{k+1} \\ y_{k+1} \end{pmatrix} = \begin{pmatrix} L - (\tilde{x}_{k+1} - L) \\ \tilde{y}_{k+1} \end{pmatrix} \tag{7.24}$$

and

$$v_{k+1} = \begin{pmatrix} v_{k+1,x} \\ v_{k+1,y} \end{pmatrix} = \begin{pmatrix} -\tilde{v}_{k+1,x} \\ \tilde{v}_{k+1,y} \end{pmatrix}. \tag{7.25}$$

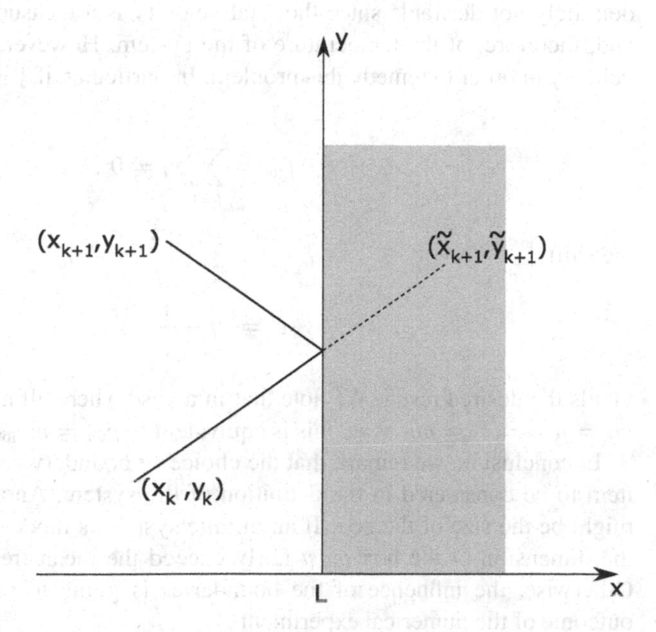

Fig. 7.2 Illustration of the reflection principle for a box of finite dimension with reflecting boundaries

Here, L denotes the length of the box and $\tilde{x}_{k+1}$, $\tilde{y}_{k+1}$, $\tilde{v}_{k+1,x}$ and $\tilde{v}_{k+1,y}$ are the positions and velocities one would have obtained in the absence of the boundary, see Fig. 7.2.

(ii) The system is not confined. Then the situation is entirely different. Of course, one could approximate the infinite volume by a large but finite volume. In such a case the influence of a constraint to finite size is usually not negligible. The induced errors are referred to as *finite volume effects*. A very popular choice are so called *periodic boundary conditions* which means that a finite system is surrounded by an infinite number of completely identical replicas of the system, where the forces are allowed to act across the boundaries. Because of this, calculating the force on one particle requires the evaluation of an infinite sum. This is numerically not manageable and we have to find ways to truncate the sum. For instance, it might be a good approximation to restrict the sum to nearest-neighbor cells. However, the applicability of such an approach highly depends on the properties of the system under investigation and, in particular, on the range of the interaction potential. In case of a LENNARD-JONES potential the quantity defining the range of the interaction potential is ϵ, see Fig. 7.1.

If a particle leaves the box, it enters the box at the same time on the opposite side. More generally, due to the requirement of identical replicas, we have for all observables $O(r)$ that $O(r + nK) = O(r)$, where r lies within the central box, K is a lattice vector pointing to one of the neighboring cells and $n \in \mathbb{Z}$.

There is another crucial point concerning periodic boundary conditions. In case of a closed system, the system is definitely at rest. However, if periodic boundary conditions are imposed it is possible that the particles move with constant velocity from one cell to another, which, in our case, resembles circling trajectories. This is definitely not desirable since the total velocity is a measure of the kinetic energy and, therefore, of the temperature of the system. However, one can *shift* the total velocity in order to remedy this problem. In particular, if

$$v_{\mathrm{tot}} = \sum_{i=1}^{N} v_i \neq 0 , \tag{7.26}$$

the shift

$$v_i' = v_i - \frac{1}{N} v_{\mathrm{tot}} , \tag{7.27}$$

yields the desired result. We note that in a case where all masses are identical, i.e. $m_1 = m_2 = \ldots = m_N \equiv m$, this is equivalent to $p_{\mathrm{tot}} = m v_{\mathrm{tot}} = 0$.

In conclusion, we remark that the choice of boundary conditions is not the only item to be considered in the definition of the system. Another quite crucial point might be the size of the box. If an infinite system is modeled using finite systems, the dimension of the box *must* fairly exceed the mean free path of the particles. Otherwise, the influence of the boundaries is going to perturb significantly the outcome of the numerical experiment.

Initialization and Equilibration

We remember from statistical physics [1–4] that every degree of freedom in the system (7.1) contributes just $k_B T/2$ to the total kinetic energy because of the equipartition theorem. Here k_B is BOLTZMANN's constant and T is the temperature. If we regard N particles, which move in a d-dimensional space, we have $d(N-1)$ degrees of freedom, if we demand that $v_{tot} = 0$. Hence, we have

$$E_{kin} = \frac{1}{2}\sum_{i=1}^{N} m_i v_i^2 = \frac{d(N-1)}{2} k_B T \,, \tag{7.28}$$

which gives a relation from which we can determine the temperature of the system:

$$k_B T = \frac{1}{d(N-1)} \sum_{i=1}^{N} m_i v_i^2 \,. \tag{7.29}$$

However, in many applications the system is supposed to be simulated at a *given* temperature, i.e. the temperature T is an input rather than an output parameter and is supposed to stay constant during the simulation. We can control the temperature by rescaling the velocities and this might be necessary at several times during the simulation in order to guarantee a constant temperature. We define

$$v_i' = \lambda v_i \,, \tag{7.30}$$

where λ is a rescaling parameter. The temperature associated with the velocities v_i' is given by

$$k_B T' = \frac{\lambda^2}{d(N-1)} \sum_{i=1}^{N} m_i v_i^2 \,. \tag{7.31}$$

This allows to determine how to choose λ in order to obtain a certain temperature T':

$$\lambda = \sqrt{\frac{d(N-1) k_B T'}{2E_{kin}}} \,. \tag{7.32}$$

We note that if the total velocity, which is the sum of all velocities v_i, is zero, the total velocity corresponding to the rescaled velocities v_i' is also equal to zero since

$$\sum_{i=1}^{N} v_i' = \lambda \sum_{i=1}^{N} v_i = 0 \,. \tag{7.33}$$

This ensures that rescaling of the velocities does not induce a bias.

The choice of the initial conditions highly influences the time the system needs to reach thermal equilibrium. For instance, if a gas is to be simulated at a given temperature T it might be advantageous to choose the initial velocities according to a MAXWELL-BOLTZMANN distribution. The MAXWELL-BOLTZMANN distribution states that the probability [more precisely: the pdf (*probability density function* describing the probability, see Appendix E)] that a particle with mass m has velocity v is proportional to

$$p(|v|) \propto |v|^2 \exp\left(-\frac{m|v|^2}{2k_B T}\right) . \tag{7.34}$$

Another intriguing question is how to check whether or not thermal equilibrium has been reached. In statistical mechanics one is usually confronted with expectation values of observables $O(t)$ as a function of time. The expectation value $\langle O \rangle$ is defined as

$$\langle O \rangle = \lim_{\tau \to \infty} \frac{1}{\tau} \int_0^\tau dt O(t) . \tag{7.35}$$

Since $O(t)$ is not known analytically one replaces the mean value by its arithmetic mean

$$\langle O \rangle \approx \overline{O} = \frac{1}{n} \sum_{j=k+1}^{k+n} O(t_j) . \tag{7.36}$$

If n and k are sufficiently large, the average value can be regarded as converged. In particular, one has to choose n reasonably large and then find k in such a way, that for all values $k' \geq k$ the same result for $\overline{O}$ is obtained. Hence, equilibrium has been reached after k time-steps and it is now possible to 'measure' the observables by calculating their mean values. A more detailed discussion of such a procedure, as, for instance, the influence of time correlations or a discussion of more advanced techniques is postponed to Chap. 19.

There is one last point: In many cases the *natural units* of the physical system might be disadvantageous because they are likely to induce numerical instabilities. In such cases a common technique is to switch to *rescaled variables* by introducing new units, which are characteristic quantities for the system and all physical quantities are expressed in these new units. For instance one might introduce the length L of the box as the unit of space. The new spatial coordinates would then be given by

$$r' = \frac{r}{L} . \tag{7.37}$$

Hence all coordinates take on values within the interval $r' \in [0, 1]$. However, one cannot introduce an arbitrary set of characteristic quantities due to the physical relations they have to obey. For instance, one might introduce a characteristic energy E_0, a characteristic length λ, and a characteristic mass m. In this case the characteristic temperature $\tilde{T}$ is determined via

$$\tilde{T} = \frac{E_0}{k_B} . \tag{7.38}$$

Moreover the characteristic time τ is fixed to the value

$$\tau = \sqrt{\frac{m\lambda^2}{E_0}} , \tag{7.39}$$

which results from the relation between the kinetic energy and the velocity.

To illustrate a molecular dynamics simulation we study a set of $N = 100$ particles of mass $m = 1$ which are subject to a LENNARD-JONES potential (7.4) characterized by $\epsilon = \sigma = 1$ and to a gravitational force mg, $g = 9.81$. At initialization the particles are placed in a 10×10 lattice starting with the lower left hand edge at $x = 10.5$ and $y = 10$. The particles are equally spaced with $\Delta x = \Delta y = \epsilon$. This initial configuration is shown in Fig. 7.3a. Furthermore, the left hand side, the right hand side, and the bottom of the confinement ($L = 30$) are described by reflecting boundary conditions, Eqs. (7.24) and (7.25). The confinement is open at the top, i.e. it extends to infinity. The time step is given by $\Delta t = 10^{-3}$. Figure 7.3b–d demonstrate how the system developed after 1200, 1800, and 3000 time steps, respectively.

This chapter closes our discussion of the numerics of initial value problems. In the following chapters we will introduce some of the basic concepts developed to solve boundary value problems with numerical methods.

Summary

This chapter dealt with the classical dynamics of many particles (not necessarily identical particles) which are confined in a box of finite dimension or which are allowed to roam freely in infinite space. The particles are subject to a particle-particle interaction and to an external force. The discussion was restricted to classical molecular dynamics. From NEWTON's equations of motion for N interacting particles numerical methods were developed which allowed the simulation of the particles' dynamics. Based on the symplectic EULER method the STÖRMER-VERLET algorithm was derived. Another approach was based on the central rectangular rule and resulted in the *leap-frog* algorithm. Finally, the *velocity* VERLET algorithm was introduced. All three methods do have their merits. The first gives very accurate results for the particles' positions but calculates inaccurate

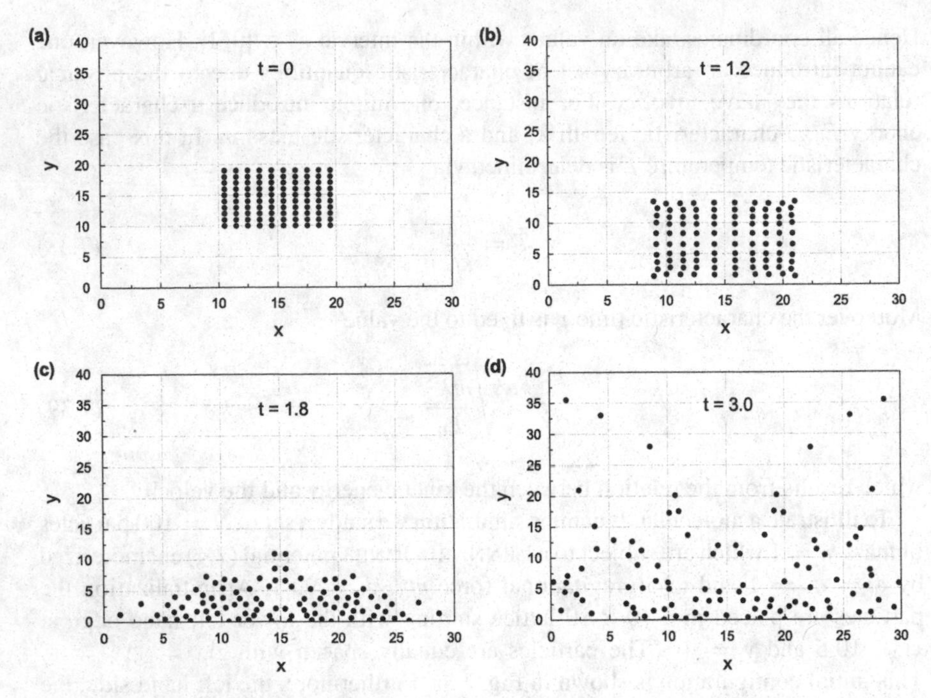

Fig. 7.3 **(a)** Initial configuration: the particles are placed in a 10×10 equally spaced lattice starting with $x = 10.5$ and $y = 10.0$; $g = 9.81$. The initial velocities are equal to zero. **(b)** Configuration after 1200 time steps. **(c)** Configuration after 1800 time steps. **(d)** Configuration after 3000 time steps

velocities. It has the advantage that it is time reversible. The other two methods lack this property but give very accurate estimates of the particles' velocities. The final part of this chapter was dedicated to the discussion of various subtleties of the numerical implementation of these algorithms as there were: (i) definition of boundary conditions, (ii) initialization of the algorithm, (iii) equilibration to a given temperature, (iv) ensuring constant temperature throughout the simulation, and (v) transformation to rescaled variables.

Problems

1. We investigate the pendulum of Chap. 1 and write its equation of motion as

$$\ddot{x} + \omega^2 x = 0 ,$$

with $\omega = \sqrt{g/\ell}$. The STÖRMER-VERLET algorithm is applied to simulate the pendulum's motion and to compare the numerical results with the exact solution.

Demonstrate that the result is very sensitive to the choice of the time step Δt and, in particular, of the product $\omega \Delta t$. Note that in this particular case the STÖRMER-VERLET algorithm can also be studied analytically! What happens for the choice $\omega \Delta t = 1$ or $\omega \Delta t \geq 2$? Which conclusions can be drawn from this example for a proper choice of the time discretization?

Try the other two methods to simulate the pendulum's dynamics.

2. Write a molecular dynamics code with the help of the following instructions. You can use either the leap-frog or the velocity VERLET algorithm. We consider the following system:

   - There are $N = 100$ particles in a two-dimensional box with side length $L = 30$. The boundaries at the bottom, at the left- and at right-hand side are considered as reflecting, as in Fig. 7.2. The top of the box is regarded as open (no periodic boundary condition or reflecting boundary is imposed).
   - The particles interact through a LENNARD-JONES potential of the form (7.4) where ϵ and σ define the interaction.
   - Furthermore, a gravitational force $F_{\text{ext}} = -mge_y$ acts on each particle, where m is the particle's mass, g is the acceleration due to gravity, and e_y denotes the unit vector in y-direction.
   - As an initial condition, the particles can be placed within the box on a regular lattice, where the distance between the particles is the characteristic distance according to the LENNARD-JONES potential, i.e. ϵ. The form and position of this lattice is arbitrary. This is illustrated in Fig. 7.3.

   We measure the velocities and the positions of all particles. Since the particle's velocities and positions are to be analyzed with the help of an extra program, the data are written to external files (it is not necessary to save *all* time steps!).

   Perform the following analysis:

   - Determine the temperature T from the kinetic energy as discussed in in this chapter. Note that in this particular case we do not demand that $v_{tot} = 0$!
   - Try different initial conditions. For instance, set the initial velocity equal to zero and stack the particles in different geometric configurations (rectangle, triangle, ..., one can also use more than one configurations at the same time!). The nearest neighbor distance between the particles can be set equal to ϵ. Choose one configuration and place it at different positions in the box. What happens?
   - Set $\epsilon = \sigma = m = 1$ (we change the units) and set in the initial condition to the inter-atomic distance of $2^{\frac{1}{6}}\epsilon$. (Why?) Vary the gravitational acceleration g (different systems of units) in order to simulate different states of matter. The reference program developed solid behavior for $g \approx 0$, liquid behavior for $g \approx 0.1$ and gaseous behavior for $g > 1$. Explain this behavior!
   - Measure the particle density $\rho(h)$ as a function of the height h. You should be able to reproduce the barometric formula:

$$\rho \propto \rho_0 \exp\{-\gamma h/T\}, \quad \gamma > 0 .$$

- Determine the momentum distribution ($p_i = mv_i$) of the particles and demonstrate that it follows a MAXWELL-BOLTZMANN distribution

$$p(|v|) \propto |v|^2 \exp\{-\gamma|v|^2/T\}, \qquad \gamma > 0\,,$$

with $|v| = \sqrt{v_x^2 + v_y^2}$, the Euclidean norm.
- Illustrate the results of the simulation graphically.

References

1. Mandl, F.: Statistical Physics, 2nd edn. Wiley, New York (1988)
2. Schwabl, F.: Statistical Mechanics. Advanced Texts in Physics. Springer, Berlin/Heidelberg (2006)
3. Halley, J.W.: Statistical Mechanics. Cambridge University Press, Cambridge (2006)
4. Hardy, R.J., Binek, C.: Thermodynamics and Statistical Mechanics: An Integrated Approach. Wiley, New York (2014)
5. Hoover, W.G.: Molecular Dynamics. Springer, Berlin/Heidelberg (1986)
6. Griebel, M., Knapek, S., Zumbesch, G.: Numerical Simulation in Molecular Dynamics. Texts in Computational Science and Engeneering, vol. 5. Springer, Berlin/Heidelberg (2007)
7. Marx, D., Hutter, J.: Ab Initio Molecular Dynamics. Cambridge University Press, Cambridge (2012)
8. Gatti, F. (ed.): Molecular Quantum Dynamics. Springer, Berlin/Heidelberg (2014)
9. Jones, J.E.: On the determination of molecular fields. II. From the equation of state of a gas. Proc. R. Soc. A **106**, 463–477 (1924). doi:10.1098/rspa.1924.0082
10. Hairer, E., Lubich, C., Wanner, G.: Geometric numerical integration illustrated by the Störmer-Verlet method. Acta Numer. **12**, 399–450 (2003). doi:10.1017/S0962492902000144

Chapter 8
Numerics of Ordinary Differential Equations: Boundary Value Problems

8.1 Introduction

It is the aim of this chapter to introduce some of the basics methods developed to solve boundary value problems. Since a treatment of all available concepts is far too extensive, we will concentrate on two approaches, namely the finite difference approach and shooting methods [1–5]. Furthermore, we will strictly focus on *linear boundary value problems* defined on a finite interval $[a, b] \subset \mathbb{R}$. A boundary value problem is referred to as linear if both the differential equation and the boundary conditions are linear. Such a problem of order n is of the form

$$\begin{cases} L[y] = f(x), & x \in [a, b] , \\ U_\nu[y] = \lambda_\nu, & \nu = 1, \dots, n . \end{cases} \tag{8.1}$$

Here, $L[y]$ is a linear operator

$$L[y] = \sum_{k=0}^{n} a_k(x) y^{(k)}(x) , \tag{8.2}$$

where $y^{(k)}(x)$ denotes the k-th spatial derivative of $y(x)$, i.e. $y^{(k)} \equiv \mathrm{d}^k y(x)/\mathrm{d}x^k$ and $f(x)$ as well as the $a_k(x)$ are given functions which we assume to be continuous. Accordingly, linear boundary conditions $U_\nu[y]$ can be formulated as

$$U_\nu[y] = \sum_{k=0}^{n-1} \left[\alpha_{\nu k} y^{(k)}(a) + \beta_{\nu k} y^{(k)}(b) \right] = \lambda_\nu , \tag{8.3}$$

B.A. Stickler, E. Schachinger, *Basic Concepts in Computational Physics*,
DOI 10.1007/978-3-319-27265-8_8

where the $\alpha_{\nu k}$, $\beta_{\nu k}$ and λ_ν are given constants. The question in which cases a solution to the boundary value problem (8.1) exists and whether or not this solution will be unique [6], will not be discussed here.

Let us introduce some further conventions: The differential equation in the first line of Eq. (8.1) is referred to as *homogeneous* if the function $f(x) = 0$ for all $x \in [a, b]$. In analogy, the boundary conditions are referred to as *homogeneous* if the constants $\lambda_\nu = 0$ for all $\nu = 1, \ldots, n$. Finally, the boundary value problem (8.1) is referred to as *homogeneous* if the differential equation is homogeneous and the boundary conditions are homogeneous as well. In all other cases it is referred to as *inhomogeneous*. Moreover, the boundary conditions are said to be *decoupled* if the function values at the two different boundaries do not mix.

One of the most important types of boundary value problems in physics are linear second order boundary value problems with decoupled boundary conditions. They are of the form:

$$a_2(x)y''(x) + a_1(x)y'(x) + a_0(x)y(x) = f(x) , \quad x \in [a, b] , \tag{8.4a}$$

$$\alpha_0 y(a) + \alpha_1 y'(a) = \lambda_1, \quad |\alpha_0| + |\alpha_1| \neq 0 , \tag{8.4b}$$

$$\beta_0 y(b) + \beta_1 y'(b) = \lambda_2, \quad |\beta_0| + |\beta_1| \neq 0 . \tag{8.4c}$$

This chapter focuses mainly on problems of this kind.

In particular, for second order differential equations, boundary conditions of the form

$$y(a) = \alpha , \qquad y(b) = \beta , \tag{8.5}$$

are referred to as *boundary conditions of the first kind* or DIRICHLET *boundary conditions*. On the other hand, boundary conditions of the form

$$y'(a) = \alpha , \qquad y'(b) = \beta , \tag{8.6}$$

are referred to as *boundary conditions of the second kind* or NEUMANN *boundary conditions* and boundary conditions of the form (8.4) are referred to as *boundary conditions of the third kind* or STURM *boundary conditions*.

We note, that the particular case of decoupled boundary conditions does not include problems like

$$y(a) = y(b) \neq 0 . \tag{8.7}$$

We encountered such a condition in Sect. 7.3 where we introduced boundary conditions of this form as periodic boundary conditions.

In the following section the method of finite differences will be applied to solve boundary value problems of the form (8.4). On the other hand, shooting methods, in particular the method developed by NUMEROV (see, for instance, [7] and references therein), will be the topic of the third section.

A common alternative in the case of constant coefficients is to solve the differential equation with the help of FOURIER transform techniques. A brief introduction to the numerical implementation of the FOURIER transform is given in Appendix D.

8.2 Finite Difference Approach

For illustrative purposes, we regard a boundary value problem of the form (8.4). The extension to more complex problems might be tedious but follows the same line of arguments. We discretize the interval $[a, b]$ according to the recipe introduced in Chap. 2: the positions x_k are given by $x_k = a + (k-1)h$, where the grid-spacing h is determined via the maximum number of grid-points N as $h = (b-a)/(N-1)$. Hence, we have $x_1 = a$ and $x_N = b$. Furthermore, we use the notation $y_k \equiv y(x_k)$ for all $k = 1, \ldots, N$. It will be used for all functions which appear in Eqs. (8.4).

Let us now employ the central difference derivative (2.10c) in order to approximate

$$y_k'' \equiv y''(x_k) \approx \frac{y_{k+1} - 2y_k + y_{k-1}}{h^2} , \tag{8.8}$$

for $k = 2, \ldots, N-1$ and

$$y_k' \equiv y'(x_k) \approx \frac{y_{k+1} - y_{k-1}}{2h} . \tag{8.9}$$

The boundary points x_1 and x_N will be treated in a separate step. In order to abbreviate the notation we will rewrite the differential equation (8.4) without the indices as

$$a(x)y''(x) + b(x)y'(x) + c(x)y(x) = f(x) . \tag{8.10}$$

Equations (8.8) and (8.9) are then applied and we arrive at the difference equation

$$a_k \frac{y_{k+1} - 2y_k + y_{k-1}}{h^2} + b_k \frac{y_{k+1} - y_{k-1}}{2h} + c_k y_k = f_k , \tag{8.11}$$

where $k = 2, \ldots, N-1$. Sorting the y_k yields

$$\left(\frac{a_k}{h^2} - \frac{b_k}{2h}\right) y_{k-1} + \left(c_k - \frac{2a_k}{h^2}\right) y_k + \left(\frac{a_k}{h^2} + \frac{b_k}{2h}\right) y_{k+1} = f_k , \tag{8.12}$$

and this equation is only valid for $k = 2, \ldots, N-1$ because we defined N grid-points within the interval $[a, b]$.

A final step is necessary in which the boundary conditions are incorporated. This will then enable us to reduce the whole problem to a system of linear equations. Decoupled boundary conditions of a second order differential equation for the left-hand boundary (8.4b) are of the general form:

$$\alpha_0 y(a) + \alpha_1 y'(a) = \lambda_1, \quad |\alpha_0|, |\alpha_1| \neq 0\,. \tag{8.13}$$

In analogy, we find for the right-hand boundary (8.4c):

$$\beta_0 y(b) + \beta_1 y'(b) = \lambda_2, \quad |\beta_0|, |\beta_1| \neq 0\,. \tag{8.14}$$

We discretize $y'(a)$ as

$$y_1' \equiv y'(a) \approx \frac{y_2 - y_0}{2h}\,, \tag{8.15}$$

and set $y_1 = y(a)$. Note that the function value y_0 in Eq. (8.15) is unknown since the virtual point $x_0 = a - h$ is not within our interval $[a, b]$. Nevertheless, we use Eq. (8.15) in Eq. (8.13) and obtain:

$$\alpha_0 y_1 + \alpha_1 \frac{y_2 - y_0}{2h} = \lambda_1\,. \tag{8.16}$$

We solve now Eq. (8.16) for y_0 under the premise that $\alpha_1 \neq 0$,

$$y_0 = y_2 - \frac{2h}{\alpha_1}\left(\lambda_1 - \alpha_0 y_1\right)\,, \tag{8.17}$$

rewrite Eq. (8.12) for $k = 1$,

$$\left(\frac{a_1}{h^2} - \frac{b_1}{2h}\right) y_0 + \left(c_1 - \frac{2a_1}{h^2}\right) y_1 + \left(\frac{a_1}{h^2} + \frac{b_1}{2h}\right) y_2 = f_1\,, \tag{8.18}$$

and insert (8.17) into (8.18):

$$\left[c_1 - \frac{2a_1}{h^2} + \frac{\alpha_0}{\alpha_1}\left(\frac{2a_1}{h} - b_1\right)\right] y_1 + \frac{2a_1}{h^2} y_2 = f_1 - \frac{\lambda_1}{\alpha_1}\left(b_1 - \frac{2a_1}{h}\right)\,. \tag{8.19}$$

On the other hand, in the specific case of $\alpha_1 = 0$ we immediately obtain from Eq. (8.16):

$$y_1 = \frac{\lambda_1}{\alpha_0}\,. \tag{8.20}$$

The same strategy can be applied to incorporate the right-hand side boundary condition, Eq. (8.14): We discretize Eq. (8.14) by introducing the function value y_{N+1} at the virtual grid-point $x_{N+1} = N + h$ outside the interval $[a, b]$ via:

$$\beta_0 y_N + \beta_1 \frac{y_{N+1} - y_{N-1}}{2h} = \lambda_2 . \tag{8.21}$$

This equation is solved for y_{N+1} under the premise that $\beta_1 \neq 0$

$$y_{N+1} = y_{N-1} + \frac{2h}{\beta_1} (\lambda_2 - \beta_0 y_N) , \tag{8.22}$$

and insert this into Eq. (8.12) for $k = N$. This results in:

$$\frac{2a_N}{h^2} y_{N-1} + \left[c_N - \frac{2a_N}{h^2} - \frac{\beta_0}{\beta_1} \left(b_N + \frac{2a_N}{h} \right) \right] y_N = f_N - \frac{\lambda_2}{\beta_1} \left(b_N + \frac{2a_N}{h} \right) . \tag{8.23}$$

In the specific case $\beta_1 = 0$, the value y_N is fixed at the boundary and one obtains from Eq. (8.14):

$$y_N = \frac{\lambda_2}{\beta_0} . \tag{8.24}$$

All these manipulations reduced the boundary value problem to a system of inhomogeneous linear equations, namely Eqs. (8.12), (8.19), and (8.23). It can be written as

$$Ay = F , \tag{8.25}$$

where we introduced the vector $y = (y_1, y_2, \ldots, y_N)^T$, the vector F

$$F = \begin{pmatrix} f_1 - \frac{\lambda_1}{\alpha_1} \left(b_1 - \frac{2a_1}{h} \right) \\ f_2 \\ f_3 \\ \vdots \\ f_{N-1} \\ f_N - \frac{\lambda_2}{\beta_1} \left(b_N + \frac{2a_N}{h} \right) \end{pmatrix} , \tag{8.26}$$

and the tridiagonal matrix A:

$$A = \begin{pmatrix} B_1 & C_1 & 0 & & \cdots & 0 \\ A_2 & B_2 & C_2 & & \cdots & \vdots \\ 0 & \ddots & \ddots & \ddots & & \\ \vdots & & \ddots & \ddots & \ddots & 0 \\ & & & A_{N-1} & B_{N-1} & C_{N-1} \\ 0 & \cdots & & 0 & A_N & B_N \end{pmatrix} . \tag{8.27}$$

Here we defined

$$A_k = \begin{cases} \left(\frac{a_k}{h^2} - \frac{b_k}{2h} \right) & k = 2, \ldots, N-1 , \\ \frac{2a_N}{h^2} & k = N , \end{cases} \tag{8.28}$$

$$B_k = \begin{cases} \left[c_1 - \frac{2a_1}{h^2} + \frac{\alpha_0}{\alpha_1} \left(\frac{2a_1}{h} - b_1 \right) \right] & k = 1 , \\ \left(c_k - \frac{2a_k}{h^2} \right) & k = 2, \ldots N-1 , \\ \left[c_N - \frac{2a_N}{h^2} - \frac{\beta_0}{\beta_1} \left(b_N + \frac{2a_N}{h} \right) \right] & k = N , \end{cases} \tag{8.29}$$

and, finally,

$$C_k = \begin{cases} \frac{2a_1}{h^2} & k = 1 , \\ \left(\frac{a_k}{h^2} + \frac{b_k}{2h} \right) & k = 2, \ldots, N-1 . \end{cases} \tag{8.30}$$

The remaining task is now to solve this linear system of equations (8.25). (A brief introduction to the numerical treatment of linear systems of equations can be found in Appendix C.) Very effective methods exist for cases where the matrix A is tridiagonal [8] as it is the case here. Although we discussed the method of finite differences for the particular case of a second order differential equation with decoupled boundary conditions, the same strategy can be employed to derive similar methods for higher order boundary value problems. However, these methods will, in general, be more complex. Furthermore, we note that in cases where $\alpha_1 = \beta_1 = 0$ the function values at the boundaries y_1 and y_N are fixed and the corresponding system of linear equations reduces to $(N-2)$ equations.

Let us briefly investigate the differential equation which corresponds to the problem (8.4) together with periodic boundary conditions of the form (8.7). In this case we have to consider that

$$y_1 = y_N \,, \tag{8.31}$$

and, for a solution to exist, we have necessarily

$$a_1 = a_N, \quad b_1 = b_N, \quad \text{and} \quad c_1 = c_N \,. \tag{8.32}$$

The finite difference approximations (8.8) and (8.9) are again applied to derive Eqs. (8.12) for $k = 2, \ldots, N-1$. For $k = 2$ Eq. (8.12) becomes

$$\left(\frac{a_2}{h^2} - \frac{b_2}{2h}\right) y_1 + \left(c_2 - \frac{2a_2}{h^2}\right) y_2 + \left(\frac{a_2}{h^2} + \frac{b_2}{2h}\right) y_3 = f_2 \,, \tag{8.33}$$

and we have for $k = N-1$

$$\left(\frac{a_{N-1}}{h^2} - \frac{b_{N-1}}{2h}\right) y_{N-2} + \left(c_{N-1} - \frac{2a_{N-1}}{h^2}\right) y_{N-1} + \left(\frac{a_{N-1}}{h^2} + \frac{b_{N-1}}{2h}\right) y_N = f_{N-1} \,. \tag{8.34}$$

Since $y_1 = y_N$ this can be rewritten as

$$\left(\frac{a_{N-1}}{h^2} - \frac{b_{N-1}}{2h}\right) y_{N-2} + \left(c_{N-1} - \frac{2a_{N-1}}{h^2}\right) y_{N-1} + \left(\frac{a_{N-1}}{h^2} + \frac{b_{N-1}}{2h}\right) y_1 = f_{N-1} \,. \tag{8.35}$$

Finally, Eq. (8.12) results for $k = 1$ in

$$\left(\frac{a_1}{h^2} - \frac{b_1}{2h}\right) y_{N-1} + \left(c_1 - \frac{2a_1}{h^2}\right) y_1 + \left(\frac{a_1}{h^2} + \frac{b_1}{2h}\right) y_2 = f_1 \,, \tag{8.36}$$

where we identified $y_0 = y(x_1 - h) \equiv y(x_N - h) = y_{N-1}$. All this results in a closed system of $N-1$ equations, which is of the form (8.25)

$$Ay = F \,, \tag{8.37}$$

where $y = (y_1, y_2, \ldots, y_{N-1})^T$,

$$F = \begin{pmatrix} f_1 \\ f_2 \\ \vdots \\ f_{N-1} \end{pmatrix} , \tag{8.38}$$

and the $(N-1) \times (N-1)$ matrix A is given by

$$A = \begin{pmatrix} B_1 & C_1 & 0 & \cdots & 0 & A_1 \\ A_2 & B_2 & C_2 & & \cdots & 0 \\ 0 & \ddots & \ddots & \ddots & & \vdots \\ \vdots & & \ddots & \ddots & \ddots & 0 \\ 0 & & & A_{N-2} & B_{N-2} & C_{N-2} \\ C_{N-1} & 0 & \cdots & 0 & A_{N-1} & B_{N-1} \end{pmatrix} . \tag{8.39}$$

Here, we defined

$$A_k = \left(\frac{a_k}{h^2} - \frac{b_k}{2h} \right) , \qquad k = 1, \ldots, N-1 , \tag{8.40}$$

$$B_k = \left(c_k - \frac{2a_k}{h^2} \right) , \qquad k = 1, \ldots, N-1 , \tag{8.41}$$

and

$$C_k = \left(\frac{a_k}{h^2} + \frac{b_k}{2h} \right) , \qquad k = 1, \ldots, N-1 . \tag{8.42}$$

In contrast to the matrix (8.27) the matrix (8.39) is not tridiagonal since the matrix elements $(A)_{1,N-1}$ and $(A)_{N-1,1}$ are non-zero. Nevertheless, it was possible to reduce the boundary value problem to a system of linear equations which can be solved iteratively.

8.3 Shooting Methods

For illustrative purposes, we restrict here the discussion to a second order boundary value problem with decoupled boundary conditions of the form (8.4). The essential idea of shooting methods is to treat the boundary value problem as an initial value problem. The resulting equations can then be solved with the help of methods discussed in Chap. 5. Of course, such an approach is ill-defined because no initial conditions but only boundary conditions are given. The trick is, that one modifies the initial conditions iteratively in such a way that in the end the boundary conditions are fulfilled. Let us put this train of thoughts into a mathematical form: We rewrite the second order differential equation (8.4a) as

$$y'' = f(y, y', x) , \tag{8.43}$$

which can be reduced to a set of first order differential equations as was demonstrated in Chap. 5. We note that Eq. (8.43) is not yet well posed since the initial conditions have not been defined. The boundary condition on the left-hand side reads:

$$\alpha_0 y(a) + \alpha_1 y'(a) = \lambda_1 . \tag{8.44}$$

We now *assume* that $y'(a) = z$, where z is some number. This gives the well posed initial value problem

$$\begin{cases} y'' = f(y, y', x) , \\ y(a) = \dfrac{\lambda_1}{\alpha_0} - \dfrac{\alpha_1}{\alpha_0} z , \\ y'(a) = z , \end{cases} \tag{8.45}$$

under the assumption that $\alpha_0 \neq 0$. The solution of this problem will be written as $y(x; z)$ in order to indicate its dependence on the particular choice $y'(a) = z$. We remember, that the boundary condition at the right-hand boundary is defined as:

$$\beta_0 y(b) + \beta_1 y'(b) = \lambda_2 . \tag{8.46}$$

Let us introduce the function:

$$F(z) = \beta_0 y(b; z) + \beta_1 y'(b; z) - \lambda_2 . \tag{8.47}$$

We observe that the solution of the equation

$$F(z) = 0 , \tag{8.48}$$

gives the desired solution to the boundary value problem (8.4), because in this case the second boundary condition (8.46) is fulfilled. In practice, one tries several values of z until relation (8.48) is fulfilled. However, from a numerical point of view this method is very inefficient since usually several initial value problems have to be solved until the correct value of z is found. Nevertheless, in some cases shooting methods proved to be very useful [7].

For instance, shooting methods are particularly effective if a solution to an eigenvalue problem of the form

$$a(x)y''(x) + b(x)y'(x) + c(x)y(x) = \lambda y(x) , \tag{8.49a}$$

in combination with homogeneous boundary conditions,

$$\alpha_0 y(a) + \alpha_1 y'(a) = 0 , \tag{8.49b}$$

and

$$\beta_0 y(b) + \beta_1 y'(b) = 0 , \tag{8.49c}$$

is to be found. We note that Eq. (8.49a) has the trivial solution $y(x) \equiv 0$ for all values of λ. However a non-trivial solution will only exist for particular values of λ. These particular values will be indexed by λ_n and are referred to as *eigenvalues* of Eq. (8.49a) [9]. The corresponding functions $y_n(x)$ are referred to as *eigenfunctions*. We note that the differential equation (8.49a) in combination with the boundary conditions (8.49b) and (8.49c) define a homogeneous boundary value problem. Such a problem is commonly referred to as an *eigenvalue problem* [9]. Furthermore, we note the following property of homogeneous boundary value problems: Suppose that $y(x)$ is a solution of the boundary value problem (8.49), then $\tilde{y}(x) = \gamma y(x)$, with $\gamma = \text{const}$ will also be a solution of (8.49). Hence, the solution of a homogeneous boundary value problem is not unique but invariant under multiplication by a constant γ. Typically, the multiplicative factor γ is fixed by some additional condition, such as a normalization condition of the form

$$\int_a^b dx |y(x)|^2 = 1 . \tag{8.50}$$

We now employ this property and *choose* $y(a) = 1$. Inserting this choice into (8.49b) yields

$$y'(a) = -\frac{\alpha_0}{\alpha_1} . \tag{8.51}$$

Note that for $\alpha_0 = 0$ or $\alpha_1 = 0$, we are restricted to the choices $y'(a) = 0$ and $y(a)$ is arbitrary or $y(a) = 0$ and $y'(a)$ is arbitrary, respectively. If we assume that $a(x) \neq 0$ for all $x \in [a, b]$, we can solve the initial value problem

$$\begin{cases} y''(x) = -\dfrac{b(x)}{a(x)} y'(x) - \dfrac{c(x) - \lambda}{a(x)} y(x) , \\ y(a) = 1 , \\ y'(a) = -\dfrac{\alpha_0}{\alpha_1} . \end{cases} \tag{8.52}$$

The solutions are denoted by $y(x; \lambda)$ in order to emphasize that they will highly depend on the choice of the parameter λ. The strategy is to solve the initial value problem (8.52) for several values of λ and whenever one finds that

$$F(\lambda_n) = \beta_0 y(b; \lambda_n) + \beta_1 y'(b; \lambda_n) - \lambda_n = 0 , \tag{8.53}$$

is satisfied, an eigenvalue λ_n with the corresponding eigenfunction $y_n(x) = y(x; \lambda_n)$ of the eigenvalue problem (8.49) has been found.

However, this strategy is also very time consuming. The most common application of the shooting method is its combination with a very fast and accurate solution of initial value problems. This method is known as the NUMEROV method [7]. It is applicable whenever one is confronted with a differential equation of the form

$$y''(x) + k(x)y(x) = 0, \tag{8.54}$$

in combination with homogeneous boundary conditions. Here $k(x)$ is some function. If we are particularly interested in eigenvalue problems then $k(x)$ has the form $k(x) = q(x) - \lambda$, where $q(x)$ is some function and λ is the eigenvalue [see the discussion after Eq. (8.49)]. For instance, consider the one-dimensional stationary SCHRÖDINGER equation [10–12],

$$\psi''(x) + \frac{2m}{\hbar^2}\left[E - V(x)\right]\psi(x) = 0 , \tag{8.55}$$

where $\psi(x)$ is the wave-function, m is the mass, $\hbar$ denotes the reduced PLANCK constant, E is the energy, and $V(x)$ is some potential. In this case we identify

$$k(x) = \frac{2m}{\hbar^2}\left[E - V(x)\right] . \tag{8.56}$$

We note that Eq. (8.55) together with its boundary conditions defines an eigenvalue problem with eigenvalues E_n, the possible energies of the system. We remember from Chap. 2, Eq. (2.34), that

$$y_j'' = \frac{y_{j+1} - 2y_j + y_{j-1}}{h^2} - \frac{h^2}{12}y_j^{(4)} - \ldots = -k_j y_j . \tag{8.57}$$

Here we made use of Eq. (8.54) and introduced $k_j \equiv k(x_j)$. Furthermore, we write the fourth derivative of $y(x)$ at point $x = x_j$ as

$$y_j^{(4)} \approx \frac{y_{j+1}'' - 2y_j'' + y_{j-1}''}{h^2} = \frac{-k_{j+1}y_{j+1} + 2k_j y_j - k_{j-1}y_{j-1}}{h^2} , \tag{8.58}$$

where we employed Eq. (8.54). Truncating (8.57) after the fourth order derivative $y_j^{(4)}$, inserting relation (8.58), and solving for y_{j+1} yields

$$y_{j+1} = \frac{2\left(1 - \frac{5h^2}{12}k_j\right)y_j - \left(1 + \frac{h^2}{12}k_{j-1}\right)y_{j-1}}{1 + \frac{h^2}{12}k_{j+1}} . \tag{8.59}$$

This gives a very fast algorithm to solve the differential equation (8.54) with some initial values of the form (8.52). The remaining strategy is the same as discussed above, i.e. one *screens* the parameter λ in order to find the eigenvalues λ_n and eigenfunctions $y_n(x)$. In case of the SCHRÖDINGER equation, one can screen the energy E in order to obtain the energy eigenvalues E_n which satisfy a condition of the form (8.53).

Before we present two illustrating examples in the next two chapters let us conclude this section with two important remarks on the NUMEROV method. We note from Eq. (8.59) that in order to compute y_3 one already needs the function values y_1 and y_2. Usually, one obtains these values from the boundary conditions in combination with some additional condition for the problem at hand. Such an additional condition might be, for instance, the normalization of the function $y(x)$, like Eq. (8.50). We also emphasized that one has to run the NUMEROV algorithm several times for different *trial values* of the parameter λ. In order to reduce the computational cost of the method it is in many cases advantageous to store the function values q_l, where $k_i = q_i - \lambda$, in an array which is then regarded as an input argument of the NUMEROV algorithm.

Summary

We focused on linear boundary value problems defined on a finite interval $[a, b] \subset \mathbb{R}$. Most important for physics are second order boundary value problems with decoupled boundary conditions, i.e. the boundary conditions at the two different boundaries do not mix. The numerical treatment of the second order differential equation together with its boundary conditions concentrated either on the application of finite differences or on shooting methods. In the finite difference approach the methods developed in Chap. 2 were applied and the boundary conditions were incorporated directly. This resulted in a set of linear algebraic equations which was to be solved for each grid-point of the discretisized interval $[a, b]$. The case of periodic boundary conditions was also discussed in detail.

The shooting methods, on the other hand, try to link the decoupled boundary value problem to an initial value problem. This allowed the application of the methods discussed in Chap. 5. The idea was to start with some initial value at one of the two boundaries, solve the differential equation numerically and to modify the initial value iteratively until it agreed with the original boundary condition within some predefined error. Such a procedure is rather time consuming. Nevertheless, shooting methods, in particular its NUMEROV variation, proved to be very useful in the numerical solution of eigenvalue problems. This was demonstrated using the homogeneous boundary value problem of the one-dimensional stationary SCHRÖDINGER equation as an example.

Problems

1. Solve the linear second order boundary value problem

$$y''(x) + y(x) = x,$$

for $x \in [0, \pi/2]$ with the boundary conditions $y(0) = 0$ and $y(\pi/2) = 1$ analytically and then numerically with the help of finite differences.

2. Solve the linear, second order boundary value problem

$$y''(x) - 2\cos(2x)y(x) = 0,$$

on the interval $[-\pi/2, \pi/2]$ and with the boundary conditions $y(\pm\pi/2) = 1$. Use the finite difference method.

Comment: The solution can be expressed analytically in terms of so-called MATHIEU functions [13, 14] which might be intrinsically available from your computing environment. If this happens to be the case, it might be a good idea to compare the numerical solution with the analytical result.

References

1. Collatz, L.: The Numerical Treatment of Differential Equations. Springer, Berlin/Heidelberg (1960)
2. Lapidus, L., Pinder, G.F.: Numerical Solution of Partial Differential Equations. Wiley, New York (1982)
3. Stoer, J., Bulirsch, R.: Introduction to Numerical Analysis, 2nd edn. Springer, Berlin/Heidelberg (1993)
4. Asher, U.M., Mattheij, R.M.M., Russell, R.D.: Numerical Solution of Boundary Value Problems for Ordinary Differential Eqautions. Classics in Applied Mathematics, vol. 13. Cambridge University Press, Cambridge (1995)
5. Powers, D.L.: Boundary Value Problems, 6th edn. Academic, San Diego (2009)
6. Polyanin, A.D., Zaitsev, V.F.: Boundary Value Problems, 2nd edn. Chapman & Hall/CRC, Boca Raton (2003)
7. Hairer, E., Nørsett, S.P., Wanner, G.: Solving Ordinary Differential Equations I, 2nd edn. Springer Series in Computational Mathematics, vol. 8. Springer, Berlin/Heidelberg (1993)
8. Press, W.H., Teukolsky, S.A., Vetterling, W.T., Flannery, B.P.: Numerical Recipes in C++, 2nd edn. Cambridge University Press, Cambridge (2002)
9. Courant, R., Hilbert, D.: Methods of Mathematical Physics, vol. 1. Wiley, New York (1989)
10. Baym, G.: Lectures on Quantum Mechanics. Lecture Notes and Supplements in Physics. The Benjamin/Cummings, London/Amsterdam (1969)
11. Cohen-Tannoudji, C., Diu, B., Laloë, F.: Quantum Mechanics, vol. I. Wiley, New York (1977)
12. Sakurai, J.J.: Modern Quantum Mechanics. Addison-Wesley, Menlo Park (1985)
13. Abramovitz, M., Stegun, I.A. (eds.): Handbook of Mathemathical Functions. Dover, New York (1965)
14. Olver, F.W.J., Lozier, D.W., Boisvert, R.F., Clark, C.W.: NIST Handbook of Mathematical Functions. Cambridge University Press, Cambridge (2010)

Chapter 9
The One-Dimensional Stationary Heat Equation

9.1 Introduction

This is the first of two chapters which illustrate the applicability of the methods introduced in Chap. 8. Within this chapter the finite difference approach is employed to solve the stationary heat equation. Let us motivate briefly this particular problem. We consider a rod of length L which is supposed to be kept at constant temperatures T_0 and T_N at its ends as illustrated in Fig. 9.1. The homogeneous heat equation is a linear partial differential equation of the form

$$\frac{\partial}{\partial t} T(x, t) = \kappa \Delta T(x, t) . \tag{9.1}$$

Here $T(x, t)$ is the temperature as a function of space $x \in \mathbb{R}^3$ and time $t \in \mathbb{R}$, $\Delta = \nabla^2 = \partial_x^2 + \partial_y^2 + \partial_z^2$ is the LAPLACE operator, and $\kappa = \text{const}$ is the thermal diffusivity.

We remark, that Eq. (9.1) is a partial differential equation together with initial and boundary conditions. Moreover, we note in passing that the heat equation is equivalent to the diffusion equation [1]

$$\frac{\partial}{\partial t} \rho(x, t) = D \Delta \rho(x, t) , \tag{9.2}$$

with particle density $\rho(x, t)$ and the diffusion coefficient $D = \text{const}$. Here we restrict ourselves to a simplified situation in order to test the validity of the finite difference approach discussed in Sect. 8.2. The general solution of the heat or diffusion equation will be discussed in Sect. 11.3. (The problem of the one-dimensional heat equation was studied in all conceivable detail by J. R. CANNON [2].)

If we assume that the cylindrical surface of the rod is perfectly isolated, we can restrict the problem to a one-dimensional problem. Furthermore, we assume that the

B.A. Stickler, E. Schachinger, *Basic Concepts in Computational Physics*,
DOI 10.1007/978-3-319-27265-8_9

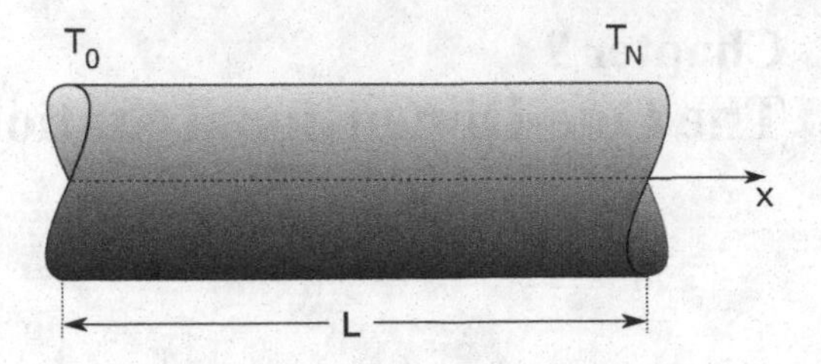

Fig. 9.1 We consider a rod of length L. Its ends are kept at constant temperatures T_0 and T_N, respectively

steady-state has been reached, i.e. $\frac{\partial}{\partial t}T(x,t) = 0$. Hence, the remaining boundary value problem is of the form

$$\begin{cases} \frac{d^2}{dx^2}T(x) = 0, & x \in [0, L] , \\ T(0) = T_0 , & \\ T(L) = T_N . & \end{cases} \tag{9.3}$$

The solution can easily be found analytically and one obtains

$$T(x) = T_0 + (T_N - T_0)\frac{x}{L} . \tag{9.4}$$

In the following section we will apply the approach of finite differences to the boundary value problem (9.3) as discussed in Sect. 8.2.

9.2 Finite Differences

We discretize the interval $[0, L]$ according Chap. 2 by the introduction of $N+1$ grid-points $x_n = nh$, with $h = L/N$, $x_0 = 0$, and $x_N = L$. Furthermore, $T_n \equiv T(x_n)$ and, in particular, we refer to the boundary conditions (9.3) as T_0 and T_N, respectively.

On the basis of this discretization, we approximate Eq. (9.3) by

$$\frac{T_{n+1} - 2T_n + T_{n-1}}{h^2} = 0 , \tag{9.5}$$

or equivalently

$$T_{n+1} - 2T_n + T_{n-1} = 0 . \tag{9.6}$$

We can rewrite this as a matrix equation,

$$AT = F , \tag{9.7}$$

where the boundary conditions have already been included. In Eq. (9.7) the vector $T = (T_1, T_2, \ldots, T_{N-1})^T$. Furthermore, the tridiagonal matrix A is given by

$$A = \begin{pmatrix} -2 & 1 & 0 & \ldots & 0 \\ 1 & -2 & 1 & 0 \ldots & 0 \\ 0 & 1 & -2 & 1 & \\ \vdots & & \ddots & \ddots & \ddots \\ 0 & \ldots & & 1 & -2 \end{pmatrix}, \tag{9.8}$$

and the vector F by

$$F = \begin{pmatrix} -T_0 \\ 0 \\ \vdots \\ 0 \\ -T_N \end{pmatrix}. \tag{9.9}$$

It is an easy task to solve Eq. (9.7) analytically. It follows from Eq. (9.6) that

$$T_{n+1} = 2T_n - T_{n-1}, \qquad n = 1, \ldots, N-1 . \tag{9.10}$$

We insert $n = 1, 2, 3$ in order to obtain

$$T_2 = 2T_1 - T_0 , \tag{9.11}$$

$$\begin{aligned} T_3 &= 2T_2 - T_1 , \\ &= 3T_1 - 2T_0 , \end{aligned} \tag{9.12}$$

$$\begin{aligned} T_4 &= 2T_3 - T_2 , \\ &= 4T_1 - 3T_0 . \end{aligned} \tag{9.13}$$

We recognize the pattern and conclude that T_n has the general form

$$T_n = nT_1 - (n-1)T_0 , \tag{9.14}$$

which we prove by complete induction:

$$\begin{aligned} T_{n+1} &= 2T_n - T_{n-1} \\ &= 2[nT_1 - (n-1)T_0] - [(n-1)T_1 - (n-2)T_0] \\ &= (n+1)T_1 - nT_0 . \end{aligned} \tag{9.15}$$

Hence, expression (9.14) is valid for all $n = 1, \ldots, N$. However, since T_N is kept constant according to the boundary condition, we can determine T_1 from

$$T_N = NT_1 - NT_0 + T_0 , \tag{9.16}$$

which yields

$$T_1 = \frac{T_N - T_0}{N} + T_0 . \tag{9.17}$$

Inserting (9.17) into (9.14) gives

$$\begin{aligned} T_n &= T_0 + (T_N - T_0)\frac{n}{N} \\ &= T_0 + (T_N - T_0)\frac{nh}{L} , \end{aligned} \tag{9.18}$$

which is exactly the discretized version of Eq. (9.4). Hence the finite difference approach to the boundary value problem (9.3) is exact and independent of the grid-spacing h. This is not surprising since we proved already in Chap. 2 that finite difference derivatives are exact for linear functions.

9.3 A Second Scenario

We consider the inhomogeneous heat equation

$$\frac{\partial}{\partial t}T(x,t) = \kappa \Delta T(x,t) - \Gamma(x,t) . \tag{9.19}$$

Here $\Gamma(x,t) \equiv \Gamma(x)$ is some heat source or heat drain, which is assumed to be independent of time t. Again, we consider the one dimensional, stationary case, i.e.

$$\frac{d^2}{dx^2}T(x) = \frac{1}{\kappa}\Gamma(x) , \tag{9.20}$$

with the same boundary conditions as in Eq. (9.4). Furthermore, we assume $\Gamma(x)$ to be of the form

$$\Gamma(x) = \frac{\Theta}{\ell}\exp\left[-\frac{\left(x - \frac{L}{2}\right)^2}{\ell^2}\right] , \tag{9.21}$$

i.e. $\Gamma(x)$ has the form of a GAUSS peak which is centered at $x = L/2$ and has a width determined by the parameter ℓ and a maximum height given by the constant Θ. Such a situation might occur, for instance, when the rod is heated with some kind of a heat gun or cooled by a cold spot. (In cases where the diffusion equation (9.2) is used to describe the random motion of electrons in some device, one can imagine, that the density of electrons ρ is constant at the contacts at $x = 0$ and $x = L$. The source/drain term $\Gamma(x)$ then accounts for a constant generation or recombination rate of electrons, for instance, through incoming light or intrinsic traps, respectively [3].)

Furthermore, we note that in the limiting case $\ell \to 0$ we have

$$\lim_{\ell \to 0} \Gamma(x) \propto \Theta \delta\left(x - \frac{L}{2}\right), \tag{9.22}$$

where $\delta(\cdot)$ is the DIRAC δ-distribution; in this case the spatial extension of the source/drain term $\Gamma(x)$ is infinitesimal.

We now employ the results of Sect. 8.2 and rewrite the system of equations in the familiar form[1]

$$AT = F, \tag{9.23}$$

where A has already been defined in Eq. (9.8), $T = (T_1, T_2, \ldots, T_{N-1})^T$, and F is given by

$$F = \frac{h^2}{\kappa} \begin{pmatrix} \Gamma_1 - \frac{\kappa}{h^2} T_0 \\ \Gamma_2 \\ \vdots \\ \Gamma_{N-2} \\ \Gamma_{N-1} - \frac{\kappa}{h^2} T_N \end{pmatrix}. \tag{9.24}$$

Here we used the notation $\Gamma_n \equiv \Gamma(x_n)$.

The system is solved numerically quite easily using methods discussed by PRESS et al. [4] for the solution of sets of algebraic equations of the kind (9.24) with tridiagonal matrix A. We chose $L = 10$, $\kappa = 1$, $\Theta = -0.4$, $\ell = 1$, $T_0 = 0$ and $T_N = 2$. The resulting temperature profiles $T(x)$ (solid line) for different values of N can be found in Figs. 9.2, 9.3, and 9.4 as well as the respective form of the function $\Gamma(x)$ (dashed line). With increasing number of steps we see, as it was to be expected, a refinement of the temperature profile. Its maximum does not quite agree with the minimum of $\Gamma(x)$, it is shifted slightly towards the end of the rod because of the boundary conditions, i.e. $T_0 < T_N$.

[1] We note that Eq. (9.20) can also be solved with the help of FOURIER transforms, see Appendix D.

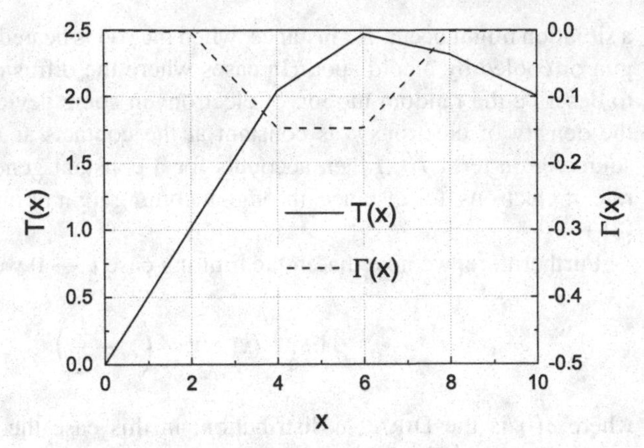

Fig. 9.2 Temperature profile $T(x)$ (*solid line, left hand scale*) and the source function $\Gamma(x)$ (*dashed line, right hand scale*) for $N = 5$

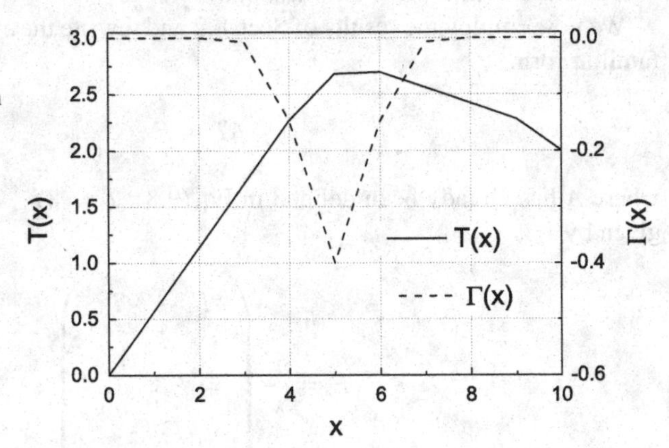

Fig. 9.3 Temperature profile $T(x)$ (*solid line, left hand scale*) and the source function $\Gamma(x)$ (*dashed line, right hand scale*) for $N = 10$

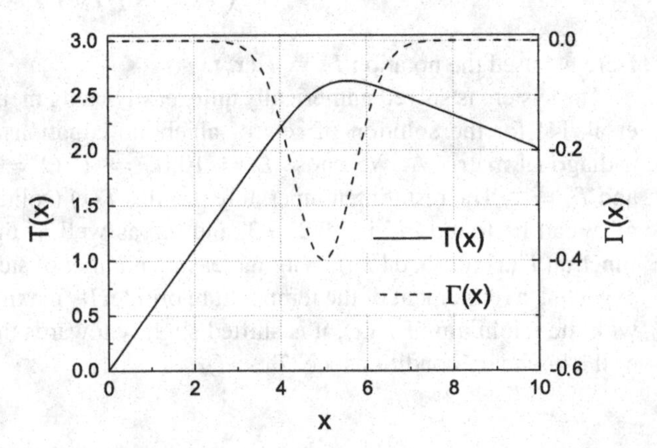

Fig. 9.4 Temperature profile $T(x)$ (*solid line, left hand scale*) and the source function $\Gamma(x)$ (*dashed line, right hand scale*) for $N = 100$

Summary

The methods of Sect. 8.2 were applied to find the numerical solution of the stationary heat equation with DIRICHLET boundary conditions. We studied the particular case of an isolated rod of length L. This reduced the dimensionality of the differential equation to one. The length of the rod was then divided into N discrete grid-points. Using finite differences the one-dimensional ordinary second order differential equation which described this particular problem was transformed into a set of linear algebraic equations which determined the temperatures at each grid-point. This set of algebraic equations was characterized by a tridiagonal coefficient matrix. Solutions have been studied without and with a heat source which was described as a 'point' source characterized by a Gaussian of given width and amplitude. In the first case analytic solutions were easily derived. They described a linear temperature profile increasing (decreasing) from T_0 to T_N. In the latter case solutions were generated numerically using specific algorithms designed for sets of algebraic equations with a tridiagonal coefficient matrix A.

Problems

1. Calculate the stationary temperature profile across the cylindrical rod of Fig. 9.1 which is exposed to a heat sink centered around $x = L/2$. This heat sink is described by a function $\Gamma(x)$ which is of rectangular shape of width a and depth θ. Both ends of the rod are kept at constant temperatures T_0 and T_N, respectively.
2. Investigate the three cases $T_0 > T_N$, $T_0 < T_N$, $T_0 = T_N > 0$, and study the influence of the width a of the heat sink on the temperature profile.
3. Consider the one-dimensional drift-diffusion equation [5]

$$\frac{\partial}{\partial t}\rho(x,t) = -D_1\frac{\partial}{\partial x}\rho(x,t) + D_2\frac{\partial^2}{\partial x^2}\rho(x,t) + \Gamma(x,t)$$

where D_1 is the drift constant and D_2 the diffusion constant. For instance, if we want to model the electron density in an electronic device, the drift constant would be in the simplest case $D_1 = \mu E$, where μ is the charge carrier mobility and E is the x-component of the electric field.

Discretize the above equation for the stationary case and solve it numerically for different values of D_1. The boundary conditions are $\rho(0) > 0$ and $\rho(L) > \rho(0)$, where L denotes the length of the device. Solve it numerically and analytically for $\Gamma(x) = 0$ and compare the results. Investigate also a scenario with a generation rate Γ given by an exponential function like $\Gamma(x) = \Gamma_0 \exp(-\lambda x)$ with $\lambda > 0$.

References

1. Cussler, E.L.: Diffusion. Cambridge University Press, Cambridge (2009)
2. Cannon, J.R.: The One-Dimensional Heat Equation. Encyclopedia of Mathematics and Its Applications. Cambridge University Press, Cambridge (1985)
3. Glicksman, M.E.: Diffusion in Solids. Wiley, New York (1999)
4. Press, W.H., Teukolsky, S.A., Vetterling, W.T., Flannery, B.P.: Numerical Recipes in C++, 2nd edn. Cambridge University Press, Cambridge (2002)
5. Jüngel, A.: Drift-Diffusion Equations. Lecture Notes in Physics, vol. 773. Springer, Berlin/Heidelberg (2009)

Chapter 10
The One-Dimensional Stationary SCHRÖDINGER Equation

10.1 Introduction

The numerical solution of the stationary SCHRÖDINGER equation is discussed to illustrate the application of NUMEROV's shooting method introduced in Sect. 8.3.

We start the discussion with a brief survey of basic quantum mechanics. Of course, this chapter is not supposed to give a self-contained introduction into this field and the reader not familiar with quantum mechanics should, therefore, regard the following discussion from a purely mathematical point of view. For more in-depth reading on quantum mechanics we refer to the books [1–4] to name a few.

A quantum-mechanical wave-function $\Psi \equiv \Psi(x,t) \in \mathbb{C}$ is a function of time $t \in \mathbb{R}^+$ and space $x \in \mathbb{R}^3$ and obeys the SCHRÖDINGER equation:

$$i\hbar \frac{\mathrm{d}}{\mathrm{d}t}\Psi = H\Psi \,. \tag{10.1}$$

Here, $\hbar = h/(2\pi)$ is the reduced PLANCK constant, i is the imaginary unit, and H is the HAMILTON operator or Hamiltonian. If $H \neq H(t)$, i.e. the Hamiltonian is independent of time t, we can employ a product ansatz

$$\Psi(x,t) = \exp\left(-\frac{i}{\hbar}Et\right)\psi(x) \,, \tag{10.2}$$

B.A. Stickler, E. Schachinger, *Basic Concepts in Computational Physics*,
DOI 10.1007/978-3-319-27265-8_10

where E is the energy and $\psi(x)$ is the time-independent part of the wave-function. This ansatz transforms Eq. (10.1) into

$$i\hbar \frac{\mathrm{d}}{\mathrm{d}t}\left[\exp\left(-\frac{i}{\hbar}Et\right)\psi(x)\right] = i\hbar\left(-\frac{i}{\hbar}E\right)\exp\left(-\frac{i}{\hbar}Et\right)\psi(x)$$
$$= \exp\left(-\frac{i}{\hbar}Et\right)H\psi(x) , \tag{10.3}$$

and $\psi(x)$ is determined by the eigenvalue problem [5]

$$H\psi = E\psi , \tag{10.4}$$

augmented by appropriate boundary conditions. We already came across Eq. (10.4) when we discussed shooting methods in Sect. 8.3.

The one-particle Hamiltonian is of the general form

$$H = T + V = \frac{P^2}{2m} + V , \tag{10.5}$$

with the kinetic energy operator T, the potential operator V, the momentum operator P, and the particle's mass m. If the system is not exposed to an external magnetic field, P can be expressed in position space by

$$P = -i\hbar\nabla , \tag{10.6}$$

and the potential operator V by $V(x)$. Thus we get for Eq. (10.5):

$$H = -\frac{\hbar^2}{2m}\Delta + V(x) . \tag{10.7}$$

Hence, we have to solve the linear, second order partial differential equation:

$$-\frac{\hbar^2}{2m}\Delta\psi(x) + V(x)\psi(x) = E\psi(x) . \tag{10.8}$$

This equation will certainly not have solutions for arbitrary values of the energy E. The particular values $E = E_n$[1] for which it has a solution are referred to as *eigenenergies* and the corresponding solution $\psi_n(x)$ is referred to as *eigenfunction*

[1]It depends on the problem on hand whether or not the index n will be continuous or discrete. For simplicity, we assume here n to be discrete.

to the eigenenergy E_n [1–3, 5]. To emphasize this point we rewrite Eq. (10.8) as:

$$-\frac{\hbar^2}{2m}\Delta\psi_n(x) + V(x)\psi_n(x) = E_n\psi_n(x)\,, \qquad \forall n \in \mathbb{N}\,. \tag{10.9}$$

It is the purpose of this chapter to develop a numerical procedure which will, in the end, allow to calculate numerically the eigenvalues E_n and eigenfunctions $\psi_n(x)$ as solutions of this equation.

We proceed in our analysis by defining the scalar product between two functions $\chi(x)$ and $\varphi(x)$

$$\langle\chi|\varphi\rangle = \int \mathrm{d}x\, \chi^*(x)\varphi(x)\,, \tag{10.10}$$

where $\chi^*(x)$ denotes the complex conjugate of $\chi(x)$. The corresponding L^2-norm reads

$$|\chi| = \sqrt{\langle\chi|\chi\rangle} = \sqrt{\int \mathrm{d}x |\chi(x)|^2}\,. \tag{10.11}$$

The expectation value of an operator O in the quantum mechanical state Ψ is given by

$$\langle O\rangle = \frac{\langle\Psi|O|\Psi\rangle}{\langle\Psi|\Psi\rangle} = \frac{\int \mathrm{d}x \Psi^*(x) O \Psi(x)}{\int \mathrm{d}x |\Psi(x)|^2}\,. \tag{10.12}$$

It follows from Eq. (10.4) that the energy is the expectation value of the Hamiltonian H

$$\langle H\rangle = \frac{\int \mathrm{d}x \Psi^*(x) H \Psi(x)}{\int \mathrm{d}x |\Psi(x)|^2} = E\,. \tag{10.13}$$

We quote now some important properties; a detailed discussion can be found in any textbook on quantum mechanics.

- The expectation value $\langle O\rangle$ of a Hermitian operator O, $O^\dagger = O$, is real, i.e. $\langle O\rangle = \langle O\rangle^*$. Here $O^\dagger$ denotes the adjoint of O, i.e. $O^\dagger = (O^*)^T$.
- Every real expectation value can be described by a Hermitian operator.
- All observables can be described by Hermitian operators, in particular, the Hamiltonian *has* to be a *Hermitian* operator to ensure that the eigenenergies E_n are real, $E_n \in \mathbb{R}$.

- It follows from the hermiticity of H that the eigenfunctions $\psi_n(x)$ form a complete, orthogonal basis in HILBERT space [5]. Furthermore, the functions can be normalized and the relation

$$\langle \psi_n | \psi_m \rangle = \delta_{nm} , \tag{10.14}$$

holds, with δ_{nm} the KRONECKER-δ.

Thus, with the help of Eq. (10.14) we rewrite the expectation value (10.12) of a Hermitian operator O as:

$$\langle O \rangle = \int \mathrm{d}x\, \Psi^*(x) O \Psi(x) . \tag{10.15}$$

In a next step we define the wave function $\Psi_n(x, t)$ following the ansatz (10.2)

$$\Psi_n(x, t) = \exp\left(-\frac{i}{\hbar} E_n t\right) \psi_n(x) , \tag{10.16}$$

and the total wave-function $\Psi(x, t)$ is then a superposition of wave-functions $\Psi_n(x, t)$

$$\Psi(x, t) = \sum_n c_n \Psi_n(x, t) , \tag{10.17}$$

because the $\Psi_n(x, t)$ constitute a complete, orthogonal basis.[2] Furthermore, we demand $\Psi(x, t)$ to be normalized for all t. Employing Eq. (10.14) in Eq. (10.17) yields

$$\int \mathrm{d}x |\Psi(x, t)|^2 = \sum_n |c_n|^2 = 1 . \tag{10.18}$$

We quote BORN's interpretation of the squared modulus of the total wave-function (referred to as BORN's rule):

$$|\Psi(x, t)|^2 \mathrm{d}x = \text{The probability that the particle described by the wave-function } \Psi(x, t) \text{ can be found at time } t \text{ within a volume } dx \text{ around the point } x. \tag{10.19}$$

[2]This is only possible because the SCHRÖDINGER equation is linear.

This interpretation justifies the requirement of a normalization of the wave-function $\Psi(x,t)$

$$\int dx|\Psi(x,t)|^2 \stackrel{!}{=} 1 , \tag{10.20}$$

because, by definition, the particle has to be found somewhere anytime.

Suppose we start with an initial state $\chi(x) = \Psi(x, t = 0)$. Since the functions $\psi_n(x)$ form a complete basis in HILBERT space, $\chi(x)$ may be written with the help of Eq. (10.17) as

$$\chi(x) = \sum_n c_n \psi_n(x) . \tag{10.21}$$

We deduce from Eq. (10.14) that

$$\langle \psi_m | \chi \rangle = \sum_n c_n \int dx \psi_m^*(x) \psi_n(x) = c_m . \tag{10.22}$$

Consequently, $|c_m|^2$ is the probability that the particle was initially in state m. This allows us to interpret Eq. (10.17) in the following way: The coefficients c_m determine the composition of the initial state. The exponential factor describes an oscillation and we note that different eigenfunctions, which correspond to different eigenenergies, oscillate with different frequencies. This can, for instance, induce the diffluence of a wave packet.

10.2 A Simple Example: The Particle in a Box

We concentrate on the one-dimensional SCHRÖDINGER equation and discuss a simple problem which will then be solved numerically in Sect. 10.3. We rewrite the one-dimensional SCHRÖDINGER equation (10.7), with $x \in \mathbb{R}$, as

$$-\frac{\hbar^2}{2m}\frac{d^2}{dx^2}\psi_n(x) + V(x)\psi_n(x) = E_n\psi_n(x) , \tag{10.23}$$

and specify

$$V(x) = \begin{cases} 0 & 0 \le x \le L, \\ \infty & \text{elsewhere,} \end{cases} \tag{10.24}$$

together with the boundary conditions

$$\psi_n(0) = \psi_n(L) = 0 , \tag{10.25}$$

and the normalization condition

$$\int \mathrm{d}x |\psi_n(x)|^2 = \int_0^L \mathrm{d}x |\psi_n(x)|^2 = 1 . \tag{10.26}$$

We note that the boundary conditions are dictated by the particular form of the potential (10.24) which requires that $\psi_n(x) = 0$ for $x \notin [0, L]$. This problem is commonly referred to as the *particle in a one-dimensional box*.

Let us introduce dimensionless variables in order to simplify the numerics of Eq. (10.23). We define new variables

$$s = \frac{x}{L}, \qquad \varepsilon_n = \frac{E_n}{\overline{E}} , \tag{10.27}$$

where L is the length scale and $\overline{E}$ is the energy scale. The energy scale $\overline{E}$ is fully determined by the relation

$$\overline{E} = \frac{\hbar^2}{mL^2} . \tag{10.28}$$

We note that $s \in [0, 1]$, hence the rescaled wave-function is given by

$$\varphi_n(s) = \sqrt{L}\psi_n(x) , \tag{10.29}$$

which satisfies the normalization condition

$$\int_0^L \mathrm{d}x |\psi_n(x)|^2 = \int_0^1 \mathrm{d}s |\varphi_n(s)|^2 = 1 . \tag{10.30}$$

The rescaled SCHRÖDINGER equation can be obtained by multiplying Eq. (10.23) with $1/\overline{E}$:

$$\begin{aligned} -\frac{\hbar^2}{2m\overline{E}} \frac{\mathrm{d}^2}{\mathrm{d}x^2} \psi_n(x) + \frac{V(x)}{\overline{E}} \psi_n(x) &= -\frac{L^2}{2} \frac{\mathrm{d}^2}{\mathrm{d}x^2} \psi_n(x) + v(s)\psi_n(x) \\ &= -\frac{1}{2} \frac{\mathrm{d}^2}{\mathrm{d}s^2} \psi_n(x) + v(s)\psi_n(x) \\ &= \frac{E_n}{\overline{E}} \psi_n(x) \\ &= \varepsilon_n \psi_n(x) . \end{aligned} \tag{10.31}$$

Here we introduced the rescaled potential $v(s)$

$$v(s) = \begin{cases} 0 & 0 \leq s \leq 1, \\ \infty & \text{elsewhere.} \end{cases} \tag{10.32}$$

Hence, the rescaled wave-function (10.29) is a solution of the differential equation:

$$-\frac{1}{2}\frac{\mathrm{d}^2}{\mathrm{d}s^2}\varphi_n(s) + v(s)\varphi_n(s) = \varepsilon_n\varphi_n(s)\ . \tag{10.33}$$

The form (10.32) of the potential implies that $\varphi_n(s) = 0$ for all $s \notin [0, 1]$ and the complete boundary value problem is defined as:

$$\begin{cases} -\frac{1}{2}\frac{\mathrm{d}^2}{\mathrm{d}s^2}\varphi_n(s) = \varepsilon_n\varphi_n(s)\ , & s \in [0, 1]\ , \\ \varphi_n(0) = 0\ , \\ \varphi_n(1) = 0\ . \end{cases} \tag{10.34}$$

It is an easy task to solve this boundary value problem analytically. For $s \in [0, 1]$ we choose the ansatz

$$\varphi_n(s) = A_n \sin(k_n s) + B_n \cos(k_n s)\ , \tag{10.35}$$

where A_n and B_n are some constants and k_n is given by

$$k_n = \sqrt{2\varepsilon_n}\ . \tag{10.36}$$

From the boundary conditions we obtain

$$\varphi_n(0) = B_n = 0\ , \tag{10.37}$$

and

$$\varphi_n(1) = A_n \sin(k_n) = 0\ . \tag{10.38}$$

Thus,

$$k_n = n\pi \,, \tag{10.39}$$

and the eigenenergies ε_n are quantized:

$$\varepsilon_n = \frac{n^2\pi^2}{2} \,. \tag{10.40}$$

The corresponding eigenfunctions $\varphi_n(s)$ are then given by:

$$\varphi_n(s) = \begin{cases} A_n \sin(n\pi s) & s \in [0,1] \,, \\ 0 & \text{elsewhere.} \end{cases} \tag{10.41}$$

The constants A_n are determined from the normalization condition (10.30)[3]

$$\begin{aligned} \int_0^1 \mathrm{d}s |\varphi_n(s)|^2 &= A_n^2 \int_0^1 \mathrm{d}s \sin^2(n\pi s) \\ &= \frac{A_n^2}{2} \\ &\stackrel{!}{=} 1 \,, \end{aligned} \tag{10.43}$$

and:

$$A_n = \sqrt{2} \,. \tag{10.44}$$

Finally, we apply the relations (10.27), (10.28), and (10.29) and obtain

$$\psi_n(x) = \frac{1}{\sqrt{L}} \varphi_n\left(\frac{x}{L}\right) = \begin{cases} \sqrt{\frac{2}{L}} \sin\left(\frac{n\pi x}{L}\right) & x \in [0,L] \,, \\ 0 & \text{elsewhere,} \end{cases} \tag{10.45}$$

[3] Here we make use of

$$\int \mathrm{d}u \sin^2(u) = \frac{1}{2} \left[u - \cos(u)\sin(u)\right] \,. \tag{10.42}$$

and

$$E_n = \varepsilon_n \overline{E} = \frac{\hbar^2 \pi^2 n^2}{2mL^2} . \tag{10.46}$$

In most cases expectation values of some observables are to be determined. We might, for instance, be interested in the expectation value $\langle x \rangle$ of the position operator x or its variance $\mathrm{var}\,(x) = \left\langle (x - \langle x \rangle)^2 \right\rangle$ (see also Appendix E). It follows from Eq. (10.27) that

$$\langle x \rangle = L \langle s \rangle , \quad \text{and} \quad \left\langle (x - \langle x \rangle)^2 \right\rangle = L^2 \left\langle (s - \langle s \rangle)^2 \right\rangle = L^2 \mathrm{var}\,(s) . \tag{10.47}$$

Definition (10.15) gives together with solution (10.41) the expectation value $\langle s \rangle$:

$$\langle s \rangle = 2 \int_0^1 \mathrm{d}s \sin^2(n\pi s) s = \frac{1}{2} . \tag{10.48}$$

Thus, the expectation value of the position operator is independent of the quantum number n. Furthermore, we obtain for $\langle s^2 \rangle$:

$$\left\langle s^2 \right\rangle = 2 \int_0^1 \mathrm{d}s \sin^2(n\pi s) s^2 = \frac{1}{3} - \frac{1}{2n^2\pi^2} . \tag{10.49}$$

Hence, the variance $\mathrm{var}\,(s)$ is determined by

$$\left\langle (s - \langle s \rangle)^2 \right\rangle = \left\langle s^2 \right\rangle - \langle s \rangle^2 = \frac{1}{3} - \frac{1}{2n^2\pi^2} - \frac{1}{4} = \frac{1}{12}\left(1 - \frac{6}{n^2\pi^2}\right) . \tag{10.50}$$

We note that the variance increases with increasing n.

In the next section these results are reproduced numerically by the NUMEROV shooting method. (See Sect. 8.3 and, for instance, Ref. [6].) Moreover, this numerical formulation will allow us to find solutions for more complex potentials $V(x)$.

10.3 Numerical Solution

The following discussion is based on the scaled SCHRÖDINGER equation (10.33), but we consider now a more general potential of the form

$$v(s) = \begin{cases} \tilde{v}(s) & 0 \le s \le 1 , \\ \infty & \text{elsewhere,} \end{cases} \tag{10.51}$$

which results in the boundary value problem:

$$\begin{cases} -\frac{1}{2}\frac{\mathrm{d}^2}{\mathrm{d}s^2}\varphi_n(s) + \tilde{v}(s)\varphi_n(s) = \varepsilon_n\varphi_n(s) \quad s \in [0,1], \quad n \in \mathbb{N}\,, \\ \varphi_n(0) = 0\,, \\ \varphi_n(1) = 0\,. \end{cases} \tag{10.52}$$

As our numerical treatment will be based on shooting methods, discussed in Sect. 8.3, the second order differential equation in Eq. (10.52) will be transformed into a form which corresponds to Eq. (8.54), namely:

$$\varphi_n''(s) + 2\left[\varepsilon_n - \tilde{v}(s)\right]\varphi_n(s) = 0\,. \tag{10.53}$$

The interval $[0,1]$ is discretized using $N+1$ grid-points $s_\ell = \ell/N$, $\ell = 0,1,2,\ldots,N$ $(h = 1/N)$ and we denote with $\varphi_n(s_\ell)$ and $\tilde{v}(s_\ell) \equiv v_\ell$ the values of $\varphi_n(s)$ and $\tilde{v}(s)$ at those grid-points. Thus, Eq. (8.59) can immediately be applied and we get:

$$\varphi_n(s_{\ell+1}) = \frac{2\left[1 - \frac{5}{6N^2}\left(\varepsilon_n - \tilde{v}_\ell\right)\right]\varphi_n(s_\ell) - \left[1 + \frac{1}{6N^2}\left(\varepsilon_n - \tilde{v}_{\ell-1}\right)\right]\varphi_n(s_{\ell-1})}{1 + \frac{1}{6N^2}\left(\varepsilon_n - \tilde{v}_{\ell+1}\right)}\,. \tag{10.54}$$

We use the initial conditions $\varphi_n(s_0) = 0$ and $\varphi_n'(s_0) = 1$, which is always possible, since (10.52) is a homogeneous boundary value problem. This gives

$$\varphi_n'(s_0) \approx \frac{\varphi_n(s_1) - \varphi_n(s_{-1})}{2h} = 1 \quad \Rightarrow \quad \varphi_n(s_1) = \frac{2}{N}\,. \tag{10.55}$$

The normalization of the wave-function (10.30) is then approximated with the help of the forward rectangular rule (3.9):

$$\int_0^1 \mathrm{d}s|\varphi_n(s)|^2 \approx h\sum_{\ell=0}^{N}\left[\varphi_n(s_\ell)\right]^2 \stackrel{!}{=} 1\,. \tag{10.56}$$

Consistently, we approximate the expectation value $\langle s\rangle$ via

$$\int_0^1 \mathrm{d}s\, s|\varphi_n(s)|^2 \approx h\sum_{\ell=0}^{N}\left[\varphi_n(s_\ell)\right]^2 s_\ell\,. \tag{10.57}$$

Table 10.1 Comparison between analytic and numerical eigenenergies for the particle in a box for $N = 100$

n	ε_n-analytic	ε_n-numeric
1	4.934802	4.934802
2	19.739209	19.739208
3	44.413219	44.413205
4	78.956835	78.956753
5	123.370055	123.369742
6	177.652879	177.651943
7	241.805308	241.802947
8	315.827341	315.822077
9	399.718978	399.708300
10	493.480220	493.460113

The NUMEROV shooting algorithm is then defined by the following steps:

1. Choose two trial energies ε_a and ε_b and define the required accuracy η.
2. Calculate $\varphi(s_N; \varepsilon_a) \equiv \varphi_a$ and $\varphi(s_N; \varepsilon_b) \equiv \varphi_b$ using Eq. (10.54).
3. If $\varphi_a \varphi_b > 0$, choose new values for ε_a or ε_b and go to step 1.
4. If $\varphi_a \varphi_b < 0$, calculate $\varepsilon_c = (\varepsilon_a + \varepsilon_b)/2$ and determine $\varphi(s_N; \varepsilon_c) \equiv \varphi_c$ using Eq. (10.54).
5. If $\varphi_a \varphi_c < 0$, set $\varepsilon_b = \varepsilon_c$ and go to step 4.
6. If $\varphi_c \varphi_b < 0$, set $\varepsilon_a = \varepsilon_c$ and go to step 4.
7. Terminate the loop when $|\varepsilon_a - \varepsilon_b| < \eta$.

These steps have been carried out for 100 grid-points, a potential $\tilde{v} = 0$, and a required accuracy of $\eta = 10^{-10}$. The first ten eigenenergies are given in Table 10.1 and are compared with analytic results (10.40).

In addition, Fig. 10.1 presents the first five eigenvalues ε_n (right hand scale) as horizontal straight lines. Aligned with these eigenvalues we find on the right hand side of this figure the corresponding normalized eigenfunctions calculated using $N = 100$ grid-points. The agreement with the analytic result of Eq. (10.41) is excellent.

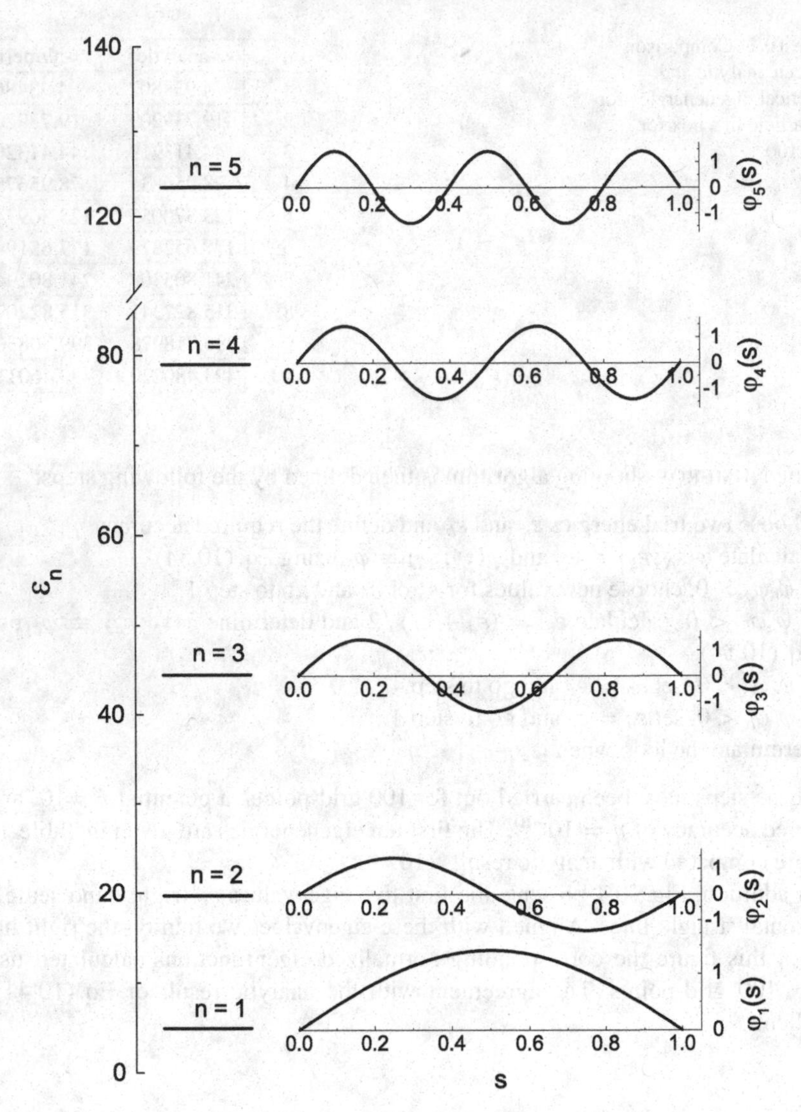

Fig. 10.1 The first five numerically determined eigenvalues ε_n of Table 10.1 are presented as *horizontal lines (left hand scale)*. Aligned with these eigenvalues are the corresponding eigenfunctions $\varphi_n(s)$ vs s for $N = 100$ (*right hand scales*)

10.4 Another Case

Here we discuss briefly some results achieved with the help of NUMEROV's shooting algorithm. In particular, we discuss the particle in the box for three different potentials $\tilde{v}(s)$

$$\tilde{v}_1(s) = 50\cos(\pi s)\,, \quad \tilde{v}_2(s) = 50\exp\left[-\frac{\left(s-\frac{1}{2}\right)^2}{0.08}\right]\,, \quad \tilde{v}_3(s) = 50s\,. \tag{10.58}$$

The potentials are illustrated in Fig. 10.2. All calculations were carried out with $N = 100$ grid-points and an accuracy $\eta = 10^{-10}$. The first five eigenenergies ε_n are shown in Figs. 10.3, 10.4, and 10.5, respectively, as horizontal lines (left hand scale). The numerically determined normalized eigenfunctions $\varphi_n(s)$ vs s (solid lines) are presented on the right hand side of these figures and are aligned with their respective eigenvalues. They are also compared with the eigenfunctions (dotted lines) of the particle in a box, i.e. $\tilde{v}(s) = 0$. In all cases the eigenfunctions reflect the symmetry of the various potentials $\tilde{v}(s)$ which becomes particularly transparent in Fig. 10.3 for the potential $\tilde{v}_1(s)$. The eigenfunctions develop an additional node in comparison to the eigenfunctions calculated for $\tilde{v}(s) = 0$. In the other two cases only the very first eigenfunctions $n \geq 3$ appear to be affected by the potential. Moreover, in all three cases, the respective eigenvalues are shifted towards higher values which is consistent with a general result of quantum mechanical perturbation theory.

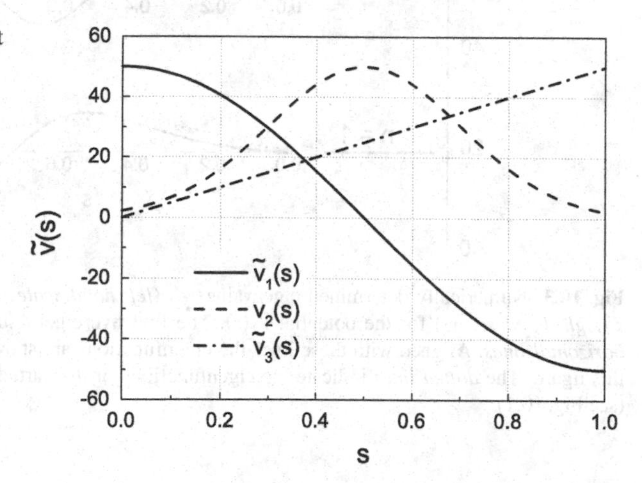

Fig. 10.2 The three different potentials $\tilde{v}(s)$ defined in Eq. (10.58)

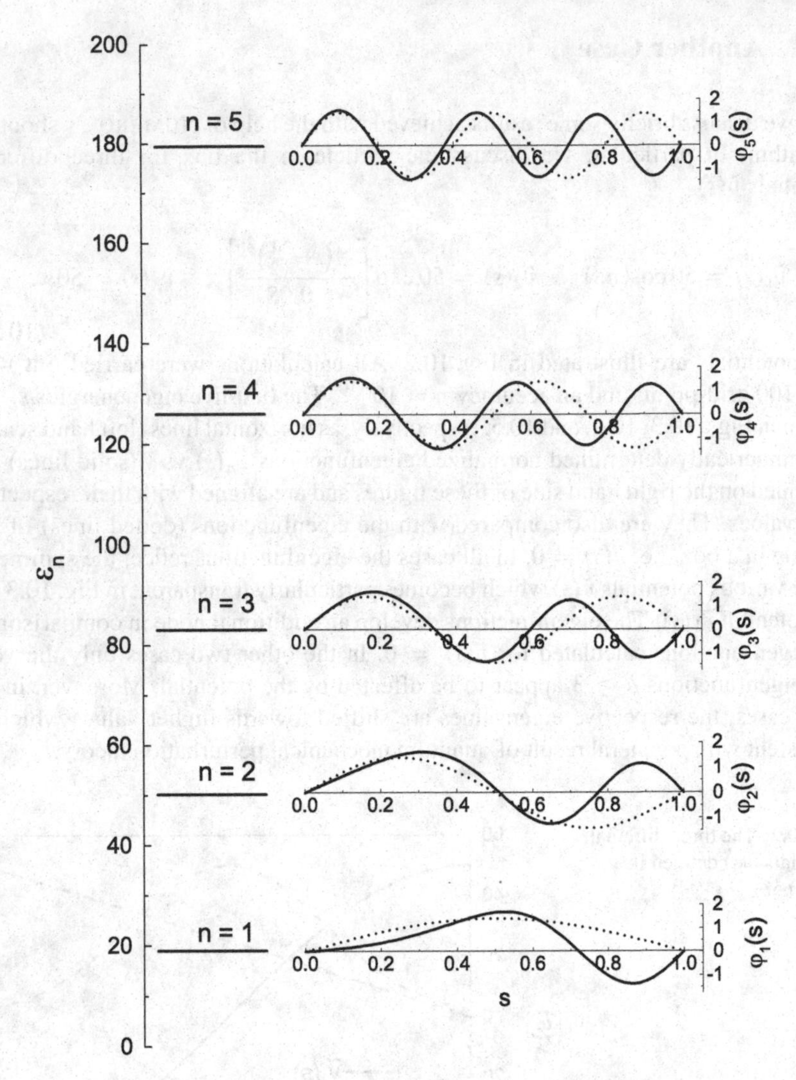

Fig. 10.3 Numerically determined eigenvalues ε_n (*left hand scale*) and eigenfunctions $\varphi_n(s)$ vs s (*right hand scales*) for the potential $\widetilde{v}_1(s)$. The first five eigenvalues are presented as straight *horizontal lines*. Aligned with these lines the eigenfunctions are shown on the *right hand side* of this figure. The *dotted lines* indicate the eigenfunctions of the particle in the box with $\widetilde{v}(s) = 0$ (see Fig. 10.1)

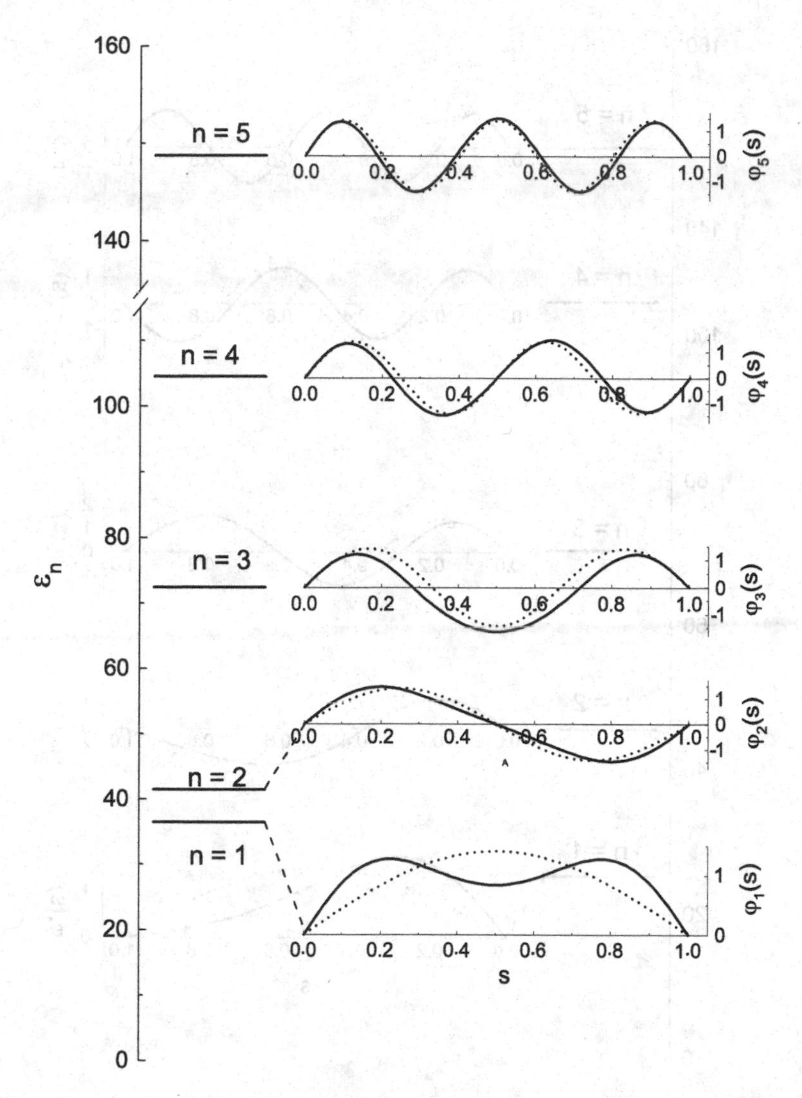

Fig. 10.4 Numerically determined eigenvalues ε_n (*left hand scale*) and eigenfunctions $\varphi_n(s)$ vs s (*right hand scales*) for the potential $\tilde{v}_2(s)$. The first five eigenvalues are presented as straight *horizontal lines*. Aligned with these lines the eigenfunctions are shown on the *right hand side* of this figure. The *dotted lines* indicate the eigenfunctions of the particle in the box with $\tilde{v}(s) = 0$ (see Fig. 10.1)

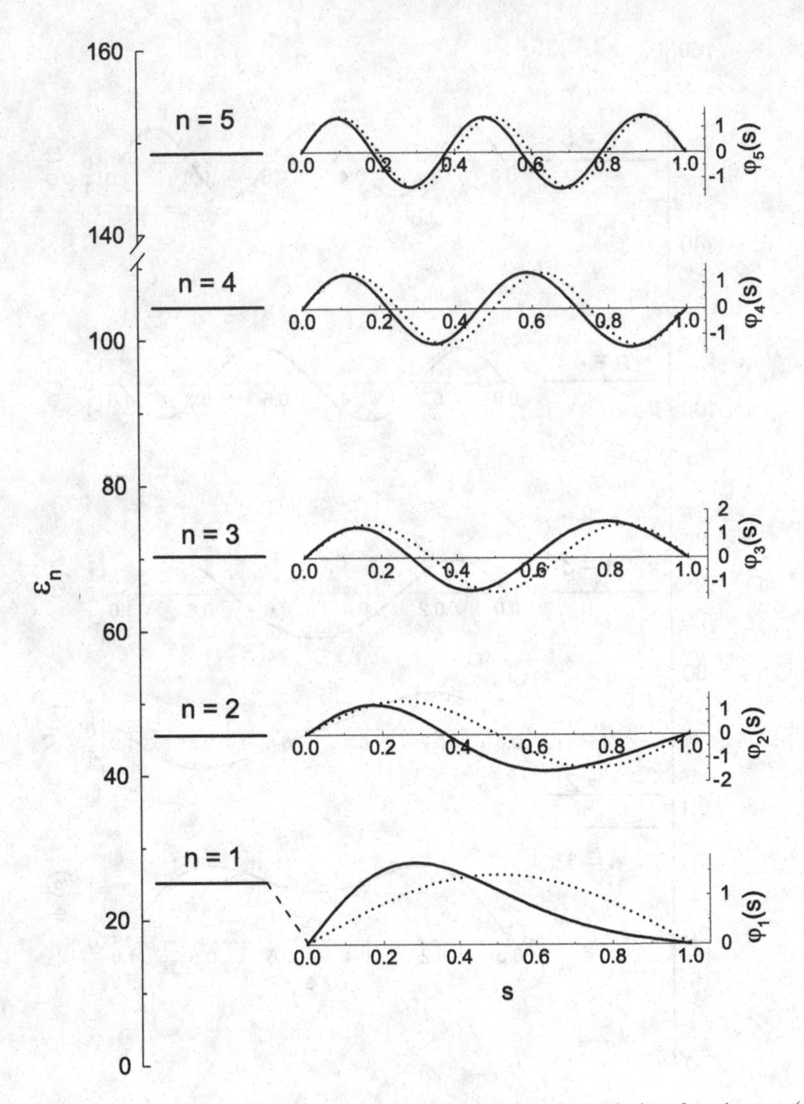

Fig. 10.5 Numerically determined eigenvalues ε_n (*left hand scale*) and eigenfunctions $\varphi_n(s)$ vs s (*right hand scales*) for the potential $\tilde{v}_3(s)$. The first five eigenvalues are presented as straight *horizontal lines*. Aligned with these lines the eigenfunctions are shown on the *right hand side* of this figure. The *dotted lines* indicate the eigenfunctions of the particle in the box with $\tilde{v}(s) = 0$ (see Fig. 10.1)

Summary

The quantum-mechanical problem of a particle in a box was described by a homogeneous boundary value problem which could be solved analytically if the box' potential $\tilde{v}(s) = 0$. On the other hand, NUMEROV's shooting algorithm was particularly designed to treat effectively homogeneous boundary value problems. Consequently, the problem of the particle in the box was used to design a NUMEROV shooting algorithm which was then tested against the analytic results. The agreement between numerics and analytical results turned out to be excellent and proved the quality of the method. For illustrative purposes the problem of the particle in the box was then solved numerically for three different, more complex, box-potentials $\tilde{v}(s)$.

Problems

1. Solve the one-dimensional stationary SCHRÖDINGER equation in an infinitely deep potential well by employing the shooting method according to NUMEROV of Sect. 8.3. The total potential $v(s)$ is assumed to be of the form (10.51). Choose different potentials $\tilde{v}(s)$ within the well.

   You can check your code by reproducing the results presented in Sects. 10.3 and 10.4. In addition, determine numerically the expectation value $\langle x \rangle$ and the variance var (x) of the position operator x for the first five eigenfunctions. This can be achieved by employing the rectangular rule of Chap. 3, as illustrated in Eq. (10.57).
2. Solve the SCHRÖDINGER equation for some potential $\tilde{v}(s)$ of your choice and plot the first five eigenfunctions. This potential should not be equal to one of the potentials discussed in this chapter. Again, calculate $\langle x \rangle$ and var (x) for the first five eigenfunctions.
3. Solve the stationary SCHRÖDINGER equation (10.4) for the harmonic potential $V(x) = m\omega^2 x^2/2$. The algorithm discussed in this chapter can be applied by choosing the box length L sufficiently large, so that the harmonic oscillator potential is well within the box for all energies of interest.
4. Solve the SCHRÖDINGER equation for a double well potential which can be obtained by adding two mutually displaced harmonic potentials.

References

1. Baym, G.: Lectures on Quantum Mechanics. Lecture Notes and Supplements in Physics. The Benjamin/Cummings Publ. Comp., Inc., London/Amsterdam (1969)
2. Cohen-Tannoudji, C., Diu, B., Laloë, F.: Quantum Mechanics, vol. I. Wiley, New York (1977)
3. Sakurai, J.J.: Modern Quantum Mechanics. Addison-Wesley, Menlo Park (1985)

4. Ballentine, L.E.: Quantum Mechanics. World Scientific, Hackensack (1998)
5. Courant, R., Hilbert, D.: Methods of Mathematical Physics, vol. 1. Wiley, New York (1989)
6. Hairer, E., Nørsett, S.P., Wanner, G.: Solving Ordinary Differential Equations I. Springer Series in Computational Mathematics, vol. 8, 2nd edn. Springer, Berlin/Heidelberg (1993)

Chapter 11
Partial Differential Equations

11.1 Introduction

This section discusses some fundamental aspects of the numerics of partial differential equations and it will be restricted to methods already encountered in previous chapters, i.e. on finite difference methods. These are particularly useful to find solutions of linear partial differential equations (PDEs). Nonlinear PDEs, such as the NAVIER-STOKES equations, require more advanced techniques as there are finite element methods or finite volume methods for conservation laws. A detailed discussion of a wide spectrum of methods can be found in many textbooks on the numerics of PDEs the interested reader is referred to [1–7].

Since we already introduced the concepts of finite difference derivatives in Chap. 2 and their application to boundary value problems of ordinary differential equations in Sect. 8.2, we concentrate mainly on the application of these methods to specific types of PDEs. In detail, we investigate the POISSON equation as an example for *elliptic PDEs*, the time dependent heat equation as an example for *parabolic PDEs*, and the wave equation as an example for *hyperbolic PDEs*. The concepts presented here are, of course, also applicable to other problems. However, in contrast to the numerics of ordinary differential equations, there exists no general recipe for the solution of PDEs.

Another important point to note is that, as in the theory of ordinary differential equations, the problem is only fully determined when initial and/or boundary conditions have been defined. For instance, in the case of the POISSON equation only boundary conditions are required, while for the time-dependent heat equation initial conditions are required as well. In general, pure boundary value problems are easier from a numerical point of view because the question whether or not the algorithm is stable does not play such an important role. For combined boundary

B.A. Stickler, E. Schachinger, *Basic Concepts in Computational Physics*,
DOI 10.1007/978-3-319-27265-8_11

and initial value problems it is essential to check carefully that the discretization of the time axis is not in conflict with the discretization of the space domain. This is of particular importance in the numerical treatment of hyperbolic PDEs, where the so called COURANT-FRIEDRICHS-LEWY (CFL) condition determines the stability of the algorithm. We shall come back to this point in Sects. 11.3 and 11.4. Finally, we conclude this chapter with a discussion of the numerical solution of the time-dependent SCHRÖDINGER equation.

11.2 The POISSON Equation

We consider the POISSON equation as a model for an elliptic PDE [8, 9]. Nevertheless, we review briefly some basics of electrodynamics [10, 11]. The force $F(r, t)$ as a function of position $r \in \mathbb{R}^3$ and time $t \in \mathbb{R}^+$ acting on a particle with charge q, which moves with velocity v within an electromagnetic field described by the electric field $E(r, t)$ and the magnetic field $B(r, t)$, is determined from:

$$F(r, t) = q\left[E(r, t) + v \times B(r, t)\right] . \tag{11.1}$$

We consider here the electrostatic case which is characterized by a zero magnetic field $[B(r, t) = 0]$ and a time independent electric field. The electric field E itself is described by the equation

$$\nabla \cdot E(r) = \frac{1}{\epsilon_0}\rho(r) , \tag{11.2}$$

where the charge density $\rho(r, t)$ acts as the source of the electric field $E(r, t)$. Here ϵ_0 is the dielectric permittivity of vacuum. Furthermore, the electric field E is connected to the electrostatic potential $\varphi(r)$ via

$$E(r) = -\nabla\varphi(r) . \tag{11.3}$$

Thus, Eq. (11.2) is reformulated as:

$$\Delta\varphi(r) = -\frac{\rho(r)}{\epsilon_0} . \tag{11.4}$$

This equation is referred to as the POISSON equation and in the particular case of $\rho(r) = 0$ it is referred to as the LAPLACE equation [12].

We focus now on the numerical solution of the two dimensional POISSON equation (11.4) on a rectangular domain $\Omega = [0, L_x]\times[0, L_y]$ together with boundary conditions $\varphi(x, y) = g(x, y)$ on $\partial\Omega$. In detail, we want to solve the two-dimensional boundary value problem

$$\begin{cases} \frac{\partial^2}{\partial x^2}\varphi(x, y) + \frac{\partial^2}{\partial y^2}\varphi(x, y) = -\rho(x, y) \,, & (x, y) \in \Omega \,, \\ \varphi(x, y) = g(x, y) \,, & (x, y) \in \partial\Omega \,, \end{cases} \tag{11.5}$$

where we absorbed ϵ_0 into the charge density $\rho(x, y)$. Note that a treatment of the three dimensional case can be carried out in analogue.

We employ a finite difference approximation to the derivatives which appear in Eq. (11.5) (see Chap. 2) and we define grid-points in x and y direction via

$$x_i = x_0 + ih_x, \quad i = 0, 1, 2, \ldots, n \,, \tag{11.6a}$$

$$y_j = y_0 + jh_y, \quad j = 0, 1, 2, \ldots, m \,, \tag{11.6b}$$

where h_x and h_y denote the grid-spacing in x- and y-direction, respectively. As discussed in Chap. 2 we consider only equally spaced grid-points. An extension to non-uniform grids is straight forward.

We define the function values on the grid-points as

$$\varphi_{i,j} \equiv \varphi(x_i, y_j) \,, \tag{11.7}$$

and similarly $\rho_{i,j} \equiv \rho(x_i, y_j)$. Consequently, we find the finite difference approximation of Eq. (11.5):

$$\frac{\varphi_{i-1,j} - 2\varphi_{i,j} + \varphi_{i+1,j}}{h_x^2} + \frac{\varphi_{i,j-1} - 2\varphi_{i,j} + \varphi_{i,j+1}}{h_y^2} = -\rho_{i,j} \,. \tag{11.8}$$

The boundary conditions (11.5) can be written as

$$\varphi_{0,j} = g_{0,j} \,, \qquad j = 0, 1, \ldots, m \,, \tag{11.9a}$$

$$\varphi_{n,j} = g_{n,j} \,, \qquad j = 0, 1, \ldots, m \,, \tag{11.9b}$$

$$\varphi_{i,0} = g_{i,0} \,, \qquad i = 1, 2, \ldots, n-1 \,, \tag{11.9c}$$

$$\varphi_{i,m} = g_{i,m} \,, \qquad i = 1, 2, \ldots, n-1 \,. \tag{11.9d}$$

Equation (11.8) is multiplied by $-h_x^2 h_y^2/2$ and we obtain after rearranging terms

$$\left(h_x^2 + h_y^2\right)\varphi_{i,j} - \frac{1}{2}\left[h_y^2\left(\varphi_{i-1,j} + \varphi_{i+1,j}\right) + h_x^2\left(\varphi_{i,j-1} + \varphi_{i,j+1}\right)\right] = \frac{(h_x h_y)^2}{2}\rho_{i,j} \,, \tag{11.10}$$

for $i = 1, \ldots, n$ and $j = 1, \ldots, m$. There are different strategies how this set of equations might be solved. The common strategy is to employ the assignments

$$\begin{aligned} \varphi_{1,1} &\to \varphi_1 \,, \\ \varphi_{1,2} &\to \varphi_2 \,, \\ &\vdots \\ \varphi_{n,m} &\to \varphi_\ell \,, \end{aligned} \tag{11.11}$$

where $\ell = nm$. Equation (11.10) is then rewritten as a matrix equation with a vector of unknowns $\varphi = (\varphi_1, \varphi_2, \ldots, \varphi_\ell)^T$ according to Eq. (11.11). The boundary conditions are to be included in the matrix. This matrix equation is then solved either by direct or iterative methods as they are discussed in Appendix C.

It is our plan to solve Eq. (11.10) iteratively. This requires the introduction of a superscript iteration index t and $\varphi_{i,j}^t$ denotes the function value $\varphi(x_i, y_j)$ after t-iteration steps. There are two different implementations of an iterative solution, namely the GAUSS-SEIDEL or the JACOBI method (Appendix C). They differ only in the update procedure of the function values $\varphi_{i,j}^t$ at the grid-points. The basic idea is to develop an update algorithm which expresses the function values $\varphi_{i,j}^t$ with the help of function values at already updated grid-points and of function values $\varphi_{i,j}^{t-1}$ determined in the preceding iteration step [Appendix, Eq. (C.27)].

We formulate this iteration rule as

$$\begin{aligned} \varphi_{i,j}^{t+1} = {} & \frac{(h_x h_y)^2}{2(h_x^2 + h_y^2)}\rho_{i,j} + \frac{1}{2(h_x^2 + h_y^2)}\left[h_y^2\left(\varphi_{i-1,j}^{t+1} + \varphi_{i+1,j}^{t}\right)\right. \\ & \left. + h_x^2\left(\varphi_{i,j-1}^{t+1} + \varphi_{i,j+1}^{t}\right)\right] , \end{aligned} \tag{11.12}$$

where we abstained from incorporating a relaxation parameter (see Appendix C). Note that by using the iteration rule (11.12) the boundary conditions have to be accounted for in an additional step.

Let us specify the boundary conditions for a concrete problem: We we want to determine the electrostatic potential of an electric monopole, dipole, and quadrupole, respectively. They are placed inside a grounded box of dimensions L_x and L_y. Thus, we have to impose DIRICHLET boundary conditions $\varphi(0, y) = \varphi(L_x, y) = 0$ in x-direction and $\varphi(x, 0) = \varphi(x, L_y) = 0$ in y-direction. In this particular case the boundary conditions can be made part of Eq. (11.12) by restricting the loop over the x-grid (y-grid) to $i = 2, \ldots, N-1$ which leaves the boundary points $\varphi(0, y)$ [$\varphi(x, 0)$] and $\varphi(L_x, y)$ [$\varphi(x, L_y)$] unchanged. Furthermore we set $L_x = L_y = 10$, the number of grid-points on both axes to $n = m = 100$, and define the domains:

$$\Omega_1 = \left(x_{\frac{n}{2}-10}, x_{\frac{n}{2}}\right] \times \left(y_{\frac{m}{2}-10}, y_{\frac{m}{2}}\right] , \tag{11.13a}$$

$$\Omega_2 = \left(x_{\frac{n}{2}}, x_{\frac{n}{2}+10}\right] \times \left(y_{\frac{m}{2}-10}, y_{\frac{m}{2}}\right] , \tag{11.13b}$$

$$\Omega_3 = \left(x_{\frac{n}{2}-10}, x_{\frac{n}{2}}\right] \times \left(y_{\frac{m}{2}}, y_{\frac{m}{2}+10}\right] , \tag{11.13c}$$

$$\Omega_4 = \left(x_{\frac{n}{2}}, x_{\frac{n}{2}+10}\right] \times \left(y_{\frac{m}{2}}, y_{\frac{m}{2}+10}\right] . \tag{11.13d}$$

The charge density $\rho(x, y)$ is described by three different scenarios, namely the electric monopole

$$\rho_1(x, y) = \begin{cases} 50 & (x, y) \in \Omega_1 \cup \Omega_2 \cup \Omega_3 \cup \Omega_4 , \\ 0 & \text{elsewhere,} \end{cases} \tag{11.14a}$$

the electric dipole

$$\rho_2(x, y) = \begin{cases} 50 & (x, y) \in \Omega_1 \cup \Omega_2 , \\ -50 & (x, y) \in \Omega_3 \cup \Omega_4 , \\ 0 & \text{elsewhere,} \end{cases} \tag{11.14b}$$

and the electric quadrupole:

$$\rho_3(x, y) = \begin{cases} 50 & (x, y) \in \Omega_1 \cup \Omega_4 , \\ -50 & (x, y) \in \Omega_2 \cup \Omega_3 , \\ 0 & \text{elsewhere.} \end{cases} \tag{11.14c}$$

These charge densities are illustrated in Fig. 11.1.

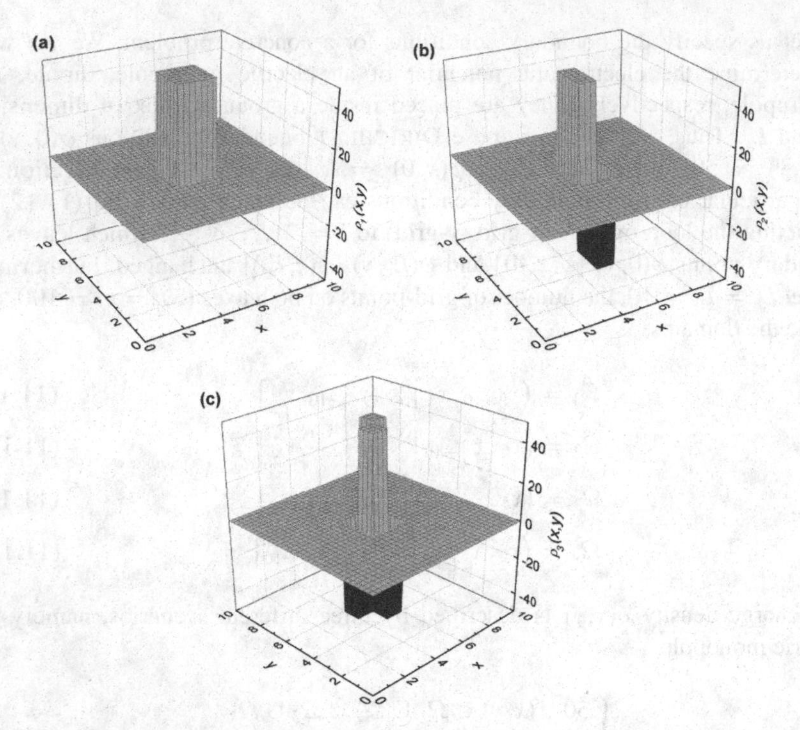

Fig. 11.1 The electric monopole, dipole, and quadrupole charge densities **(a)** $\rho_1(x, y)$, **(b)** $\rho_2(x, y)$, and **(c)** $\rho_3(x, y)$, respectively, as defined in Eq. (11.14)

The solution of Eq. (11.12) is regarded to be converged if the potential $\varphi(x, y)$ does not change significantly between two consecutive iteration steps, i.e.

$$\max_{i,j} \left(|\varphi^t_{i,j} - \varphi^{t-1}_{i,j}| \right) < \eta \,, \tag{11.15}$$

where $\eta = 10^{-4}$ is the required accuracy. A criterion to check the relative change can be formulated in a similar fashion. The resulting potential profiles $\varphi(x, y)$ are presented in Fig. 11.2. They reflect perfectly the symmetries of the charge densities $\rho_1(x, y)$, $\rho_2(x, y)$, and $\rho_3(x, y)$, respectively. Finally, standard finite difference methods can be applied to calculate, based on Eq. (11.3), the electric field $E(x, y)$ from the potential profiles $\varphi(x, y)$.[1]

[1]We note that the electrostatic potentials that we calculated here numerically can also be determined analytically with the method of mirror charges [10].

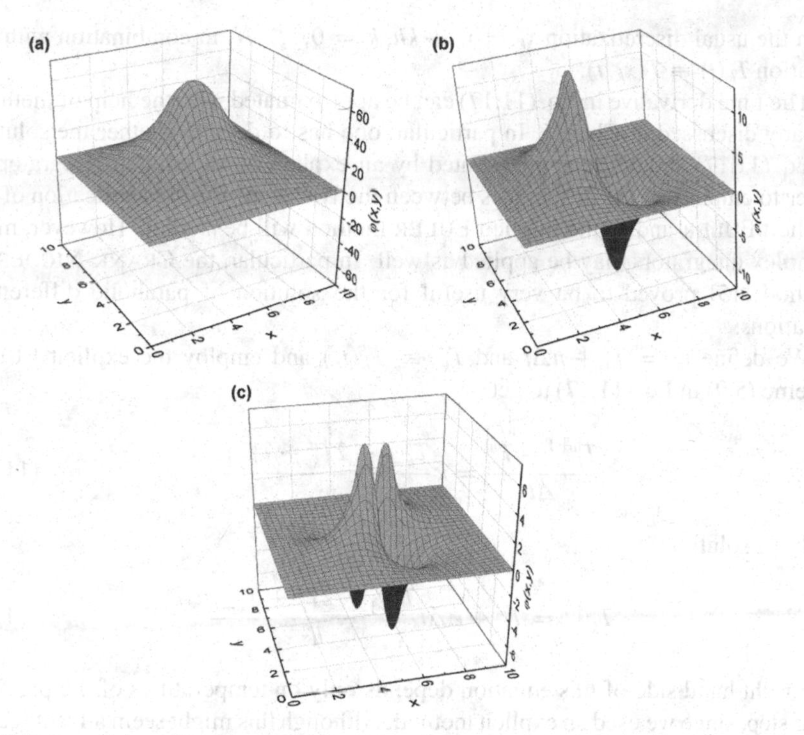

Fig. 11.2 Potential profile $\varphi(x, y)$ obtained for charge density **(a)** $\rho_1(x, y)$, **(b)** $\rho_2(x, y)$, and **(c)** $\rho_3(x, y)$

11.3 The Time-Dependent Heat Equation

We discuss here the numerical solution of the time-dependent heat equation [13, 14] which is a representative of parabolic PDEs. This equation has already been introduced in Sect. 9.1, Eq. (9.1), and is, reduced to the one-dimensional case, of the form

$$\frac{\partial}{\partial t} T(x, t) = \kappa \frac{\partial^2}{\partial x^2} T(x, t) , \tag{11.16}$$

with the thermal diffusivity κ. It is augmented by appropriate boundary and initial conditions. Again, we will not discuss the extension to higher dimensions since it is straight forward, however, maybe tedious in the general case. We approximate the right hand side of Eq. (11.16) with the help of the central finite difference approximation (Sect. 2.2) and obtain

$$\frac{\partial}{\partial t} T_k(t) = \kappa \frac{T_{k-1}(t) - 2T_k(t) + T_{k+1}(t)}{h^2} , \tag{11.17}$$

with the usual discretization $x_k = x_0 + kh$, $k = 0, \dots, N$, in combination with the notation $T_k(t) \equiv T(x_k, t)$.

The time derivative in Eq. (11.17) can be approximated with the help of methods already discussed in Chap. 5. In particular, one has to decide whether the solution of Eq. (11.17) should be approximated by an explicit or an implicit integrator. In order to emphasize the differences between the two methods, the application of the explicit EULER and of the implicit EULER method will be studied. However, more complex integrators may be applied as well. In particular, the CRANK-NICOLSON method [15] proved to be very useful for the solution of parabolic differential equations.

We define $t_n = t_0 + n\Delta t$ and $T_k^n \equiv T_k(t_n)$ and employ the explicit EULER scheme (5.9) in Eq. (11.17) to get

$$\frac{T_k^{n+1} - T_k^n}{\Delta t} = \kappa \frac{T_{k-1}^n - 2T_k^n + T_{k+1}^n}{h^2}, \tag{11.18}$$

with the solution:

$$T_k^{n+1} = T_k^n + \kappa \Delta t \frac{T_{k-1}^n - 2T_k^n + T_{k+1}^n}{h^2}. \tag{11.19}$$

The right hand side of this equation depends only on temperatures of the previous time step, since we used an explicit method. Although this might seem advantageous on a first glance, it turns out that the above scheme is not stable for arbitrary choices of Δt and h. In particular, it is possible to prove that the above discretization is stable only for

$$\frac{\kappa \Delta t}{h^2} \le \frac{1}{2}. \tag{11.20}$$

A detailed discussion and proof of this property can be found in any advanced textbook on numerics of PDEs [1–5].

On the other hand, if we apply the implicit EULER method (5.10) to solve Eq. (11.17) we obtain

$$\frac{T_k^{n+1} - T_k^n}{\Delta t} = \kappa \frac{T_{k-1}^{n+1} - 2T_k^{n+1} + T_{k+1}^{n+1}}{h^2}, \tag{11.21}$$

which is unconditionally stable. However, Eq. (11.21) is an implicit equation, i.e. the function values T_{k+1}^{n+1} and T_{k-1}^{n+1} are required in order to evaluate T_k^{n+1}. Hence, Eq. (11.21) has to be solved as a system of linear equations. This system may be written as

$$AT^{n+1} = T^n + F, \tag{11.22}$$

with the vector $T^n = (T_0^n, T_1^n, \dots, T_N^n)^T$ and the tridiagonal matrix A:

$$A = \begin{pmatrix} \ddots & \ddots & \ddots & & \\ & -\frac{\kappa\Delta t}{h^2} & 1 + \frac{2\kappa\Delta t}{h^2} & -\frac{\kappa\Delta t}{h^2} & \\ & & \ddots & \ddots & \ddots \end{pmatrix} . \tag{11.23}$$

The boundary conditions are incorporated in the matrix A and in the vector F. (See Sects. 9.2 and 9.3.) The linear system of equations (11.22) can be solved numerically using a direct or an iterative method. Employing an iterative method, imposes a third index t on the function values of the temperature T which accounts for the iteration step.

Let us give a brief numerical example. We consider the time-dependent homogeneous heat equation (11.16) on a finite interval $[0, L]$ together with the boundary conditions of Sect. 9.1:

$$T(0) = T_0, \qquad T(L) = T_N . \tag{11.24}$$

In addition we introduce the initial condition

$$T(x, 0) = 0, \qquad x \in [0, L] . \tag{11.25}$$

Figure 11.3 presents the time evolution of $T(x, t)$ at six different time-steps as it was obtained with the explicit EULER method (11.19). Here we chose $T_0 = 0$, $T_N = 2$, $N = 20$, $L = 10$, $\kappa = 1$ as well as $\Delta t \approx 0.5$. Note that for this choice of parameters, the condition (11.20) is fulfilled since $h \approx 1.05$ and therefore

$$\frac{\kappa\Delta t}{h^2} \approx 0.45 \le \frac{1}{2} . \tag{11.26}$$

Figure 11.4 corresponds to Fig. 11.3 but now Δt was chosen to be approximately 0.7 and:

$$\frac{\kappa\Delta t}{h^2} \approx 0.63 > \frac{1}{2} . \tag{11.27}$$

Thus, the stability criterion was violated and the solutions became unstable. Finally, Fig. 11.5 presents results obtained with the same parameters as for Fig. 11.4 but with the help of the implicit EULER method (11.21). Obviously, this procedure provides a stable solution of the problem.

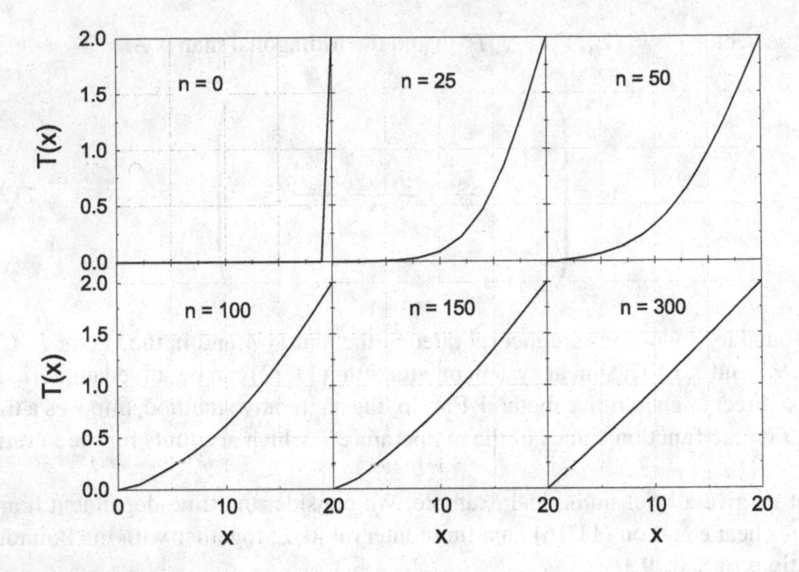

Fig. 11.3 Solutions of the time-dependent heat equation $T(x)$ vs x generated by the explicit EULER method. The stability criterion (11.20) is fulfilled. Results after $n = 25, 50, 100, 150$, and 300 time steps are presented. $n = 0$ represents the initial conditions

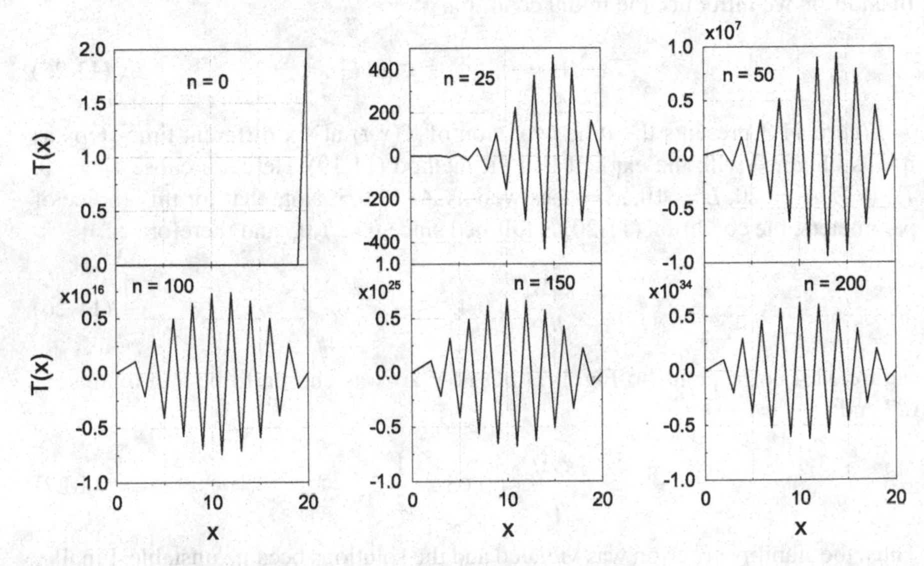

Fig. 11.4 Solutions of the time-dependent heat equation $T(x)$ vs x generated by the explicit EULER method. The stability criterion (11.20) is not fulfilled and, therefore, the solution is apparently unstable. Results after $n = 25, 50, 100, 150$, and 200 time steps are presented. $n = 0$ represents the initial conditions

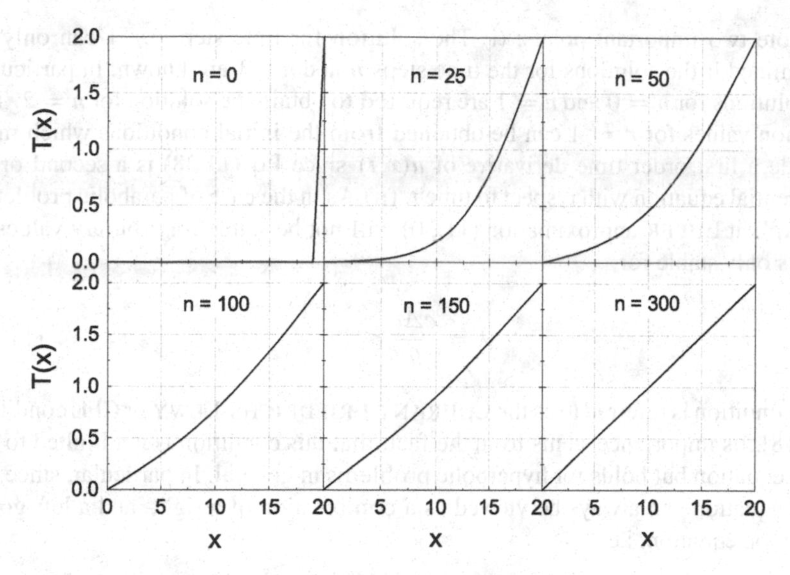

Fig. 11.5 Solutions of the time-dependent heat equation $T(x)$ vs x generated by the implicit EULER method. Results after $n = 25, 50, 100, 150$, and 300 time steps are presented. $n = 0$ represents the initial conditions

11.4 The Wave Equation

As a model hyperbolic PDE we consider briefly the wave equation [16]. Again, we regard only the one-dimensional case:

$$\frac{\partial^2}{\partial t^2}u(x,t) = c^2\frac{\partial^2}{\partial x^2}u(x,t) . \tag{11.28}$$

Here, c is the speed at which the wave $u(x,t)$ propagates. Equation (11.28) is to be augmented by appropriate boundary and initial conditions. A finite difference approach similar to the one discussed in Sect. 11.3 will be employed and the discussion will be restricted to the explicit EULER approximation. Consequently, Eq. (11.28) is replaced by

$$\frac{u_k^{n-1} - 2u_k^n + u_k^{n+1}}{\Delta t^2} = c^2\frac{u_{k-1}^n - 2u_k^n + u_{k+1}^n}{h^2} . \tag{11.29}$$

We define the parameter $\lambda = \frac{c\Delta t}{h}$ and solve Eq. (11.29) for u_k^{n+1}:

$$u_k^{n+1} = 2(1-\lambda^2)u_k^n - u_k^{n-1} + \lambda^2(u_{k-1}^n + u_{k+1}^n) . \tag{11.30}$$

We note two important points: (i) The solution for time step $n + 1$ can only be determined if the solutions for the time steps n and $n - 1$ are known. In particular, the solutions for $n = 0$ and $n = 1$ are required to obtain the solution for $n = 2$. The function values for $n = 1$ can be obtained from the initial conditions which must include a first order time derivative of $u(x, t)$ since Eq. (11.28) is a second order differential equation with respect to time t. (ii) As in the case of parabolic problems, the explicit EULER approximation (11.30) will not be stable for arbitrary values of λ. It is only stable for

$$\lambda = \frac{c\Delta t}{h} \leq 1 \, . \tag{11.31}$$

This condition is referred to as the COURANT-FRIEDRICHS-LEWY or CFL condition [17, 18]. Its importance stems from the fact, that this condition is not limited to the wave equation but holds for hyperbolic problems in general. In particular, since the wave equation can always be viewed as a combination of a right- and a left-going advection equation, i.e.

$$\frac{\partial}{\partial t} u(x, t) = \pm c \frac{\partial}{\partial x} u(x, t) \, , \tag{11.32}$$

we gain the very important property that explicit time integrators applied to solve equations of the type (11.32) are only stable if relation (11.31) is obeyed.

Let us return to the discretization (11.30). Suppose we have initial conditions of the form

$$u(x, 0) = f(x), \qquad \frac{\partial}{\partial t} u(x, 0) = g(x) \, . \tag{11.33}$$

They can be approximated by

$$u_k^0 = f_k, \qquad \frac{u_k^1 - u_k^0}{\Delta t} = g_k \, , \tag{11.34}$$

and the solution of the second relation in (11.34) yields the desired function values for $n = 1$:

$$u_k^1 = u_k^0 + \Delta t g_k \, . \tag{11.35}$$

However, in many cases it is beneficial to take higher order terms into account. This can be achieved by employing a TAYLOR expansion of the form (Chap. 2):

$$\frac{u_k^1 - u_k^0}{\Delta t} = \frac{\partial}{\partial t} u(x, 0) + \frac{\Delta t}{2} \frac{\partial^2}{\partial t^2} u(x, 0) + \mathcal{O}(\Delta t^2) \, . \tag{11.36}$$

We make now use of the initial conditions (11.33), employ the wave equation (11.28), and solve for u_k^1. This gives

$$u_k^1 = u_k^0 + \Delta t g_k + \frac{\Delta t^2 c^2}{2} f_k'' + \mathcal{O}(\Delta t^3) . \tag{11.37}$$

Here we assumed that the second spatial derivative $f_k'' = \frac{\partial^2}{\partial x^2} f(x_k)$ of the initial condition $f(x)$ exists. It may then be approximated by a finite difference approach.

To be specific we consider a vibrating string of length L, which is fixed at its ends, i.e. $u(0,t) = u(L,t) = 0$. Furthermore, we assume that the string was initially at rest, i.e.

$$\frac{\partial}{\partial t} u(x, 0) = 0 , \tag{11.38}$$

and impose initial conditions

$$u(x,0) = \begin{cases} \sin\left(\frac{2\pi x}{L}\right) & x \in \left(\frac{L}{2}, L\right] , \\ 0 & \text{elsewhere.} \end{cases} \tag{11.39}$$

Figure 11.6 presents results obtained with $L = 1, c = 2, N = 100$. Δt was chosen in such a way that $\lambda = 0.5$. On the other hand, Fig. 11.7 presents calculations

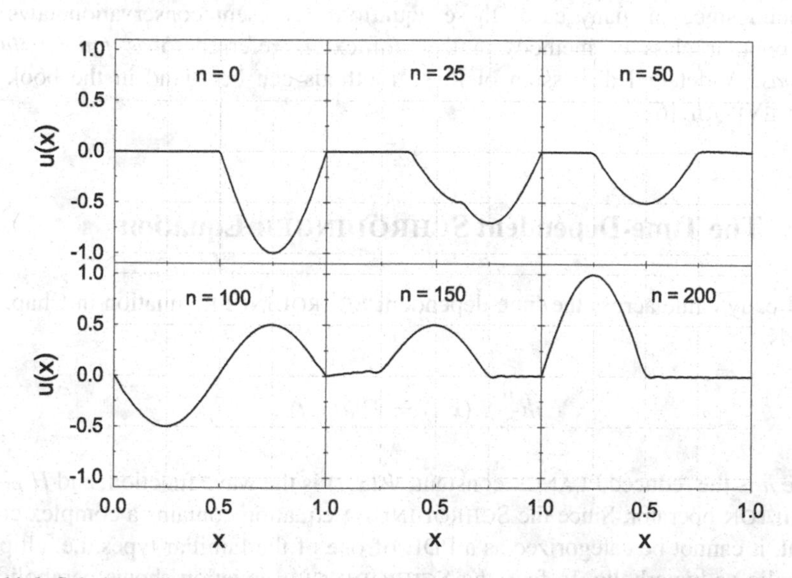

Fig. 11.6 Solutions of the wave equation $u(x)$ vs x generated by the explicit EULER method with $\lambda = 0.5$. Results after $n = 25$, 50, 100, 150, and 200 time steps are presented. $n = 0$ represents the initial conditions

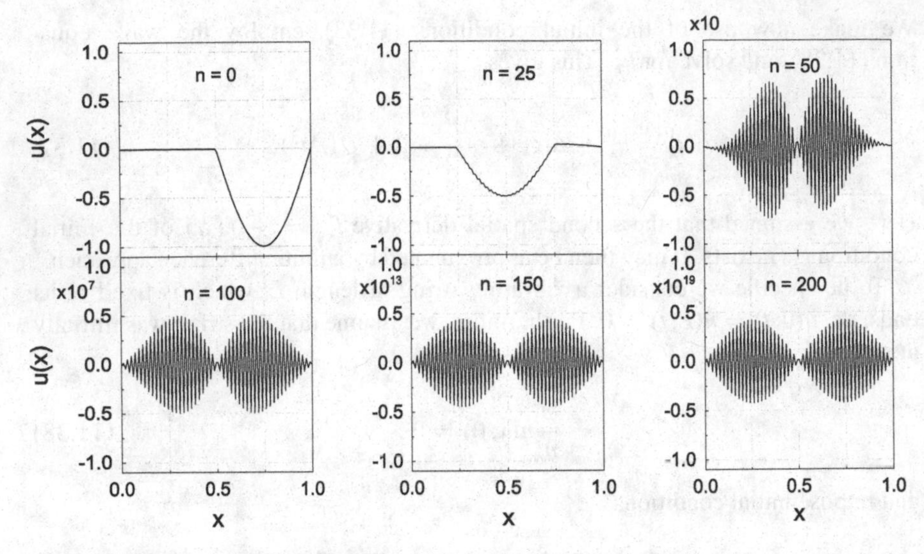

Fig. 11.7 Solutions of the wave equation $u(x)$ vs x generated by the explicit EULER method with $\lambda = 1.01$. Results after $n = 25, 50, 100, 150$, and 200 time steps are presented. $n = 0$ represents the initial conditions

performed with the same parameters but now λ was set to 1.01. Thus, the CFL condition (11.31) was violated and the solutions become unstable.

In general, the numerical solution of hyperbolic PDEs can be very difficult to obtain since in many cases these equations represent conservation laws. A very popular class of methods in this context is referred to as *finite volume methods*. A detailed discussion of these methods can be found in the book by R. J. LEVEQUE [6].

11.5 The Time-Dependent SCHRÖDINGER Equation

We already came across the time-dependent SCHRÖDINGER equation in Chap. 10. It reads

$$i\hbar\frac{\partial}{\partial t}\Psi(x,t) = H\Psi(x,t) \;, \tag{11.40}$$

where $\hbar$ is the reduced PLANCK constant, $\Psi(x,t)$ is the wave function, and H is the HAMILTON operator. Since the SCHRÖDINGER equation contains a complex coefficient, it cannot be categorized as a PDE of one of the familiar types, i.e. elliptic, parabolic or hyperbolic. In fact, the SCHRÖDINGER equation shows parabolic as well as hyperbolic behavior (it is of the form of the diffusion equation but allows for wave solutions). We discuss here briefly a very elegant method developed to numerically approximate solutions of the time-dependent SCHRÖDINGER equation, A prominent alternative method, the split operator technique, is briefly explained in Appendix D.

We note that Eq. (11.40) has the formal solution

$$\Psi(x,t) = \exp\left(-\frac{it}{\hbar}H\right)\Psi(x,0) = U(t)\Psi(x,0) , \tag{11.41}$$

where we assumed that H is independent of time t. We note that the operator $U(t)$ on the right hand side of Eq. (11.41) *propagates* the solution in time. Furthermore, it is a unitary operator and therefore preserves the norm of the wave-function $\Psi(x,t)$. $U(t)$ is usually referred to as the *unitary time-evolution operator* [19].[2]

We employ relation (11.41) and obtain

$$\Psi(x,t+\Delta t) = \exp\left[-\frac{i(t+\Delta t)}{\hbar}H\right]\Psi(x,0) = \exp\left(-\frac{i\Delta t}{\hbar}H\right)\Psi(x,t) . \tag{11.42}$$

Expanding the exponential in this equation in its series representation and truncating the series after the second term results in the approximation

$$\Psi(x,t+\Delta t) \approx \left(1 - \frac{i\Delta t}{\hbar}H\right)\Psi(x,t) . \tag{11.43}$$

Again, we introduce grid-spacing $x_k = k\Delta x$, $k \in \mathbb{N}$ and the correspondingly indexed functions $\Psi_k^n \equiv \Psi(x_k, n\Delta t)$ which results in

$$\Psi_k^{n+1} = \left(1 - \frac{i\Delta t}{\hbar}H\right)\Psi_k^n . \tag{11.44}$$

Using Eq. (10.23) for the Hamiltonian in its position space representation in the one-dimensional case and by approximating the second derivative with the help of the central difference approximation we arrive at

$$\Psi_k^{n+1} = \Psi_k^n - \frac{i\Delta t}{\hbar}\left(-\frac{\hbar^2}{2m}\frac{\Psi_{k-1}^n - 2\Psi_k^n + \Psi_{k+1}^n}{\Delta x^2} + V_k\Psi_k^n\right) , \tag{11.45}$$

where we defined $V_k \equiv V(x_k)$.

The iteration scheme (11.45) resembles the explicit EULER approximation (11.18) of the heat equation with the difference that we have here an imaginary coefficient. An implicit procedure for the time-dependent SCHRÖDINGER

[2] We remember that unitary means that $UU^\dagger = U^\dagger U = \mathbb{1}$.

equation (11.40) can be obtained by inversion of Eq. (11.42):

$$\Psi(x,t) = U^{\dagger}(\Delta t)\Psi(x,t+\Delta t) = \exp\left(\frac{i\Delta t}{\hbar}H\right)\Psi(x,t+\Delta t)\,. \tag{11.46}$$

A series expansion of the exponential results in the desired relation:

$$\Psi_k^n = \left(1 + \frac{i\Delta t}{\hbar}H\right)\Psi_k^{n+1}\,. \tag{11.47}$$

We emphasize that the unitarity of the time-evolution operator is of fundamental importance since it preserves the norm of the wave-function. However, in truncating the series representation of the unitary time evolution operator $U(\Delta t)$ we certainly violate the unitarity of $U(\Delta t)$. This problem can be remedied by imposing unitarity of the time evolution as an additional requirement. This requirement can be incorporated by normalizing the wave-function after each time step.

We demonstrate now that the CRANK-NICOLSON scheme [15] can be applied successfully to solve Eq. (11.40) numerically for a particular potential. The CRANK-NICOLSON scheme can be obtained by realizing that

$$\begin{aligned} U(\Delta t) &= \exp\left(-\frac{i\Delta t}{\hbar}H\right) \\ &= \exp\left(-\frac{i\Delta t}{2\hbar}H\right)\exp\left(-\frac{i\Delta t}{2\hbar}H\right) \\ &= \exp\left(\frac{i\Delta t}{2\hbar}H\right)^{-1}\exp\left(-\frac{i\Delta t}{2\hbar}H\right) \\ &= \left[U^{\dagger}\left(\frac{\Delta t}{2}\right)\right]^{-1}U\left(\frac{\Delta t}{2}\right)\,. \end{aligned} \tag{11.48}$$

Hence, we obtain from Eq. (11.45)

$$U^{\dagger}\left(\frac{\Delta t}{2}\right)\Psi_k^{n+1} = U\left(\frac{\Delta t}{2}\right)\Psi_k^n\,, \tag{11.49}$$

or by expanding U in a series and truncating after the second term

$$\left(1 + \frac{i\Delta t}{2\hbar}H\right)\Psi_k^{n+1} = \left(1 - \frac{i\Delta t}{2\hbar}H\right)\Psi_k^n\,. \tag{11.50}$$

Inserting the finite difference approximation of the Hamiltonian H and rearranging terms yields

$$\left[1+\frac{i\Delta t}{2\hbar}\left(\frac{\hbar^2}{m\Delta x^2}+V_k\right)\right]\Psi_k^{n+1}-\frac{i\Delta t\hbar}{4m\Delta x^2}\left(\Psi_{k-1}^{n+1}+\Psi_{k+1}^{n+1}\right)=\hat{\Omega}_k^n\,, \tag{11.51}$$

where we defined $\hat{\Omega}_k^n$ as

$$\hat{\Omega}_k^n=\left[1-\frac{i\Delta t}{2\hbar}\left(\frac{\hbar^2}{m\Delta x^2}+V_k\right)\right]\Psi_k^n+\frac{i\Delta t\hbar}{4m\Delta x^2}\left(\Psi_{k-1}^n+\Psi_{k+1}^n\right)\,. \tag{11.52}$$

Both sides of Eq. (11.51) are now multiplied by $i4m\Delta x^2/(\hbar\Delta t)$ and this gives

$$\Psi_{k-1}^{n+1}+2\left(\frac{i2m\Delta x^2}{\Delta t\hbar}-1-\frac{m\Delta x^2}{\hbar^2}V_k\right)\Psi_k^{n+1}+\Psi_{k+1}^{n+1}=\Omega_k^n\,, \tag{11.53}$$

where

$$\Omega_k^n=-\Psi_{k-1}^n+2\left(\frac{i2m\Delta x^2}{\Delta t\hbar}+1+\frac{m\Delta x^2}{\hbar^2}V_k\right)\Psi_k^n-\Psi_{k+1}^n\,. \tag{11.54}$$

We recognize that Eq. (11.53) establishes a system of linear equations and rewrite it in matrix form:

$$A\Psi^{n+1}=\Omega^n\,. \tag{11.55}$$

Here, we defined the vectors $\Psi^n=\left(\Psi_0^n,\Psi_1^n,\dots,\Psi_N^n\right)^T$, $\Omega^n=\left(\Omega_0^n,\Omega_1^n,\dots,\Omega_N^n\right)^T$ and the tridiagonal matrix

$$A=\begin{pmatrix} \ddots & \ddots & \ddots & & \\ & 1 & \Gamma_k & 1 & \\ & & \ddots & \ddots & \ddots \end{pmatrix}\,, \tag{11.56}$$

with Γ_k for $k=1,2,\dots,N$ given by

$$\Gamma_k=2\left(\frac{i2m\Delta x^2}{\Delta t\hbar}-1-\frac{m\Delta x^2}{\hbar^2}V_k\right)\,, \tag{11.57}$$

according to Eq. (11.53).

The system (11.56) is solved iteratively. However, in this case we employ a more elegant ansatz which is allowed for tridiagonal matrices. We set

$$\Psi_{k+1}^{n+1} = a_k \Psi_k^{n+1} + b_k^n \,, \tag{11.58}$$

and apply it to Eq. (11.53). After rearranging terms we arrive at:

$$2\left(1 + \frac{m\Delta x^2}{\hbar^2} V_k - \frac{i2m\Delta x^2}{\Delta t\,\hbar} - \frac{a_k}{2}\right) \Psi_k^{n+1} = \Psi_{k-1}^{n+1} + b_k^n - \Omega_k^n \,. \tag{11.59}$$

We define

$$\alpha_k = 2\left(1 + \frac{m\Delta x^2}{\hbar^2} V_k - \frac{i2m\Delta x^2}{\Delta t\,\hbar} - \frac{a_k}{2}\right) , \tag{11.60}$$

and obtain from Eq. (11.59)

$$\Psi_k^{n+1} = \frac{1}{\alpha_k} \Psi_{k-1}^{n+1} + \frac{b_k^n - \Omega_k^n}{\alpha_k} \,. \tag{11.61}$$

However, due to the ansatz (11.58) we also have

$$\Psi_k^{n+1} = a_{k-1} \Psi_{k-1}^{n+1} + b_{k-1}^n \,, \tag{11.62}$$

which results in the relations

$$a_{k-1} = \frac{1}{\alpha_k} \,, \tag{11.63}$$

and

$$b_{k-1}^n = \frac{b_k^n - \Omega_k^n}{\alpha_k} = \left(b_k^n - \Omega_k^n\right) a_{k-1} \,. \tag{11.64}$$

Equation (11.63) leads to the recursion relation

$$a_k = 2\left(1 + \frac{m\Delta x^2}{\hbar^2} V_k - \frac{i2m\Delta x^2}{\Delta t\,\hbar}\right) - \frac{1}{a_{k-1}} \,, \tag{11.65}$$

and we derive from Eq. (11.64):

$$b_k^n = \frac{b_{k-1}^n}{a_{k-1}} + \Omega_k^n \,. \tag{11.66}$$

The remaining question is how to choose a_0 and b_0^n. We impose the boundary conditions $\Psi_0^n = 0$ and $\Psi_N^n = 0$ and derive from Eq. (11.53):

$$\Omega_1^n = 2\left(\frac{i2m\Delta x^2}{\Delta t\hbar} - 1 - \frac{m\Delta x^2}{\hbar^2}V_1\right)\Psi_1^{n+1} + \Psi_2^{n+1} . \tag{11.67}$$

A comparison of this equation with the ansatz (11.58), i.e. $\Psi_2^{n+1} = a_1\Psi_1^{n+1} + b_1^n$, reveals that

$$a_1 = 2\left(1 + \frac{m\Delta x^2}{\hbar^2}V_1 - \frac{i2m\Delta x^2}{\Delta t\hbar}\right) , \tag{11.68}$$

and

$$b_1^n = \Omega_1^n . \tag{11.69}$$

These expressions are equivalent to $a_0 = \infty$ and it is, thus, impossible to calculate Ψ_1^{n+1} from Ψ_0^{n+1}. However, we can determine the function values Ψ_k^{n+1} via a backward recursion

$$\Psi_k^{n+1} = \frac{1}{a_k}\left(\Psi_{k+1}^{n+1} - b_k^n\right) , \tag{11.70}$$

which is initialized with the boundary condition $\Psi_N^{n+1} = 0$. We can now summarize the algorithm:

1. Choose the initial conditions Ψ_k^0, $k = 0, 1, \ldots, N$ which satisfy the boundary conditions $\Psi_0^0 = 0$ and $\Psi_N^0 = 0$.
2. Set

$$a_1 = 2\left(1 + \frac{m\Delta x^2}{\hbar^2}V_1 - \frac{i2m\Delta x^2}{\Delta t\hbar}\right) , \tag{11.71}$$

   and calculate for $k = 2, \ldots, N-1$

$$a_k = 2\left(1 + \frac{m\Delta x^2}{\hbar^2}V_k - \frac{i2m\Delta x^2}{\Delta t\hbar}\right) - \frac{1}{a_{k-1}} . \tag{11.72}$$

3. Start the time loop: $n = 0, 1, \ldots, M$, with M the maximum number of time steps.
4. Calculate for $k = 1, 2, \ldots, N-1$

$$\Omega_k^n = -\Psi_{k-1}^n + 2\left(\frac{i2m\Delta x^2}{\Delta t\hbar} + 1 + \frac{m\Delta x^2}{\hbar^2}V_k\right)\Psi_k^n - \Psi_{k+1}^n . \tag{11.73}$$

5. Set

$$b_1^n = \Omega_1^n , \tag{11.74}$$

and calculate for $k = 2, \ldots, N-1$

$$b_k^n = \frac{b_{k-1}^n}{a_{k-1}} + \Omega_k^n . \tag{11.75}$$

6. Calculate for $k = N-1, N-2, \ldots, 1$

$$\Psi_k^{n+1} = \frac{1}{a_k}\left(\Psi_{k+1}^{n+1} - b_k^n\right) , \tag{11.76}$$

where the boundary conditions $\Psi_0^n = \Psi_N^n = 0$ are to be considered.
7. Set $n = n + 1$ and go to step 4.

The application of this algorithm is now elucidated with the help of a specific example, the quantum mechanical tunneling effect. The initial condition is described by a Gauss wave packet

$$\Psi(x, 0) = \exp(iqx) \exp\left[-\frac{(x - x_0)^2}{2\sigma^2}\right] , \tag{11.77}$$

centered at $x = x_0$ which propagates in positive x-direction with momentum q. This wave-function is not yet normalized. Furthermore, we regard the single potential barrier

$$V_1(x) = \begin{cases} V_0 & x \in [a, b] , \\ 0 & \text{elsewhere,} \end{cases} \tag{11.78}$$

or the double potential barrier

$$V_2(x) = \begin{cases} V_0 & x \in [a, b] \cup [c, d] , \\ 0 & \text{elsewhere.} \end{cases} \tag{11.79}$$

The scales and parameters are chosen in the following way: $L = 500$, $\Delta x = 1$, $\Delta t = 0.1$, $m = \hbar = 1$, $x_0 = 200$, $q = 2$, $\sigma = 20$, $V_0 = 0.7$, $a = 250$, $b = 260$, $c = 300$, and $d = 310$. Figure 11.8 presents the time evolution of the square modulus of the resulting wave-function $|\Psi(x, n\Delta t)|^2$ vs x (solid line, left hand scale) at different time steps $n = 500$, 1000, and 1500. The time step $n = 0$ corresponds to the initial

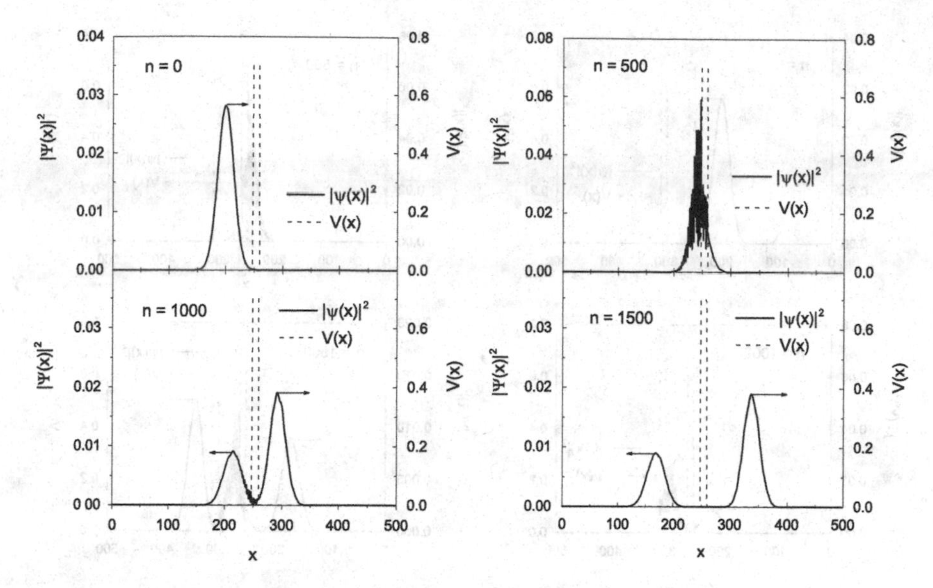

Fig. 11.8 Time evolution of the square modulus of the wave-function $|\psi(x)|^2$ vs x (*solid line, left hand scale*). The potential $V(x) = V_1(x)$ is also plotted vs x (*dashed line, right hand scale*). We present the results for $n = 500$, 1000, and 1500 time steps. The graph labeled by $n = 0$ represents the initial configuration

condition. The potential $V_1(x)$ vs x is also plotted (dashed line, right hand scale). Figure 11.9 corresponds to Fig. 11.8 but now the potential is described by $V_2(x)$ and additional time steps for $n = 2000$ and 2500 have been added.

In both figures a typical quantum mechanical effect which is referred to as *tunneling* can be observed. In particular, there exists a finite probability that the potential barrier can be crossed, although, from a classical point of view, the particle's energy is not sufficient to overcome the barrier. A detailed discussion of this effect and its technological importance can be found in any standard textbook on quantum mechanics [19–21].

In conclusion we remark that a very prominent method to solve numerically the time-dependent SCHRÖDINGER equation is based on the FOURIER transformation. The numerical implementation of the FOURIER transformation as well as its application to the SCHRÖDINGER equation is briefly discussed in Appendix D.

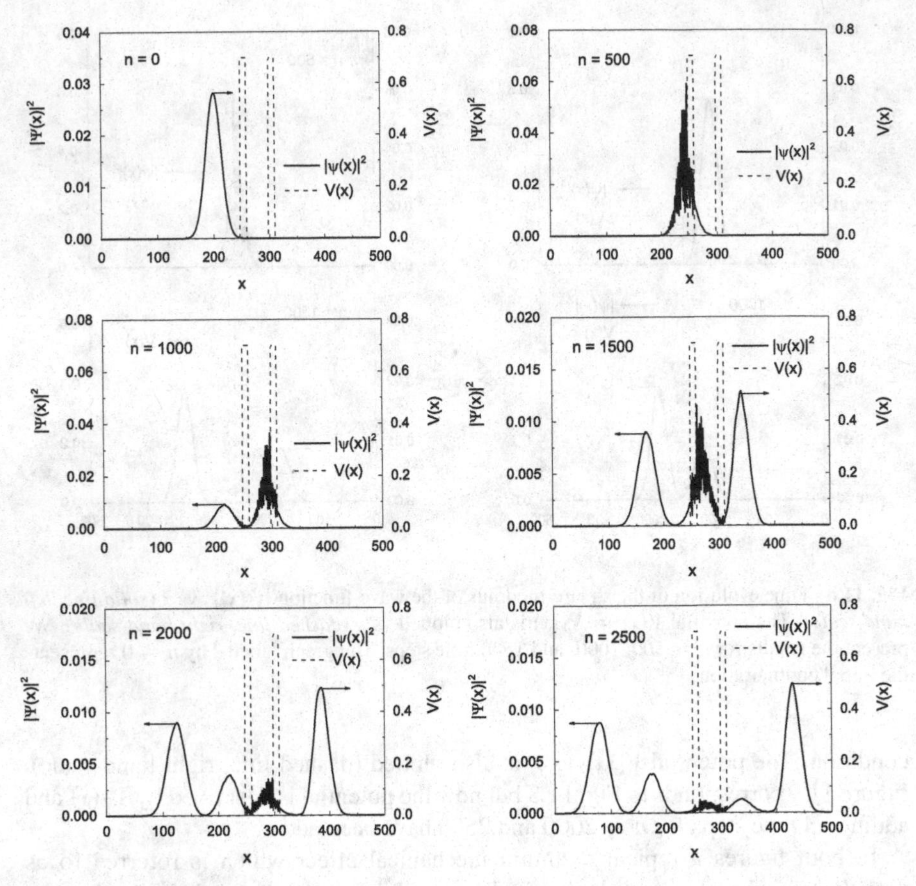

Fig. 11.9 Time evolution of the square modulus of the wave-function $|\psi(x)|^2$ vs x (*solid line, left hand scale*). The potential $V(x) = V_2(x)$ is also plotted vs x (*dashed line, right hand scale*). We present the results for $n = 500$, 1000, 1500, 2000, and 2500 time steps. The graph labeled by $n = 0$ represents the initial configuration

Summary

This chapter was about linear PDEs and how to find solutions numerically. The dominating theme was the application of the various finite difference methods. The two-dimensional POISSON equation served as an example for an elliptic PDE. The algorithm to solve this equation developed here was based on the central difference derivative. Parabolic PDEs were represented by the time-dependent one-dimensional heat equation. The numerical solution proved to be possible by either using an explicit or an implicit EULER scheme. For the explicit EULER scheme the appropriate choice of time and space discretization proved to be essential for the stability of the algorithm. The one-dimensional wave equation was introduced as an example of a hyperbolic PDE. The solution was found by employing an explicit

EULER approximation. Again time and space discretization had to follow a specific stability criterion, the COURANT-FRIEDRICHS-LEWY condition. Finally, the one-dimensional time-dependent SCHRÖDINGER equation was studied. It does not fit into any of the above categories. The algorithm to find a numerical solution was developed here on the basis of a CRANK-NICOLSON scheme and it was tested with the quantum mechanical tunneling effect.

Problems

1. Write a program which solves the two-dimensional POISSON equation for an arbitrary charge density distribution $\rho(x, y)$. Use the numerical method discussed in Sect. 11.2.
   a. Impose DIRICHLET boundary conditions $\varphi(x, 0) = \varphi(x, L_y) = \varphi(0, y) = \varphi(L_x, y) = 0$ as described in Sect. 11.2. Test the program by first reproducing Fig. 11.2.
   b. Solve the POISSON equation for different charge densities of your choice.
   c. Calculate the electric field $E(x, y)$ with the help of Eq. (11.3).
2. Calculate the time evolution of the temperature distribution $T(x, t)$ along a cylindrical rod described in Sect. 9.3. The rod is kept at constant temperatures T_0 and T_N at its ends. The parameters used in Sect. 9.3 stay unchanged. Study also the case of a heat sink as suggested in the Problems section of Chap. 9.
3. Calculate the time evolution of the square modulus of the wave-function $|\psi(x)|^2$ vs x for a potential $V_1(x)$ according to Eq. (11.78) with $V_0 < 0$ (quantum well). In a second step, modify the potential according to

$$V(x) = \begin{cases} V_1 & x \in [a, b] \cup [c, d] \\ V_2 & x \in [b, c] \\ 0 & \text{elsewhere}, \end{cases}$$

   with $V_1 > 0$, $V_2 < 0$, and $|V_1| < |V_2|$.

References

1. Lapidus, L., Pinder, G.F.: Numerical Solution of Partial Differential Equations. Wiley, New York (1982)
2. Morton, K.W., Mayers, D.F.: Numerical Solution of Partial Differential Equations. Cambridge University Press, Cambridge (2005)
3. Li, T., Qin, T.: Physics and Partial Differential Equations, vol. 1. Cambridge University Press, Cambridge (2012)

4. Li, T., Qin, T.: Physics and Partial Differential Equations, vol. 2. Cambridge University Press, Cambridge (2014)
5. Lui, S.H.: Numerical Analysis of Partial Differential Equations. Wiley, New York (2012)
6. LeVeque, R.J.: Finite Volume Methods for Hyperbolic Problems. Cambridge Texts in Applied Mathematics. Cambridge University Press, Cambridge (2002)
7. Gockenbach, M.S.: Understanding and Implementing the Finite Element Method. Cambridge University Press, Cambridge (2006)
8. Selvadurai, A.: Partial Differential Equations in Mechanics, vol. 2. Springer, Berlin/Heidelberg (2000)
9. Sleeman, B.D.: Partial differential equations, poisson equation. In: Dubitzky, W., Wolkenhauer, O., Cho, K.H., Yokota, H. (eds.) Encyclopedia of Systems Biology, pp. 1635–1638. Springer, Berlin/Heidelberg (2013)
10. Jackson, J.D.: Classical Electrodynamics, 3rd edn. Wiley, New York (1998)
11. Greiner, W.: Classical Electrodynamics. Springer, Berlin/Heidelberg (1998)
12. Polyanin, A.D.: Handbook of Linear Partial Differential Equations for Engineers and Scientists. Chapman & Hall/CRC, Boca Raton (2002)
13. Cannon, J.R.: The One-Dimensional Heat Equation. Encyclopedia of Mathematics and Its Applications. Cambridge University Press, Cambridge (1985)
14. Carslaw, H.S., Jaeger, J.C.: Conduction of Heat in Solids, 2nd edn. Oxford University Press, Oxford (1986)
15. Crank, J., Nicolson, P.: A practical method for numerical evaluation of solutions of partial differential equations of the heat-conduction type. Proc. Camb. Philos. Soc. **43**, 50–67 (1947). doi:10.1017/S0305004100023197
16. Zwillinger, D.: Handbook of Differential Equations, 3nd edn. Academic, San Diego (1997)
17. Courant, R., Friedrichs, K., Lewy, H.: On the partial difference equations of mathematical physics. IBM J. Res. Dev. **11**, 215–234 (1967 [1928])
18. Bakhvalov, N.S.: Courant-friedrichs-lewy condition. In: Hazewinkel, M. (ed.) Encyclopaedia of Mathematics. Springer, Berlin/Heidelberg (1994)
19. Sakurai, J.J.: Modern Quantum Mechanics. Addison-Wesley, Menlo Park (1985)
20. Baym, G.: Lectures on Quantum Mechanics. Lecture Notes and Supplements in Physics. The Benjamin/Cummings Publ. Comp., Inc., London/Amsterdam (1969)
21. Cohen-Tannoudji, C., Diu, B., Laloë, F.: Quantum Mechanics, vol. I. Wiley, New York (1977)

Part II
Stochastic Methods

Chapter 12
Pseudo-random Number Generators

12.1 Introduction

Stochastic methods in Computational Physics are based on the availability of random numbers and on the concepts of probability theory. (Readers not familiar with the basic concepts of probability theory are highly encouraged to study Appendix E before proceeding.) The required random numbers are provided by numerical random number generators and, thus, we have to speak, more precisely, of pseudo-random numbers. Let us now motivate the problem and discuss some preliminary items.

A popular example of randomness in physical systems is certainly the outcome of a dice-throw or the drawing of lotto numbers. Even though the outcome of a dice-throw is completely determined by the initial conditions, it is effectively unpredictable because the initial conditions cannot be determined accurately enough. A probabilistic description, which assigns the *random variables* $1, 2, 3, 4, 5$, and 6 a probability of $1/6$, respectively, is much more convenient and promising. It has to be kept in mind, of course, that all predictions obtained on the basis of such an approach are also clearly probabilistic in their nature.

Another example is Brownian motion or diffusion which describes the random motion of dust particles on a fluid surface. It is in this case particularly obvious that a description with the help of a *stochastic differential equation*, such as the LANGEVIN equation [1], is completely sufficient and more to the point than a description based on the dynamics of a large number of interacting particles.

Stochastic methods are not confined to physics: They are applied very successfully in many other fields of expertise, like biology [2], economics [3, 4], medicine [5], etc. Finally, an interesting and purely mathematical application can be found in the evaluation of integrals as an alternative to the methods discussed in Chap. 3. This method is referred to as Monte-Carlo integration and will be addressed in Chap. 14 together with a basic introduction to stochastics and its applications in physics.

B.A. Stickler, E. Schachinger, *Basic Concepts in Computational Physics*,
DOI 10.1007/978-3-319-27265-8_12

From the numerical point of view, there is one common denominator to all these applications: Random numbers are an essential tool and, consequently, so are random number generators. Thus, a closer inspection of randomness in general and the generation of random numbers or sequences in particular is required. We have to explain what we understand by randomness and how it can be measured. Moreover, based on this discussion we have to formulate requirements to be imposed on the random number generators to deliver useful random numbers.

Although, we might have an intuitive picture of randomness it is hard to formulate without relying on mathematics. For instance, consider the sequence s_1 which consists of N elements:

$$s_1 = 0, 0, 0, 0, 0, \ldots, \qquad N \text{ elements.} \tag{12.1}$$

Is it random? The question cannot be answered without further information. Suppose, the numbers in sequence s_1 were drawn from some set $\mathscr{S}$. If this set is of the form

$$\mathscr{S}_1 = \{0\}, \tag{12.2}$$

then the above sequence s_1 is certainly not random since there is only one possible outcome. However, suppose the numbers of the sequence s_1 were drawn from the set $\mathscr{S}_2$

$$\mathscr{S}_2 = \{0, 1\}, \tag{12.3}$$

with the *events* 0 and 1 together with assigned probabilities $P(0)$ and $P(1)$. These probabilities describe the probability that the outcome of a measurement on the set $\mathscr{S}_2$ yields either the event 0 or 1, respectively. For instance, in the case of tossing a coin the event 0 may correspond to *heads* while 1 stands for *tails*. (To register this result is, within this context, a *measurement*.) In this case the probabilities are given by $P(0) = P(1) = 1/2$ under the premise that the coin is perfectly ideal and has not been manipulated. Even, if we know that the coin has not been manipulated, sequence (12.1) is still a possible outcome, although it is rather improbable for a large number N of measurements (repeated tosses of the coin).

A literal definition of randomness within the context of a random sequence was given by G. J. CHAITIN [6]:

> [...] a series of numbers is random if the smallest algorithm capable of specifying it to a computer has about the same number of bits of information as the series itself.

This definition seems to include the most important features of randomness which we are used to from our experience. Since, no universal trend is observable, reproducing the sequence requires the knowledge of every single constituent. Hence, one may employ the sloppy definition: *Randomness is the lack of a universal trend.*

But how can we test whether or not a given sequence really follows a certain distribution? Of course, one can simply exploit the statistical definition of probability,

Appendix, Eq. (E.4): The probability of a certain outcome is *measured* by counting the particular results. This procedure seems to be quite promising, however, it has to be kept in mind that the statistical definition of probability is only valid in the limit $m \to \infty$, where m is the number of measurements. Hence, it is fundamentally impossible to determine whether or not a sequence is random because an infinite number of elements would have to be evaluated and analyzed. More promising appears to be the calculation of moments or correlations (see Appendix, Sects. E.2 and E.10) from the sequence and to compare such a result with known values for real random numbers. These statistical tests will be discussed in Sect. 12.3. If we consider the sequence (12.1) drawn from the set $\mathscr{S}_2$, [Eq. (12.3), uniform distribution assumed] we can deduce that it is a very improbable result for large N, although it is certainly a possible outcome. Methods based on this train of thoughts are known as *hypothesis testings* and we discuss the χ^2-test as a simple representative of such tests in Sect. 12.3.

We are now in a position to clarify what we understand by a random number: We regard a random sequence drawn from the set

$$\mathscr{S}_3 = \{0, 1, 2, 3, 4, 5, 6, 7, 8, 9\} . \tag{12.4}$$

If the random number is to be uniformly distributed, we assign probabilities $P(k) = 1/10$, $k = 0, 1, \ldots, 9$ and if we would like to obtain a random number out of the interval

$$\Omega_1 = [0, 1) , \tag{12.5}$$

we may simply draw the sequence $s_2 = a_1, a_2, a_3, \ldots$ from $\mathscr{S}_3$ and compose the random number r as

$$r = 0.a_1a_2a_3 \ldots . \tag{12.6}$$

Section 12.2 is dedicated to the discussion of different methods of how to generate so called *pseudo-random numbers*. A pseudo-random number is a number generated with the help of a deterministic algorithm, however, it shows a behavior as if it were random. This implicates that its statistical properties are close to that of true random numbers. In contrast to pseudo-random numbers, *real random numbers* are truly random. A real random number can be obtained from experiments. One could, for instance, simply toss a coin and register the resulting sequence of zeros and ones. A more sophisticated method is to exploit the radioactive decay of a nucleus, which is believed to be purely stochastic. There are also more exotic ideas, such as using higher digits of π which are assumed to behave as if they were random. However, all these methods have in common that they are far too slow for computational purposes. Moreover, an *experimental* approach is obviously not *reproducible* in the sense, that a random sequence cannot be reproduced on demand, but the reproducibility of a random number sequence is essential for many applications.

This leads us to the formulation of several criteria a random number generator will have to comply with. It should

- produce pseudo-random numbers whose statistical properties are as close as possible to those of real random numbers.
- have a long period: It should generate a non-repeating sequence of random numbers which is sufficiently long for computational purposes.
- be reproducible in the sense defined above, as well as restartable from an arbitrary break-point.
- be fast and parallelizable: It should not be the limiting component in simulations.

We restrict, within this chapter, our discussion to random numbers that are uniformly distributed over a finite set. Thus, we assign to all possible outcomes of a measurement the same probability. The generation of non-uniformly distributed random numbers from uniformly distributed random numbers is not a difficult task and will be discussed in more detail in Chap. 13.

12.2 Different Methods

We discuss here different types of pseudo-random number generators [7–9] which generate a pseudo-random number r which is uniformly distributed within the interval $[0, 1)$. Hence, its *probability density function* (pdf) is given by

$$p(r) = \begin{cases} 1 & r \in [0, 1)\,, \\ 0 & \text{elsewhere}, \end{cases} \tag{12.7}$$

and from this follows the *cumulative distribution function* (cdf; see Appendix E):

$$P(r) = \int_0^r \mathrm{d}r' p(r') = \begin{cases} 0 & r < 0\,, \\ r & 0 \le r < 1\,, \\ 1 & r \ge 1\,. \end{cases} \tag{12.8}$$

We introduce here only some of the most basic concepts for pseudo-random number generators. However, in huge simulations based on random numbers standard pseudo-random number generators provided by the various compilers may not be sufficient due to their rather short period and bad statistical properties. In this case it is, therefore, recommended to consult the literature [7, 10] and use more advanced techniques in order to obtain reliable results.

Linear Congruential Generators

Linear congruential generators are the simplest and most prominent random number generators. They produce a sequence of integers $\{x_n\}$, $n \in \mathbb{N}$ following the rule

$$x_{n+1} = (ax_n + c) \bmod m , \tag{12.9}$$

where a, c and m are positive integers which obey $0 \leq a, c < m$. Furthermore, the generator is initialized by its *seed* x_0, which is also a positive (in most cases odd) integer in the range $0 \leq x_0 < m$. The seed is commonly taken from, for instance, the system time in order to avoid repetition at a restart of the sequence. In many environments it is, therefore, necessary to fix the seed artificially whenever one wants to perform reproducible tests.

We note that the sequence resulting from Eq. (12.9) is bounded to the interval $x_n = [0, m-1]$ and, hence, its maximum period is m. However, the actual period of the sequence highly depends on the choices of the parameters a, c and m as well as on the seed x_0. In general, linear congruential generators are very fast and simple, however, they have rather short periods. Moreover, they are very susceptible to correlations since the value x_{n+1} is calculated from x_n only. (This is obviously a property which does not apply to real independent random numbers and it should therefore be eliminated!) In Sect. 12.3 we will discuss some simple methods which allow to identify such correlations.

One of the most prominent choices for the parameters in Eq. (12.9) are the PARK-MILLER parameters [10]:

$$a = 7^5, \quad c = 0, \quad m = 2^{31} - 1 . \tag{12.10}$$

Note that one has to be particularly careful when choosing $c = 0$. It follows from Eq. (12.9) that if $c = 0$ and if for any n, $x_n = 0$ one obtains $x_k = 0$ for all $k > n$.

The random numbers r_n described by the pdf (12.7) are obtained via

$$r_n = \frac{x_n}{m} \in [0, 1) . \tag{12.11}$$

Let us briefly discuss two famous improvements which concentrate on the reduction of correlations and an elongation of the period: The first idea which is referred to as *shuffling* [10] includes a second *pseudo-random* step. One calculates N numbers r_n from Eqs. (12.9) and (12.11) and stores these numbers in an array. If a random number is needed by the executing program, a second random integer $k \in [1, N]$ is drawn and the k-th element is taken from this array. In order to avoid that the same random number is used again, the k-th element of the array is replaced by a new random number which, again, is calculated from (12.9) and (12.11).

The second idea to improve the linear congruential generator is simply to include more previous elements of the sequence:

$$x_{n+1} = \left(\sum_{k=0}^{\ell} a_k x_{n-k} \right) \bmod m \,, \tag{12.12}$$

where $\ell > 0$ and $a_\ell \neq 0$. Again, the periodicity depends highly on the choice of the parameters and on the seed. A specific variation of random number generators using Eq. (12.12) are the FIBONACCI generators.

FIBONACCI *Generators*

The FIBONACCI sequence is given by

$$x_{n+1} = x_n + x_{n-1}, \qquad x_0 = 0, \quad x_1 = 1 \,, \tag{12.13}$$

which results for $n \geq 1$ in

$$1, 1, 2, 3, 5, 8, 13, 21, 34, 55, 89, \ldots \,. \tag{12.14}$$

Choosing in Eq. (12.12) $m = 10$, $\ell = 1$ and $a_0 = a_1 = 1$ simply leaves the last digits of the sequence (12.14):

$$1, 1, 2, 3, 5, 8, 3, 1, 4, 5, 9, \ldots \,. \tag{12.15}$$

This suggests the definition of a pseudo-random number generator based on the FIBONACCI sequence [11]. It is of the form

$$x_{n+1} = (x_n + x_{n-1}) \bmod m \,, \tag{12.16}$$

which, according to our previous discussion, allows a periodicity exceeding m. A straightforward generalization results in the so called *lagged* FIBONACCI *generators*:

$$x_{n+1} = (x_{n-p} \otimes x_{n-q}) \bmod m \,, \tag{12.17}$$

where $p, q \in \mathbb{N}$ and the operator $\otimes$ stands for any binary operation, such as addition, subtraction, multiplication or some logical operation. Two of the most popular lagged FIBONACCI generators are the *shift register generator* and the MARSAGLIA-ZAMAN *generator*.

The shift register generator is based on the *exclusive or* (XOR; $\oplus$) operation, which acts on each bit of the numbers x_{n-p} and x_{n-q}. In particular, the recurrence relation reads

$$x_n = x_{n-p} \oplus x_{n-q} \,. \tag{12.18}$$

The XOR operation $\oplus$ is shown in the following multiplication table:

a	b	$a \oplus b$
0	0	0
1	0	1
0	1	1
1	1	0

Hence, suppose that the binary representation of x_{n-p} is of the form 01001110... and for x_{n-q} we have 11110011... Then we get from Eq. (12.18):

x_{n-p}	0	1	0	0	1	1	1	0	...
x_{n-q}	1	1	1	1	0	0	1	1	...
x_n	1	0	1	1	1	1	0	1	...

A very prominent choice is given by $p = 250$ and $q = 103$ which yields a superior periodicity of order $\mathcal{O}(10^{75})$. The algorithm is initialized with the help of, for instance, a linear congruential generator.

In contrast, the MARSAGLIA-ZAMAN generator (*subtract-with-borrow* scheme) [12] uses the subtraction operation and may be written by introducing the so called *carry bit* Δ as

$$\Delta = x_{n-p} - x_{n-q} - c_{n-1} \,, \tag{12.19}$$

where $x_i \in [0, m]$ for all i. Then,

$$x_n = \begin{cases} \Delta & \Delta \geq 0 \,, \\ \Delta + m & \Delta < 0 \,. \end{cases} \tag{12.20}$$

and c_n is obtained via

$$c_n = \begin{cases} 0 & \Delta \geq 0 \,, \\ 1 & \Delta < 0 \,. \end{cases} \tag{12.21}$$

For the particular choice $p = 10$, $q = 24$ and $m = 2^{24}$ one finds an amazingly large periodicity of order $\mathcal{O}(10^{171})$. The random numbers x_n are integers in the interval $x_n \in [0, m]$, hence dividing the random numbers by m gives $r_n \in [0, 1]$.

12.3 Quality Tests

Here, we discuss some tests to check whether or not a given, finite sequence of numbers x_n consists of uniformly distributed random numbers out of the interval $x_n \in [0, 1]$.[1]

Statistical Tests

Statistical tests are generally the most simple methods to arrive at a first idea of the quality of a pseudo-random number generator. Statistical tests are typically based on the calculation of moments or correlations. Since we regard the simplified case of uniformly distributed, uncorrelated random numbers within the interval [0, 1], the moments can be calculated immediately from (see Appendix, Sect. E.2)

$$\langle X^k \rangle = \int \mathrm{d}x\, x^k p(x) = \int_0^1 \mathrm{d}x\, x^k = \frac{1}{k+1}\,, \tag{12.22}$$

for $k \in \mathbb{N}$. These moments are approximated using the generated finite sequence of numbers $\{x_n\}_{n=1,\ldots,N}$ via

$$\langle X^k \rangle \approx \overline{x^k} = \frac{1}{N}\sum_{n=1}^{N} x_n^k\,. \tag{12.23}$$

As illustrated in Appendix, Sect. E.2, the error of this approximation is of order $\mathcal{O}\left(1/\sqrt{N}\right)$ and

$$\langle X^k \rangle = \overline{x^k} + \mathcal{O}\left(\frac{1}{\sqrt{N}}\right)\,. \tag{12.24}$$

Another method studies correlations (see Appendix, Sects. E.2 and E.10) between the random numbers of the sequence and compare it with the analytical result. We obtain for uncorrelated random numbers:

$$\langle X_n X_{n+k} \rangle = \langle X_n \rangle^2 = \frac{1}{4}\,. \tag{12.25}$$

[1]From now on we define quite generally the interval out of which random numbers x_n are drawn by $x_n \in [0, 1]$ keeping in mind that this interval depends on the actual method applied. This method determines whether zero or one is contained in the interval.

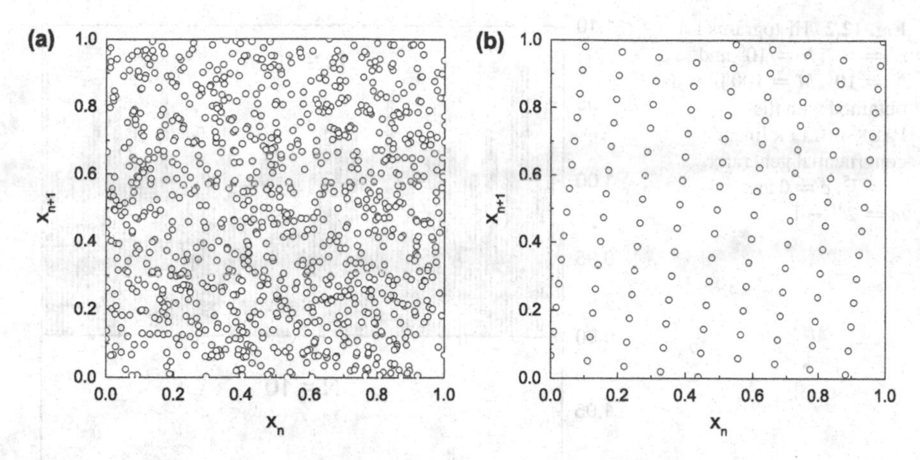

Fig. 12.1 Spectral test for a linear congruential generator. We used the PARK-MILLER parameters, **(a)** $a = 7^5$, $c = 0$, and $m = 2^{31} - 1$, **(b)** $a = 137$, $c = 0$, and $m = 2^{11}$, and plotted $N = 10^3$ subsequent pairs (x_n, x_{n+1}) of random numbers. In frame **(a)** the random numbers evolve nicely distributed within the unit square, showing no obvious correlations. On the other hand, in frame **(b)** subsequent random numbers lie on hyperplanes and, thus, develop correlations: They do not fill the unit square uniformly

Another, quite evident test, is the analysis of the symmetry of the distribution. If $X_n \in [0, 1]$ is uniformly distributed then it follows that $(1 - X_n) \in [0, 1]$ should also be uniformly distributed.

Finally, we discuss a graphical test, known as the *spectral test* [7]. The spectral test consists of plotting subsequent random numbers x_n vs x_{n+1} and of visual inspection of the result. One expects the random numbers to uniformly fill the unit-square, however, if correlations exist, particular patterns might evolve. We illustrate this method in Fig. 12.1 where it is applied to a linear congruential generator (12.9) with two different sets of parameters.

Hypothesis Testing

Basically, one could employ different hypothesis tests, such as the KOLMOGOROV-SMIRNOV test, to test random numbers. These tests are rather basic and are discussed in numerous books on statistics. In what follows we shall briefly mention the χ^2-test; for more advanced techniques we refer the reader to the literature [13, 14].

The χ^2-test tests the pdf directly. One starts by sorting the N elements of the sequence into a histogram. Suppose we would like to have M bins and, hence, the width of every bin is given by $1/M$. We now count the number of elements which lie within bin k, i.e. within the interval $[(k-1)/M, k/M]$, and denote this number by n_k. The histogram array h is given by $h = c(n_1, n_2, \ldots, n_M)^T$ where

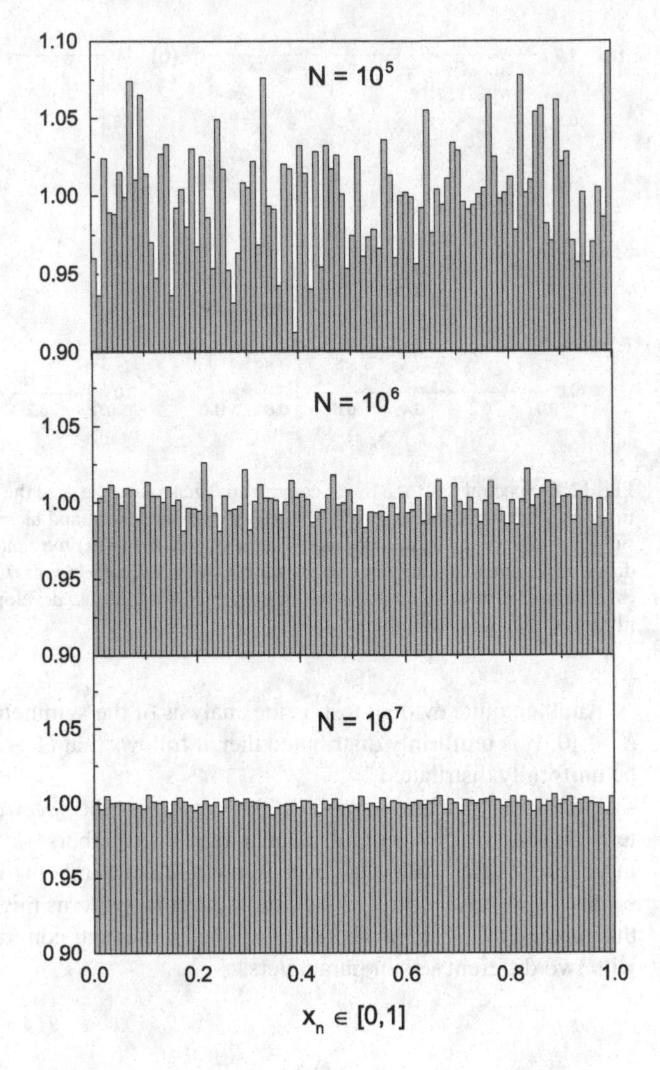

Fig. 12.2 Histograms for $N = 10^5, N = 10^6$ and $N = 10^7, M = 100$ bins as obtained with the PARK-MILLER linear congruential generator, $a = 7^5, c = 0$ and $m = 2^{31} - 1$

the constant $c = M/N$ normalizes the histogram. In Fig. 12.2 we show three different histograms for $N = 10^5, N = 10^6$ and $N = 10^7$ uniformly distributed random numbers as obtained with the PARK-MILLER linear congruential generator. In Fig. 12.3 we present a histogram for $N = 10^7$ obtained with the bad linear congruential generator defined in Fig. 12.1b. We recognize numerous empty bins which are a clear indication that the random numbers are not uniformly distributed.

Let us briefly remember some points from probability theory [15, 16]. One can show, that if numbers Q_n are normally distributed random variables, their sum

$$x = \sum_{n=1}^{\nu} Q_n^2 , \tag{12.26}$$

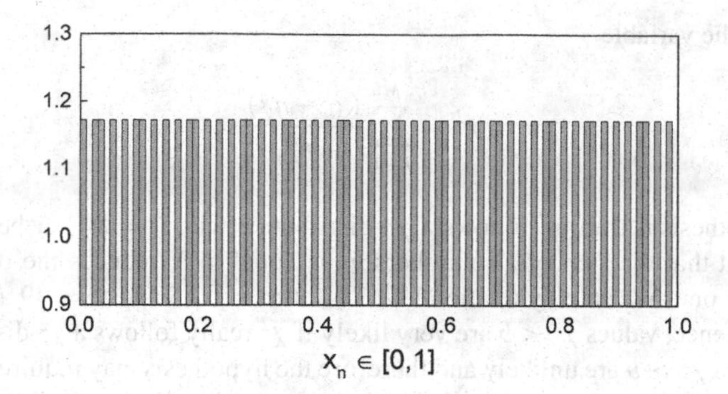

Fig. 12.3 Histogram for $N = 10^7$ and $M = 100$ bins as obtained with a linear congruential generator with parameters $a = 137$, $c = 0$ and $m = 2^{11}$

follows a χ^2-distribution where ν is the number of degrees of freedom. The pdf of the χ^2-distribution is given by

$$p(x;\nu) = \frac{x^{\frac{\nu}{2}-1}e^{-\frac{x}{2}}}{2^{\frac{\nu}{2}}\Gamma\left(\frac{\nu}{2}\right)}\,, \qquad x \geq 0\,, \tag{12.27}$$

where $\Gamma(\cdot)$ denotes the Γ-function. The probability of finding the variable x within the interval $[a,b] \subset \mathbb{R}^+$ can be calculated as

$$P(x \in [a,b];\nu) = \int_a^b \mathrm{d}x\, p(x;\nu)\,, \tag{12.28}$$

and in particular for $a = 0$ we obtain

$$P(x < b;\nu) = \int_0^b \mathrm{d}x\, p(x;\nu) = F(b;\nu)\,. \tag{12.29}$$

Here we introduced the cdf $F(b;\nu)$. Let us consider the inverse problem: the probability that $x \leq b$ is equal to α, i.e. $F(b;\nu) = \alpha$. We then calculate the upper bound b by inverting Eq. (12.29) and obtain:

$$b = F^{-1}(\alpha;\nu)\,. \tag{12.30}$$

These values are tabulated [17, 18].

We return to our particular example: the hypothesis is that the sequence $\{x_n\}$ generated by some pseudo-random number generator complies to a uniform distribution. It is a consequence of the central limit theorem (see Appendix, Sect. E.8) that the deviations from the theoretically expected values n_k^{th} obey a normal distribution.

We define the variable

$$x = \chi^2 = \sum_{k=1}^{M} \frac{(n_k - n_k^{th})^2}{n_k^{th}} . \tag{12.31}$$

If our hypothesis is true, χ^2 follows a χ^2-distribution with $\nu = M - 1$ because the requirement that the sum of all numbers n_k is equal to N reduces the degrees of freedom by one. We employ relation (12.30) for $\alpha = 0.85$ and $\nu = 99$ and obtain $b = 113$. Hence, values $\chi^2 < b$ are very likely if χ^2 really follows a χ^2 distribution, while values $\chi^2 > b$ are unlikely and therefore the hypothesis may require a review. However, it has to be emphasized that it is fundamentally impossible to *verify* a hypothesis. It can only be falsified or strengthened. We note that the resulting value of χ^2 will highly depend on the seed number of the generator as long as the maximum period has not been reached.

Summary

We first concentrated on a possible definition of randomness and on a mathematical definition of random numbers and sequences. As the generation of random numbers was the main topic of this section we moved on to describe the requirements an 'ideal' random number generator will have to obey. On the other hand, the numerics of computational physics demanded reproducible sequences of random numbers and this resulted in the notion of 'pseudo' random numbers which will be generated by deterministic methods and, thus, cannot possibly be 'ideal'. A number of rather simple but quite effective pseudo-random number generators was discussed before the question of how to test the quality (randomness) of these numbers was raised. We discussed statistical tests and demonstrated the simple spectral test using a linear congruential generator. More sophisticated is the method of quality testing. The histogram technique as a direct test for the probability density function from which the random numbers are drawn was discussed in detail. Finally, some basics of the χ^2-test have been presented.

Problems

1. Write the computer code for a linear congruential generator. This generator is described by

$$x_{j+1} = (ax_j + c) \mod m .$$

The random numbers $r_j \in [0, 1]$ can be obtained by normalizing x_j as was discussed in Sect. 12.2. Use the following parameters

a. $a = 16807\,, \quad c = 0\,, \quad m = 2^{31} - 1\,, \quad x_0 = 3141549,$
b. $a = 5\,, \quad c = 0\,, \quad m = 2^7\,, \quad x_0 = 1.$

2. Perform the following analysis:

   a. Compute the mean $\langle r \rangle$ and the variance var (r) for random numbers generated in N steps. Plot the result.
   b. Plot two successive random numbers r_k versus r_{k+1} for $k = 1, 2, \ldots, N-1$ in a two dimensional plot.
   c. Repeat the above steps for random numbers generated by your system's software.
   d. Discuss the results!

References

1. Coffey, W.T., Kalmykov, Y.P.: The Langevin Equation, 3rd edn. World Scientific Series in Contemporary Chemical Physics: Volume 27. World Scientific, Hackensack (2012)
2. Dubitzky, W., Wolkenhauer, O., Cho, K.H., Yokota, H. (eds.): Encyclopedia of Systems Biology, p. 1596. Springer, Berlin/Heidelberg (2013)
3. Tapiero, C.S.: Risk and Financial Management: Mathematical and Computational Methods. Wiley, New York (2004)
4. Lax, M., Cai, W., Xu, M.: Random Processes in Physics and Finance. Oxford Finance Series. Oxford University Press, Oxford (2013)
5. Laing, C., Lord, G.J. (eds.): Stochastic Methods in Neuroscience. Oxford University Press, Oxford (2009)
6. Chaitin, G.J.: Randomness and mathematical proof. Sci. Am. **232**, 47 (1975)
7. Knuth, D.: The Art of Computer Programming, vol. II, 3rd edn. Addison Wesley, Menlo Park (1998)
8. Gentle, J.E.: Random Number Generation and Monte Carlo Methods. Statistics and Computing. Springer, Berlin/Heidelberg (2003)
9. Ripley, B.D.: Stochastic Simulation. Wiley, New York (2006)
10. Press, W.H., Teukolsky, S.A., Vetterling, W.T., Flannery, B.P.: Numerical Recipes in C++, 2nd edn. Cambridge University Press, Cambridge (2002)
11. Knuth, D.: The Art of Computer Programming, vol. IV. Addison Wesley, Menlo Park (2011)
12. Marsaglia, G., Zaman, A.: A new class of random number generators. Ann. Appl. Prob. **1**, 462–480 (1991)
13. Iversen, G.P., Gergen, I.: Statistics. Springer Undergraduate Textbooks in Statistics. Springer, Berlin/Heidelberg (1997)
14. Keener, R.W.: Theoretical Statistics. Springer, Berlin/Heidelberg (2010)
15. Chow, Y.S., Teicher, H.: Probability Theory, 3rd edn. Springer Texts in Statistics. Springer, Berlin/Heidelberg (1997)
16. Kienke, A.: Probability Theory. Universitext. Springer, Berlin/Heidelberg (2008)
17. Abramovitz, M., Stegun, I.A. (eds.): Handbook of Mathemathical Functions. Dover, New York (1965)
18. Olver, F.W.J., Lozier, D.W., Boisvert, R.F., Clark, C.W.: NIST Handbook of Mathematical Functions. Cambridge University Press, Cambridge (2010)

Chapter 13
Random Sampling Methods

13.1 Introduction

Most applications require random number generators that follow a particular probability density function (pdf) which is not a uniform distribution on the interval [0, 1]. We present in this chapter methods to generate random numbers that follow some arbitrary pdf. As a source will serve uniformly distributed random numbers as they are generated with the help of the methods we discussed in Chap. 12.

The two most prominent techniques to generate random numbers from an arbitrary distribution, are the *inverse transformation method* and the *rejection method*. They will be discussed in Sects. 13.2 and 13.3, respectively. In addition, we comment in Sect. 13.4 briefly on sampling from piecewise defined pdfs and combined pdfs. It has to be emphasized that these methods are in many cases not sufficient and a more powerful approach is required. One of these is based on the idea of *importance sampling* and is referred to as the METROPOLIS method. It will be discussed briefly in Chap. 14.

Nevertheless, it is also possible to obtain quite easily random numbers for some specific pdfs by *direct sampling* [1]. For instance, suppose x_1, x_2 are two uniformly distributed random numbers. Hence, their pdf is given by

$$p_{\mathrm{u}}(x) = \begin{cases} 1 & x \in [0,1]\,, \\ 0 & \text{elsewhere.} \end{cases} \tag{13.1}$$

and the corresponding *cumulative distribution function* (cdf) follows:

$$P_{\mathrm{u}}(x) = \int_0^x \mathrm{d}x' p_{\mathrm{u}}(x') = \begin{cases} 0 & x < 0\,, \\ x & x \in [0,1]\,, \\ 1 & x > 1\,. \end{cases} \tag{13.2}$$

B.A. Stickler, E. Schachinger, *Basic Concepts in Computational Physics*,
DOI 10.1007/978-3-319-27265-8_13

One can prove that the new random number y

$$y = \max(x_1, x_2) \ , \tag{13.3}$$

conforms to the cdf

$$F(y) = y^2 \ , \tag{13.4}$$

and to the pdf[1]:

$$f(y) = 2y \ . \tag{13.5}$$

The consequence is an elegant method to generate random numbers z which follow the pdf

$$g(z) = kz^{k-1} \ , \tag{13.6}$$

by defining

$$z = \max(x_1, x_2, \ldots, x_k) \ . \tag{13.7}$$

Here, the random numbers x_i are uniformly distributed and can be obtained with the help of the methods introduced in Chap. 12.

Another equally elegant method can be employed to calculate random numbers which follow a normal distribution:

$$p(z) = \frac{1}{\sqrt{2\pi}} \exp\left(-\frac{z^2}{2}\right) \ . \tag{13.8}$$

Again, we act on the assumption that the random numbers x_i are uniformly distributed within the unit interval [0, 1]. We take two random numbers (x_1, x_2) and construct two random numbers (z_1, z_2) using the transformation:

$$z_1 = \cos(2\pi x_2)\sqrt{-2\ln x_1}, \qquad z_2 = \sin(2\pi x_2)\sqrt{-2\ln x_1} \ . \tag{13.9}$$

[1]This follows from the transformation of pdfs (see Chap. 14):

$$f(y) = \int_0^1 \mathrm{d}x_1 \int_0^1 \mathrm{d}x_2 \, \delta[y - \max(x_1, x_2)] = 2y.$$

It is easy to demonstrate that (z_1, z_2) follow the pdf (13.8). We introduce the joint distribution $p_u(x_1, x_2) = p_u(x_1)p_u(x_2)$ (assumption of no correlations). The transformation of probabilities [2–4] gives

$$p(z_1, z_2)\mathrm{d}z_1\mathrm{d}z_2 = p_u(x_1, x_2)\mathrm{d}x_1\mathrm{d}x_2\,, \tag{13.10}$$

or, equivalently, the JACOBIAN determinant

$$p(z_1, z_2) = \frac{\partial(x_1, x_2)}{\partial(z_1, z_2)}\,, \tag{13.11}$$

where we employed Eq. (13.1). We recognize that Eq. (13.9) is equivalent to

$$x_1 = \exp\left(-\frac{z_1^2 + z_2^2}{2}\right), \qquad x_2 = \frac{1}{2\pi}\tan^{-1}\left(\frac{z_2}{z_1}\right). \tag{13.12}$$

The JACOBIAN determinant is readily evaluated and gives[2]:

$$\begin{aligned}
\frac{\partial(x_1, x_2)}{\partial(z_1, z_2)} &= \begin{vmatrix} \frac{\partial x_1}{\partial z_1} & \frac{\partial x_1}{\partial z_2} \\ \frac{\partial x_2}{\partial z_1} & \frac{\partial x_2}{\partial z_2} \end{vmatrix} \\
&= \begin{vmatrix} -z_1 x_1 & -z_2 x_1 \\ -\frac{z_2}{2\pi(z_1^2+z_2^2)} & \frac{z_1}{2\pi(z_1^2+z_2^2)} \end{vmatrix} \\
&= \frac{x_1}{2\pi} \\
&= \frac{1}{2\pi}\exp\left(-\frac{z_1^2 + z_2^2}{2}\right) \\
&= p(z_1)p(z_2)\,.
\end{aligned} \tag{13.13}$$

This is the product of two normal distributions and, thus, z_1 and z_2 follow indeed a normal distribution.

[2] We make use of:

$$\frac{\mathrm{d}}{\mathrm{d}x}\tan^{-1}(x) = \frac{1}{1 + x^2}.$$

13.2 Inverse Transformation Method

The inverse transformation method is one of the simplest and most useful methods to sample random variables from an arbitrary pdf [1, 5–7]. Let $p(x)$, $x \in [x_{\min}, x_{\max}]$, denote the pdf from which we want to obtain our random numbers. The corresponding cdf will be denoted by

$$P(x) = \int_{x_{\min}}^{x} \mathrm{d}x' p(x') . \tag{13.14}$$

It follows immediately from the positivity and the normalization condition of pdfs (Appendix, Sect. E.5) that $P(x)$ is monotonically increasing and, furthermore, that $P(x_{\min}) = 0$ and $P(x_{\max}) = 1$. Let ξ denote some random number uniformly distributed within the interval $[0, 1]$. We obtain from the conservation of probability [2–4]

$$p_{\mathrm{u}}(\xi)\mathrm{d}\xi = p(x)\mathrm{d}x \Longrightarrow 1 = p_{\mathrm{u}}(\xi) = p(x)\left(\frac{\mathrm{d}\xi}{\mathrm{d}x}\right)^{-1} , \tag{13.15}$$

which can be solved by the choice $\xi = P(x)$, since

$$\frac{\mathrm{d}}{\mathrm{d}x}P(x) = p(x) . \tag{13.16}$$

Hence, we arrive at

$$x = P^{-1}(\xi) \tag{13.17}$$

where P^{-1} denotes the inverse of P. It is an obvious caveat of this method that it requires the inverse $P^{-1}(\xi)$ to exist and that $P(x)$ must be calculated and inverted analytically. This is, for instance, not possible in the case of the normal distribution (13.8).

Let us illustrate this method with two simple examples:

1. Suppose we want to draw random numbers which are uniformly distributed within the interval $[a, b]$. The corresponding pdf reads

$$p(x) = \frac{1}{b-a} , \tag{13.18}$$

and the cdf takes on the form

$$P(x) = \frac{x-a}{b-a} , \tag{13.19}$$

where we set, in this particular example, $x_{\min} = a$. Hence, we have

$$\xi = \frac{x-a}{b-a}\,, \tag{13.20}$$

which is uniformly distributed within [0, 1]. Consequently, we determine random numbers $x \in [a, b]$ uniformly distributed via

$$x = a + (b-a)\xi\,. \tag{13.21}$$

2. We are interested in random numbers x drawn from a pdf given by the exponential distribution:

$$p(x) = \frac{1}{\lambda}\exp\left(-\frac{x}{\lambda}\right)\,. \tag{13.22}$$

Here, $\lambda > 0$ and $x \in [0, \infty)$. These could, for instance, describe the free path x of a particle between interactions, where the mean free path $\langle x\rangle = \lambda$. From Eq. (13.17) we obtain

$$\xi = \frac{1}{\lambda}\int_0^x \mathrm{d}x' \exp\left(-\frac{x'}{\lambda}\right) = 1 - \exp\left(-\frac{x}{\lambda}\right)\,, \tag{13.23}$$

and consequently the relation

$$x = -\lambda \ln(1-\xi)\,, \tag{13.24}$$

gives random variables x which comply to the exponential distribution (13.22) if ξ follows the pdf $p_u(\xi)$ of Eq. (13.1). Moreover, it follows from the symmetry of the uniform distribution that

$$x = -\lambda \ln(\xi)\,, \tag{13.25}$$

without affecting the resulting random numbers. In Fig. 13.1 we show a histogram with random numbers drawn according to (13.25).

We pointed out already that it is certainly a caveat of this method that the cdf $P(x)$ has to be calculated and inverted analytically. Even if $P(x)$ is not analytically invertible, it is possible to employ the inverse transformation method by calculating $P(x)$ for certain grid-points x_i and then interpolating $P(x)$ piecewise between these points with the help of an invertible function. However, in many cases it is advantageous to employ the rejection method, which will be discussed next.

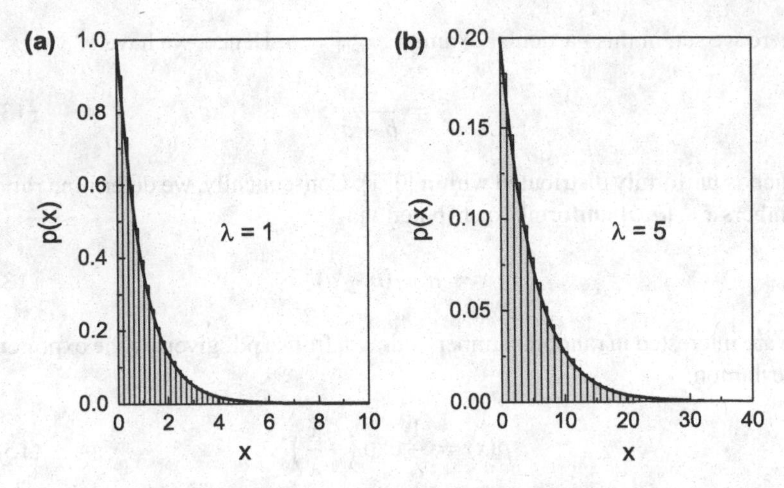

Fig. 13.1 The histogram representation of the pdf $p(x)$ vs x generated by random numbers drawn from an exponential distribution Eq. (13.22) with the help of the inverse transformation sampling method. Two different values for λ have been considered, namely **(a)** $\lambda = 1$ and **(b)** $\lambda = 5$. $N = 10^5$ random numbers have been sampled. The *solid line* corresponds to the pdf $p(x)$ according to Eq. (13.22)

13.3 Rejection Method

The rejection method is particularly suitable if the inverse transformation method fails [1, 6, 7]. One of the most prominent versions of the rejection method is the METROPOLIS algorithm. It will be introduced in Sect. 14.3.

The basic idea of the rejection method is to draw random numbers x from another, preferably analytically invertible pdf $h(x)$ and check whether or not they lie within the desired pdf $p(x)$. If this is the case the random number x is accepted, otherwise it will be rejected. This is also the basic idea of the *hit and miss* version of Monte-Carlo integration which will be discussed in Sect. 14.2.

We specify the rejection method: Let $p(x)$ denote the pdf from which we want to draw random numbers. Furthermore, let $h(x)$ be another pdf, which can easily be sampled (for instance with the help of the inverse transformation method) and which is chosen in a such a way that the inequality

$$p(x) \leq c\,h(x)\,, \tag{13.26}$$

holds for all $x \in [x_{\min}, x_{\max}]$, where $c \geq 1$ is some constant. The function $c\,h(x)$ is referred to as the *envelope* of $p(x)$ within the interval $[x_{\min}, x_{\max}]$. The strategy is clear: we sample a random variable x' (trial state) from $h(x)$ and accept it with probability $p(x)/[c\,h(x)]$. This procedure is sketched in Fig. 13.2. Let $p(A|x)$ denote the probability that a given value x is accepted and $g(x)$ denotes the probability that

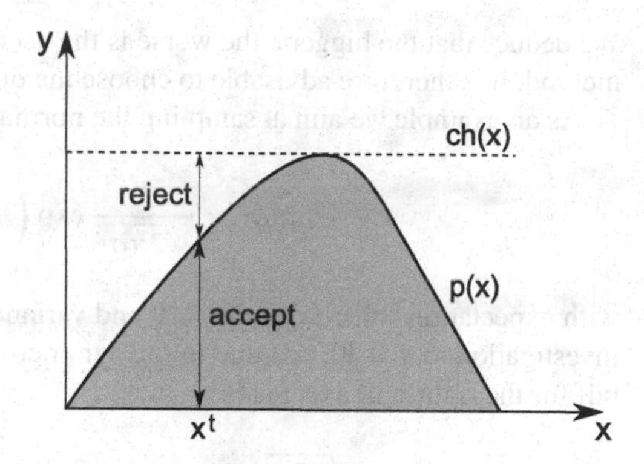

Fig. 13.2 Schematic illustration of the rejection method. The trial state x^t is accepted with probability $p(x)/[c\,h(x)]$

we produce a variable x with the help of this algorithm. Furthermore, $P(x = x^t)$ stands for the probability that a trial state x^t is generated. We have

$$
\begin{aligned}
g(x) &\propto P(x = x^t)p(A|x^t) \\
&= h(x^t)\frac{p(x^t)}{c\,h(x^t)} \\
&\propto p(x^t) \,.
\end{aligned} \tag{13.27}
$$

Hence, we indeed generate random numbers which follow the pdf $p(x)$. We may also calculate the probability $P(A)$ that an arbitrary trial state x^t is accepted. This is done with the help of the marginalization rule (E.39):

$$
\begin{aligned}
P(A) &= \int \mathrm{d}x^t p(A \wedge x^t) \\
&= \int \mathrm{d}x^t p(A|x^t)P(x = x^t) \\
&= \int \mathrm{d}x^t \frac{p(x^t)}{c\,h(x^t)} h(x^t) \\
&= \frac{1}{c}\int \mathrm{d}x^t p(x^t) \\
&= \frac{1}{c} \,.
\end{aligned} \tag{13.28}
$$

More generally, the probability $P(A)$ to accept a d-dimensional random variable is given by:

$$
P(A) = \frac{1}{c^d} \,. \tag{13.29}
$$

We deduce that the bigger c the worse is the acceptance probability of the rejection method. It is therefore advisable to choose the envelope $h(x)$ very carefully.

As an example we aim at sampling the normal distribution (E.43) for $x \in \mathbb{R}$

$$p(x) = \frac{1}{\sqrt{2\pi\sigma^2}} \exp\left(-\frac{x^2}{2\sigma^2}\right) , \tag{13.30}$$

with expectation value $\langle x \rangle \equiv x_0 = 0$ and variance σ^2. In a first step we restrict our investigation to $x \in [0, \infty)$ due to the symmetry of the pdf. The slightly modified pdf for the right-half axis reads

$$q(x) = \sqrt{\frac{2}{\pi\sigma^2}} \exp\left(-\frac{x^2}{2\sigma^2}\right) , \qquad x \in [0, \infty) , \tag{13.31}$$

where we adjusted the normalization. The complete normal distribution (13.30) is re-obtained by sampling the sign of x in an additional step. We use as an envelope $h(x)$ the exponential distribution Eq. (13.22). Furthermore, λ and c are chosen in such a way that the acceptance probability (13.28) has a maximum under the constraint (13.26). Since this is equivalent to $c \to \min$ we have to solve the optimization problem

$$c \geq \frac{q(x)}{h(x)} \to \max . \tag{13.32}$$

The resulting $c_{\min}$ is then given by

$$c_{\min} = \frac{q(x_{\text{opt}})}{h(x_{\text{opt}})} \tag{13.33}$$

and x_{opt} is the yet unknown optimal value for x. We obtain

$$\begin{aligned} \frac{\mathrm{d}}{\mathrm{d}x} \frac{q(x)}{h(x)} &= \sqrt{\frac{2\lambda^2}{\pi\sigma^2}} \frac{\mathrm{d}}{\mathrm{d}x} \exp\left(\frac{x}{\lambda} - \frac{x^2}{2\sigma^2}\right) \\ &= \sqrt{\frac{2\lambda^2}{\pi\sigma^2}} \exp\left(\frac{x}{\lambda} - \frac{x^2}{2\sigma^2}\right) \left[\frac{1}{\lambda} - \frac{x}{\sigma^2}\right] \\ &\stackrel{!}{=} 0 , \end{aligned} \tag{13.34}$$

and, therefore,

$$x_{\text{opt}} = \frac{\sigma^2}{\lambda} . \tag{13.35}$$

Consequently, we have

$$c_{\min} = \sqrt{\frac{2\lambda^2}{\pi\sigma^2}} \exp\left(\frac{\sigma^2}{2\lambda^2}\right) . \tag{13.36}$$

The above relation gives the minimum value of c for arbitrary values of λ. However, since $h(x)$ is our envelope, we can choose λ in such a way, that $c_{\min} \to \min$. This is achieved in a second step

$$\begin{aligned} \frac{\mathrm{d}}{\mathrm{d}\lambda} c_{\min} &= \sqrt{\frac{2}{\pi\sigma^2}} \exp\left(\frac{\sigma^2}{2\lambda^2}\right)\left(1 - \frac{\sigma^2}{\lambda^2}\right) \\ &\stackrel{!}{=} 0 . \end{aligned} \tag{13.37}$$

which results in the optimum value $\lambda_{\text{opt}} = \sigma$. This, finally, results together with Eq. (13.36) in:

$$c_{\min} = \sqrt{\frac{2e}{\pi}} . \tag{13.38}$$

The algorithm is executed in the following steps:

1. Draw a uniformly distributed random number $\xi \in [0, 1]$.
2. Calculate $x^t = -\lambda_{\text{opt}} \ln(\xi)$, where $\lambda_{\text{opt}} = \sigma$.
3. Draw a uniformly distributed random number $r \in [0, 1]$. If $r \leq q(x^t)/[c_{\min} h(x^t)]$, then $x = x^t$ is accepted and if $r > q(x^t)/[c_{\min} h(x^t)]$, x^t is rejected and we return to step 1.
4. If x^t was accepted, we draw a uniformly distributed random number $r \in [0, 1]$ and only if $r < 0.5$ we set $x = -x$ otherwise x stays as is.
5. We repeat steps 1–4 until the number N of desired random numbers has been reached.

Figure 13.3 shows random numbers obtained with the help of this method in a histogram representation. We calculated (a) $N = 10^3$, (b) $N = 10^4$, and (c) $N = 10^5$ random numbers for $\sigma = 1$. It is quite obvious that the original pdf (13.30) is the better approximated the bigger the number N of sampled random numbers becomes.

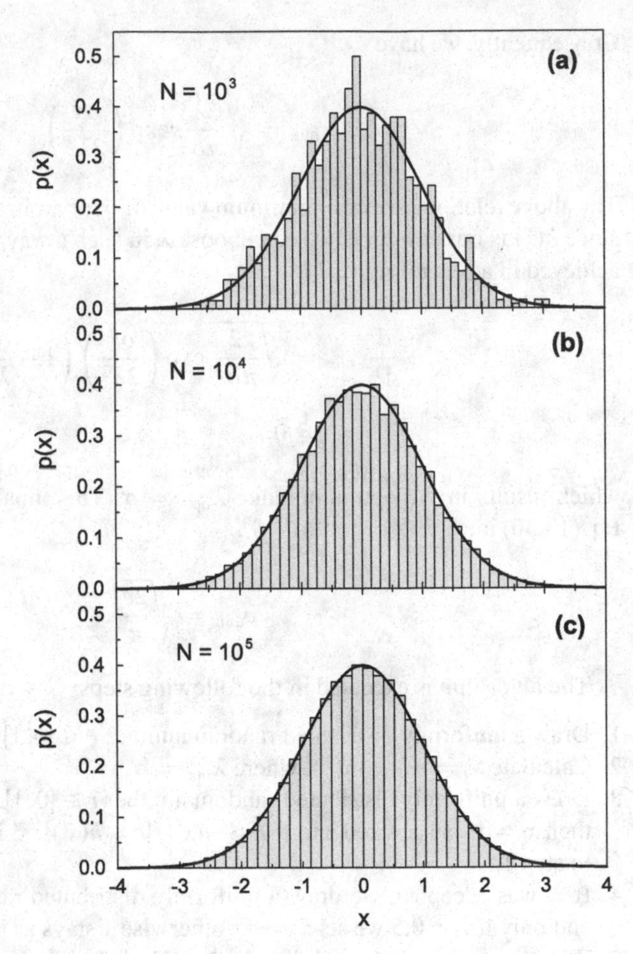

Fig. 13.3 The histogram representation of the pdf $p(x)$ vs x generated by random numbers drawn from the normal distribution Eq. (13.30) ($\sigma = 1$) with the help of the rejection method. We sampled **(a)** $N = 10^3$, **(b)** $N = 10^4$, and **(c)** $N = 10^5$ random numbers. The *solid line* represents the pdf $p(x)$ (13.30)

13.4 Probability Mixing

The method of probability mixing was developed to offer an algorithm which allows to generate random numbers by sampling piecewise defined or composite pdfs. Such a pdf is of the general form

$$p(x) = \sum_{i=1}^{N} \alpha_i f_i(x), \quad \alpha_i \neq 0 , \tag{13.39}$$

where the sub-pdfs $f_i(x)$ fulfill the normalization requirement

$$\int \mathrm{d}x' f_i(x') = 1 , \tag{13.40}$$

and are non-negative, i.e.

$$f_i(x) \geq 0 \,, \tag{13.41}$$

for all $i = 1, \ldots, N$. It follows that

$$\sum_{i=1}^{N} \alpha_i = 1 \,, \tag{13.42}$$

which ensures that

$$\int \mathrm{d}x' p(x') = 1 \,, \tag{13.43}$$

is fulfilled. The question is how to sample random numbers from such a pdf, since in most cases it might be hard to invert the sum (inverse transformation method) or find a suitable envelope (rejection method). However, this question can easily be answered: We define

$$q_i = \sum_{\ell=1}^{i} \alpha_\ell \,. \tag{13.44}$$

Thus, $q_N = 1$ and the interval [0, 1] has been divided according to:

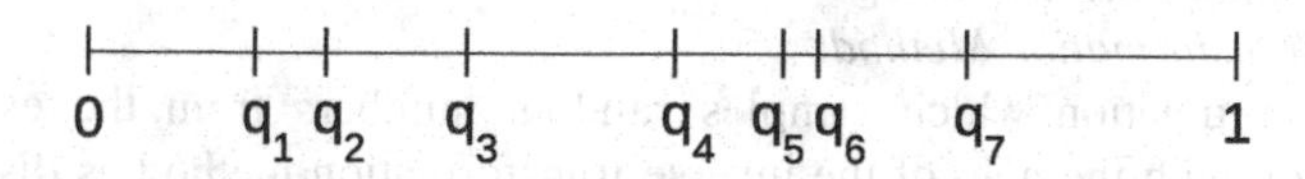

The index i of the relevant pdf is determined by the condition

$$q_{i-1} < r < q_i \,, \tag{13.45}$$

where $r \in [0, 1]$ is a uniformly distributed random number. We then draw the required random number x from the sub-pdf $f_i(x)$ with any of the methods discussed above.

This procedure is quite plausible, since the coefficients α_i give the relative weight of the sub-pdfs $f_i(x)$. In particular, α_i determines the importance of the sub-pdf $f_i(x)$. It is, therefore, a natural approach to use α_i as a measure of the probability that a random variable is to be sampled from the particular sub-pdf $f_i(x)$.

Summary

To generate random numbers is essential for many application in Computational Physics. This chapter concentrated on basic methods to generate the desired random numbers: (a) the *direct sampling method* used transformations of the uniform distribution to generate the required random numbers; (b) the *inverse transformation method* was based on the availability of an inverse cdf which was in most cases required to be calculated analytically; finally, (c) the *rejection method* which was basically a hit or miss method. It used an easily invertible pdf $h(x)$ which enveloped the desired pdf $p(x)$ completely within some interval $x \in [x_{\min}, x_{\max}]$. The effectiveness of this method depended on how 'well' the envelope $h(x)$ enclosed the original pdf $p(x)$. In a last step the method of *probability mixing* was discussed. It was an easily verifiable method which allowed to sample random numbers from composite pdfs.

Problems

Draw random numbers from the following pdfs:

1. *Direct Sampling*:
   Sample the normal distribution with $\langle x \rangle = 0$ and $\sigma = 1$ with the help of the method discussed in Sect. 13.1. Check the result by plotting the random numbers against the pdf $p(x)$ in a histogram.
2. *Inverse Transformation Method*:
   Write a function which samples random numbers from the exponential distribution with the help of the inverse transformation method as discussed in Sect. 13.2. Compare the generated random numbers to the pdf in a histogram.
3. *Rejection Method*:
   Sample the normal distribution with $\langle x \rangle = 0$ and $\sigma = 1$ with the help of the exponential distribution as discussed in Sect. 13.3. Compare the generated random numbers with the pdf in a histogram. Determine the acceptance probability numerically.
4. *Probability Mixing*:
   We choose an alternative envelope for the normal distribution with $\langle x \rangle = 0$ and $\sigma = 1$. This envelope is chosen to be constant for all $|x| < x_0$, and decays exponentially for $|x| \geq x_0$. (x_0 is a parameter of your choice.) The parameters do not need to optimize the acceptance probability. Again, plot the generated random numbers in a histogram and compare the acceptance probability with the acceptance probability of point 3.

References

1. Devroye, L.: Non-uniform Random Variate Generation. Springer, Berlin/Heidelberg (1986)
2. Chow, Y.S., Teicher, H.: Probability Theory, 3rd edn. Springer Texts in Statistics. Springer, Berlin/Heidelberg (1997)
3. Kienke, A.: Probability Theory. Universitext. Springer, Berlin/Heidelberg (2008)
4. Stroock, D.W.: Probability Theory. Cambridge University Press, Cambridge (2011)
5. Bratley, P., Fox, B.L., Schrage, L.E.: A Guide to Simulation. Springer, Berlin/Heidelberg (1987)
6. Knuth, D.: The Art of Computer Programming, vol. II, 3rd edn. Addison Wesley, Menlo Park (1998)
7. Press, W.H., Teukolsky, S.A., Vetterling, W.T., Flannery, B.P.: Numerical Recipes in C++, 2nd edn. Cambridge University Press, Cambridge (2002)

Chapter 14
A Brief Introduction to Monte-Carlo Methods

14.1 Introduction

This chapter presents a brief introduction to Monte-Carlo methods in general, and to Monte-Carlo integration as well as to the METROPOLIS-HASTINGS algorithm in particular. A detailed discussion of the fundamental concepts involved is postponed to Chap. 16. The introduction given here is not supposed to be self-contained and methods will be introduced without reference to their background.

The notion of Monte-Carlo methods, Monte-Carlo algorithms or Monte-Carlo techniques is not well defined. In particular, the term *Monte-Carlo* summarizes a wide field of methods which are based on the sampling of random numbers [1–3]. In general, the advantage of Monte-Carlo algorithms lies in their computational strength. In many cases it is simply not feasible to employ deterministic methods due to their very high computational cost. However, in many cases the use of methods based on random sampling is also motivated by the nature of the processes to be described. We mentioned in the previous chapter as a typical example the radioactive decay of some nucleus. This process is believed to be purely stochastic in nature.

The development of Monte-Carlo techniques was initialized in the 1940s by J. VON NEUMANN, S. M. ULAM and N. METROPOLIS who coined the term *Monte-Carlo methods*. One of the earliest illustrations of the principle of Monte-Carlo techniques in general, and of Monte-Carlo integration in particular is the Monte-Carlo approximation of π. The discussion which follows now includes the essential ideas of Monte-Carlo integration.

We regard the unit square characterized by the corner points $(0, 0)$, $(0, 1)$, $(1, 0)$, and $(1, 1)$. The area A_s of this square is one. We insert a quarter-circle of radius $r = 1$ which, consequently, possesses the area $A_c = \pi/4$. Suppose, we are throwing darts on this unit square in such a way that the impact points are uniformly distributed;

B.A. Stickler, E. Schachinger, *Basic Concepts in Computational Physics*,
DOI 10.1007/978-3-319-27265-8_14

then the probability P that a certain dart becomes stuck within the interior of the quarter-circle is given by

$$P = \frac{A_c}{A_s} = A_c = \frac{\pi}{4} = 0.785398\ldots. \tag{14.1}$$

From a probabilistic point of view, we have after N throws of which n hit the interior of the quarter-circle the probability:

$$P = \lim_{N\to\infty} \frac{n}{N}. \tag{14.2}$$

The strategy is clear: we draw uniformly distributed random numbers x_i, y_i from the interval $[0, 1]$. These are the intersection points of the darts. Repeating this *experiment* several times and counting the number of hits n within the quarter-circle allows us to approximate π via

$$P = \frac{\pi}{4} \approx \frac{n}{N}. \tag{14.3}$$

The resulting approximation of π will be strongly influenced by the number of experiments N as well as by the performance of the random number generator used. Table 14.1 lists computed approximations of π for different numbers of experiments N as they were obtained with the help of a linear congruential generator. Linear congruential generators have been introduced and discussed in Sect. 12.2. The parameters used to initialize the generators are given in the caption of the table. Furthermore, Fig. 14.1 illustrates the result after $N = 10^3$ experiments for both generators.

Table 14.1 Approximate values $\pi_a^{(i)}$ obtained with the method discussed in the text. The linear congruential generators are initialized by the following parameters: generator (1): $a = 7^5$, $c = 0$, $m = 2^{31} - 1$, and $x_0 = 281$ (PARK-MILLER) and generator (2): $a = 7^5$, $c = 0$, $m = 2^{11}$, and $x_0 = 281$. We also give the absolute errors $|\pi_a^{(i)} - \pi|$

N	$\pi_a^{(1)}$	$\lvert\pi_a^{(1)} - \pi\rvert$	$\pi_a^{(2)}$	$\lvert\pi_a^{(2)} - \pi\rvert$
10	2.8000	0.34159	2.8000	0.34159
10^2	2.9200	0.22159	3.1600	0.01841
10^3	3.1600	0.01841	3.1840	0.04241
10^4	3.1304	0.01119	3.1868	0.04521
10^5	3.1358	0.00579	3.1875	0.04589
10^6	3.1393	0.00229	3.1875	0.04599
10^7	3.1413	0.00028	3.1875	0.04599

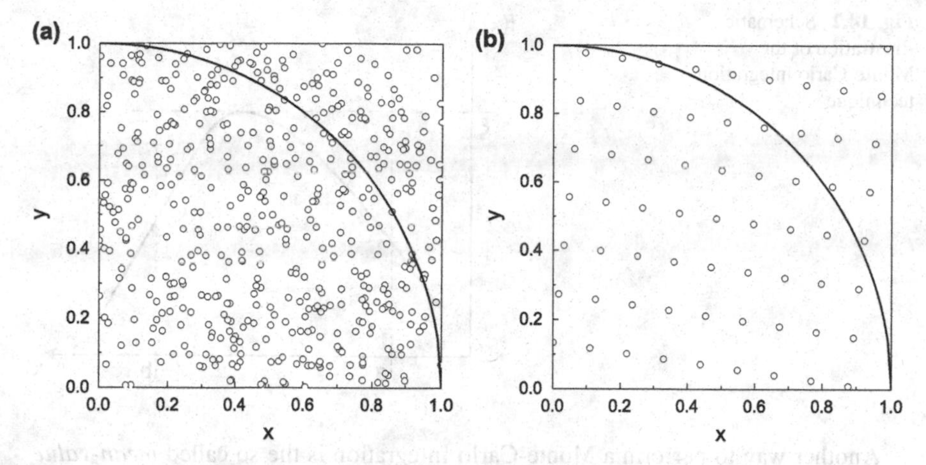

Fig. 14.1 $N = 10^3$ uniformly distributed random numbers within the unit-square. Frame **(a)** gives the results for generator (1) while frame **(b)** is for generator (2). The number of elements within the quarter-circle indicated by the *solid line* determines the value of $\pi_a^{(i)}$. The inferior result of the approximation obtained with generator (2) [frame **(b)**] originates in correlations between the x and y coordinates

14.2 Monte-Carlo Integration

We generalize the ideas formulated above and consider a function $f(x) \geq 0$ for $x \in [a, b] \subset \mathbb{R}$ where the area of interest is

$$A = \int_a^b \mathrm{d}x f(x) \,. \tag{14.4}$$

We denote

$$\xi = \max_{x \in [a,b]} f(x) \,, \tag{14.5}$$

and obtain using the above example

$$A = A_s \lim_{N \to \infty} \frac{n}{N} \,, \tag{14.6}$$

where n is the number of random points under the curve indicated schematically in Fig. 14.2. The area A_s is given by

$$A_s = (b - a)\xi \,, \tag{14.7}$$

and the random numbers $r_i = (x_i, y_i)$ are uniformly distributed within the intervals $x_i \in [a, b]$ and $y_i \in [0, \xi]$. This method is referred to as *hit and miss integration* [4].

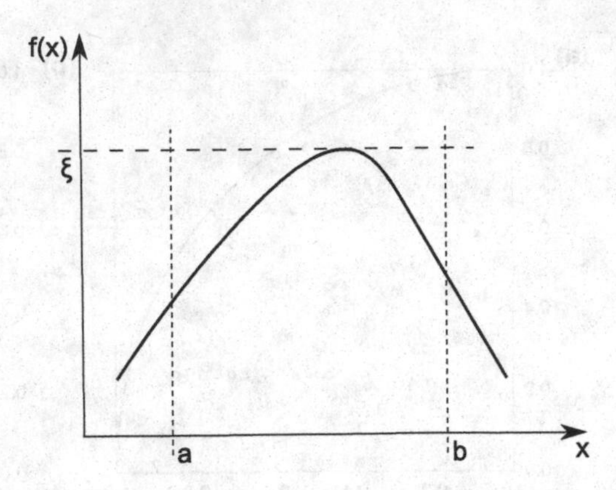

Fig. 14.2 Schematic illustration of the Monte-Carlo integration technique

Another way to perform a Monte-Carlo integration is the so called *mean-value integration*. It is essentially based on the mean value theorem of calculus which we already employed in our discussion of quadrature in Chap. 3. We restate it here for the sake of a more transparent presentation: The mean-value theorem states that if $f(x)$ is a continuous function for $x \in [a, b]$ then there exists a $z \in (a, b)$ such that

$$\int_a^b \mathrm{d}x f(x) = f(z)(b-a) \,. \tag{14.8}$$

The function value $f(z) \equiv \langle f \rangle$ is referred to as the *expectation value* or *mean value* of $f(x)$. We know from probability theory [5–7] that the expectation value can be approximated by the arithmetic mean $\overline{f}$

$$\frac{1}{b-a} \int_a^b \mathrm{d}x' f(x') \simeq \overline{f} \pm \sqrt{\frac{\overline{f^2} - \overline{f}^2}{N}} \,, \tag{14.9}$$

with the error given by the standard error, Eq. (E.14). The arithmetic mean $\overline{f}$, on the other hand, is given by

$$\overline{f} = \frac{1}{N} \sum_{i=1}^{N} f(x_i) \,, \tag{14.10}$$

and consequently

$$\overline{f^2} = \frac{1}{N} \sum_{i=1}^{N} f^2(x_i) \,. \tag{14.11}$$

Note that here the variables x_i are assumed to be uniformly distributed random numbers within the interval $[a, b]$. (This result will immediately be discussed in more detail.) However, first of all we note from the law of large numbers, Eq. (E.25), that this approach is exact in the limit $N \to \infty$:

$$\frac{1}{b-a}\int_a^b \mathrm{d}x' f(x') = \lim_{N\to\infty}\frac{1}{N}\sum_{i=1}^{N} f(x_i) . \tag{14.12}$$

Let us now consider the more general case which, in the end, will guide us to a very prominent formulation of Monte-Carlo integration. We want to estimate the expectation value

$$\langle f \rangle = \int \mathrm{d}x f(x) p(x) , \tag{14.13}$$

where $x \in \mathbb{R}^d$ and $p(x)$ is a pdf. A typical example is the calculation of the thermal expectation value in statistical physics where the pdf $p(x)$ is given by the normalized BOLTZMANN distribution

$$p(x) = \frac{1}{Z}\exp\left[-\frac{E(x)}{k_B T}\right] . \tag{14.14}$$

Here $E(x)$ denotes the energy as a function of the parameter $x \in \mathbb{R}^d$, k_B stands for BOLTZMANN's constant, T is the temperature, and the normalization factor Z is referred to as the canonical partition function [8–11].

Equation (14.13) may be rewritten as

$$\langle f \rangle = \int \mathrm{d}x f(x) p(x) = \int \mathrm{d}f f\, q(f) , \tag{14.15}$$

where we introduced the probability density $q(f)$ of f via

$$q(f) = \int \mathrm{d}x\, \delta\left[f - f(x)\right] p(x) , \tag{14.16}$$

with $\delta(\cdot)$ DIRAC's δ-distribution. Let us briefly explain how we arrived at this definition. Let the cdf $P(x)$ be defined by[1]

$$P(x) = \Pr(X \le x) = \int_{-\infty}^{x} \mathrm{d}x\, p(x) . \tag{14.17}$$

[1]Please note that according to the conventions established in Appendix E capital letters denote random variables.

We define in analogy the cdf $Q(f)$:

$$Q(f) = \Pr(F \leq f) = \Pr[f(X) \leq f] \,. \tag{14.18}$$

Note that we distinguish between the function $f(X)$ of the random variable X [which follows the pdf $p(X)$] and the particular function value $f \in \mathbb{R}$. Furthermore, the probability $\Pr[f(X) \leq f]$ can be rewritten as

$$\Pr[f(X) \leq f] = \sum_n \Pr(a_n \leq X \leq b_n) \,, \tag{14.19}$$

where the values $a_n < b_n$ are the ordered intersection points $a_1 < b_1 < a_2 < b_2 < \ldots < a_N < b_N$ chosen in such a way that

$$f(a_n) = f(b_n) = f \,, \quad \text{and} \quad f[x \in (a_n, b_n)] < f \,. \tag{14.20}$$

It is a matter of the particular form of $f(x)$ whether or not the boundary points have to be included. Equation (14.19) can be rewritten:

$$\Pr(a_n \leq X \leq b_n) = P(b_n) - P(a_n) = \int_{a_n}^{b_n} \mathrm{d}x\, p(x) \,. \tag{14.21}$$

The pdf $q(f)$ is related to the cdf $Q(f)$ via

$$q(f) = \frac{\mathrm{d}}{\mathrm{d}f} Q(f) \,, \tag{14.22}$$

and we obtain

$$\begin{aligned} q(f) &= \sum_n \frac{\mathrm{d}}{\mathrm{d}f} \Pr(a_n \leq X \leq b_n) \\ &= \sum_n \frac{\mathrm{d}}{\mathrm{d}f} \int_{a_n}^{b_n} \mathrm{d}x\, p(x) \\ &= \sum_n \left[\frac{\mathrm{d}b_n}{\mathrm{d}f} p(b_n) - \frac{\mathrm{d}a_n}{\mathrm{d}f} p(a_n) \right] \\ &= \sum_n \left[\left(\frac{\mathrm{d}f(x)}{\mathrm{d}x} \right)^{-1} p(x) \bigg|_{x=b_n} - \left(\frac{\mathrm{d}f(x)}{\mathrm{d}x} \right)^{-1} p(x) \bigg|_{x=a_n} \right] . \end{aligned} \tag{14.23}$$

However, we know from Eq. (14.20) that:

$$\left.\frac{\mathrm{d}f(x)}{\mathrm{d}x}\right|_{x=b_n} \overset{!}{>} 0 \quad \text{and} \quad \left.\frac{\mathrm{d}f(x)}{\mathrm{d}x}\right|_{x=a_n} \overset{!}{<} 0 . \tag{14.24}$$

We introduce the intersection points x_k where $x_1 < x_2 < \ldots < x_K$ and $K = 2N$ (if the boundary points are not included) for which $f(x_1) = f(x_2) = \ldots = f(x_K) = f$. Hence, Eq. (14.23) may be rewritten as

$$\begin{aligned} q(f) &= \sum_k \left|\left.\frac{\mathrm{d}f(x)}{\mathrm{d}x}\right|^{-1}\right|_{x=x_k} p(x_k) \\ &= \sum_k \frac{p(x_k)}{|f'(x_k)|} . \end{aligned} \tag{14.25}$$

We want to improve this result and remember that the DIRAC δ-distribution of an arbitrary function $g(y)$ can be expressed as [12]

$$\delta[g(y)] = \sum_i \frac{\delta(y - y_i)}{|g'(y_i)|} , \tag{14.26}$$

where the y_i are the zeros of $g(y)$, i.e. $g(y_i) = 0$. Hence, we arrive at the final form of Eq. (14.23)[2]:

$$q(f) = \int \mathrm{d}x \delta\left[f - f(x)\right] p(x) . \tag{14.27}$$

We note, furthermore, that:

$$\int \mathrm{d}f\, q(f) = \int \mathrm{d}f \int \mathrm{d}x \delta\left[f - f(x)\right] p(x) = \int \mathrm{d}x\, p(x) = 1 . \tag{14.28}$$

[2]We give an example. Suppose $f(x) = \exp(x)$. Then we deduce that

$$\delta\left[f - \exp(x)\right] = \frac{\delta(x - \ln f)}{f} ,$$

and, consequently,

$$q(f) = \frac{p(\ln f)}{f} .$$

A second example was given in Chap. 13 where we derived the pdf (13.5).

As a result, the variance of f, var (f), can be expressed as

$$\mathrm{var}\,(f) = \int \mathrm{d}x\,[f(x) - \langle f \rangle]^2\, p(x) = \int \mathrm{d}f\,[f - \langle f \rangle]^2\, q(f)\,. \tag{14.29}$$

Let us define in a next step the arithmetic mean of $f(X)$

$$\mathscr{F} = \frac{1}{N}\sum_{i=1}^{N} f(x_i) = \frac{1}{N}\sum_{i=1}^{N} f_i\,, \tag{14.30}$$

calculated with the help of N random numbers. Hence, we have

$$\langle \mathscr{F} \rangle = \langle f \rangle\,, \tag{14.31}$$

and

$$\mathrm{var}\,(\mathscr{F}) = \frac{\mathrm{var}\,(f)}{N}\,, \tag{14.32}$$

according to Appendix E. It follows from the central limit theorem, Appendix Sect. E.8, that for large values of N, the pdf of $\mathscr{F}$, $p(\mathscr{F})$ converges to a normal distribution (E.43) with $\langle \mathscr{F} \rangle$ and var $(\mathscr{F})$:

$$p(\mathscr{F}) \approx \mathscr{N}\left(\mathscr{F}\,\middle|\,\langle f \rangle, \frac{\mathrm{var}\,(f)}{N}\right)\,. \tag{14.33}$$

Based on this property $\langle f \rangle$ can be estimated from:

$$\langle f \rangle = \mathscr{F} \pm \sqrt{\frac{\mathrm{var}\,(f)}{N}} = \frac{1}{N}\sum_{i=1}^{N} f(x_i) \pm \sqrt{\frac{\mathrm{var}\,(f)}{N}}\,. \tag{14.34}$$

Here, the random numbers x_i are sampled from the pdf $p(x)$. This method is the most prominent formulation of Monte-Carlo integration.

We shall briefly discuss some properties of this method. We deduce from Eq. (14.34) that the error scales like $N^{-\frac{1}{2}}$. In contrast to the integration methods we discussed in Chap. 3, N is no longer the number of grid-points but the number of random numbers sampled.[3] In principle, the error scaling is worse than in the case of classical integrators. For instance, in the case of the central rectangular rule (Sect. 3.2) we had an error scaling of N^{-2} when summed over the whole interval. However, we obtained this result for the one-dimensional case, in higher dimensions we will certainly need much more grid-points. On the other hand, in Eq. (14.34) N

[3]Nevertheless, there is certainly some conceptual similarity between grid-points and random numbers within this context.

corresponds to the number of d-dimensional random numbers x. Hence, Monte-Carlo integration can be of advantage whenever one has to deal with complicated, high dimensional integrals. In contrast, restricted to one dimension it is in most cases not an improvement of the methods discussed already.

Monte-Carlo integration can also be of advantage whenever the integrand $f(x)$ is not well behaved. In such a case a very fine grid would be required to compute a reasonable estimate of the true value of the integral. Monte-Carlo integration offers a very convenient alternative due to its conceptual simplicity [13].

It is certainly a drawback of Monte-Carlo integration in its formulation (14.34), that the error is also proportional to $\sqrt{\text{var}(f)}$ which is a yet unknown quantity. One has to approximate it with an adequate estimator, for instance with the help of the sampling variance. Moreover, if the variance $\text{var}(f)$ diverges, the central limit theorem does not hold and the procedure (14.34) is no longer justified and will fail for sure.

Closely related to the problem of how to determine $\text{var}(f)$, is the question of how many random numbers should be drawn. In most cases an iterative approach is the most promising strategy. In a first step N random numbers are drawn and the integral is computed using Eq. (14.34). Then another set of N random numbers is sampled and Eq. (14.34) is reevaluated now using all $2N$ random numbers. If the change in the resulting estimate of the integral is less than some given tolerance ϵ, the loop is terminated otherwise another set of N random numbers is added.

We mention that this form of Monte-Carlo integration can be improved particularly by sampling only from points which dominantly contribute to the integral. This method is referred to as *importance sampling* [13–16] and will be discussed in more detail later on.

14.3 The METROPOLIS Algorithm: An Introduction

The METROPOLIS algorithm is a more sophisticated method to produce random numbers from given distributions. In fact, the METROPOLIS algorithm is a special form of the rejection method (Sect. 13.3). This section introduces the algorithm on a very basic level which will, in the end, allow a first glance at an interesting model out of statistical physics, namely the ISING model. It will be discussed in Chap. 15 and a more detailed discussion of the METROPOLIS algorithm will be postponed to Sect. 16.4.

The METROPOLIS algorithm is particularly useful to treat problems in statistical physics where thermodynamic expectation values of some observable O are of interest [8–11]. They are defined as

$$\langle O \rangle = \int \mathrm{d}x O(x) q(x) , \tag{14.35}$$

where x is a set of parameters and $q(x)$ is the BOLTZMANN distribution (14.14). The set of parameters x could be, for instance, the position- and momentum-space coordinates of N different particles. In most cases x is a high dimensional object which makes classical numerical integration (Chap. 3) cumbersome. Instead Monte-Carlo integration is employed and the integral (14.35) is approximated with the help of Eq. (14.34) by

$$\langle O \rangle \approx \frac{1}{N} \sum_{i=1}^{N} O(x_i) \pm \sqrt{\frac{\mathrm{var}\,(O)}{N}}\,, \tag{14.36}$$

where the uncorrelated random numbers x_i, $i = 1, 2, \ldots, N$ are sampled from the pdf, Eq. (14.14). We recognize immediately the problem: we need to know the exact functional form of $q(x)$ if we want to apply either the inverse transformation method or the rejection method discussed in Chap. 13. However, the partition function Z itself is determined by an integral which can be approximated using Eq. (14.36). We set

$$q(x) = \frac{p(x)}{Z}\,, \tag{14.37}$$

and

$$Z = \int \mathrm{d}x\, p(x) \tag{14.38}$$

follows from the normalization of $q(x)$. The METROPOLIS algorithm was designed to avoid precisely this problem. We concentrate on a pdf which is of the form (14.37), but $q(x)$ must not necessarily be described by a normalized BOLTZMANN distribution, Eq. (14.14). Thus, $p(x)$ is arbitrary but it ensures that

$$\int \mathrm{d}x\, q(x) = 1 \Longleftrightarrow \int \mathrm{d}x\, p(x) = Z\,, \tag{14.39}$$

and $q(x) \geq 0$ for all x. In other words, $q(x)$ is a pdf. Suppose we already have a sequence $x_0, x_1, \ldots, x_n = \{x_n\}$ of parameters which indeed follows the pdf $q(x)$.[4] We now add to the last element of this sequence x_n a small perturbation δ and set

$$x_t = x_n + \delta\,. \tag{14.40}$$

Note that the perturbation δ is of the same dimension as the vector x. Similar to the rejection method we seek for a criterion which helps us to decide whether or not the test value x_t can be accepted as the next element of the sequence $\{x_n\}$.

[4]The question of how one can obtain such a sequence will be discussed in Sect. 16.3.

The METROPOLIS method proposes an acceptance probability of the form

$$\Pr(A|x_t, x_n) = \begin{cases} 1 & \text{if} \quad \dfrac{q(x_t)}{q(x_n)} \geq 1 , \\ \dfrac{q(x_t)}{q(x_n)} & \text{otherwise.} \end{cases} \tag{14.41}$$

Hence, if $\Pr(A|x_t, x_n) = 1$, we set $x_{n+1} = x_t$, and if $\Pr(A|x_t, x_n) < 1$, we draw a random number $r \in [0, 1]$ and accept x_t if $r \leq \Pr(A|x_t, x_n)$ and reject x_t otherwise. We note that in this formulation the knowledge of the normalization factor Z is no longer required since it follows from Eq. (14.37) that

$$\frac{q(x_t)}{q(x_n)} = \frac{p(x_t)}{p(x_n)} . \tag{14.42}$$

Consequently we rewrite Eq. (14.41) as

$$\Pr(A|x_t, x_n) = \min\left(\frac{p(x_t)}{p(x_n)}, 1\right) = p(x_t|x_n) , \tag{14.43}$$

where we introduced in the last step a more compact notation.

A discussion of the underlying concepts and why the choice (14.41) indeed samples random numbers according to the pdf $q(x)$ requires some basic knowledge of stochastics in general and of MARKOV-chains in particular. This is the reason why we postponed this discussion to Chap. 16. Nevertheless, there is a particular property, referred to as *detailed balance* which requires our attention because it is crucial for the METROPOLIS algorithm: Let $p(x_t|x_n)$ denote the pdf for the probability that a random number x_t is generated from the random number x_n as defined in Eq. (14.43). Then the condition of detailed balance is defined as

$$p(x_t|x_n)q(x_n) = p(x_n|x_t)q(x_t) . \tag{14.44}$$

In words: The probability $p(x_t|x_n)$ that a random number x_t is generated from a random number x_n times the probability $q(x_n)$ that the random number x_n occurred at all is equal to the probability $p(x_n|x_t)$ that the random number x_n is generated from x_t times the probability $q(x_t)$ that x_t occurred. Detailed balance is motivated by physics and is a condition of thermodynamic equilibrium.

Let us briefly demonstrate that the METROPOLIS algorithm (14.43) satisfies detailed balance: We distinguish three different cases: (i) Suppose that $p(x_t|x_n) = p(x_n|x_t) = 1$. From Eq. (14.43) we note that this is only possible if $p(x_t) = p(x_n)$ and therefore $q(x_t) = q(x_n)$ which is already Eq. (14.44) for this particular case. (ii)

We assume that $p(x_t|x_n) = 1$ but $p(x_n|x_t) \neq 1$. It follows from Eq. (14.43) that

$$\begin{aligned} p(x_n|x_t)q(x_t) &= \frac{p(x_n)}{p(x_t)}q(x_t) \\ &= q(x_n) \,. \end{aligned} \tag{14.45}$$

This corresponds to Eq. (14.44) for $p(x_t|x_n) = 1$. Note that we made use of definition (14.37) in order to achieve this result. (iii) Finally, we find for $p(x_n|x_t) = 1$ and $p(x_t|x_n) \neq 1$ that

$$\begin{aligned} p(x_t|x_n)q(x_n) &= \frac{p(x_t)}{p(x_n)}q(x_n) \\ &= q(x_t) \,, \end{aligned} \tag{14.46}$$

which, again, is Eq. (14.44). Hence, the METROPOLIS algorithm (14.43) indeed obeys detailed balance.

So far the question of how to choose the initialization point x_0 of the sequence stayed unanswered. This is clearly not a trivial problem and it is strongly related to one of the major disadvantages of the METROPOLIS algorithm, namely that subsequent random numbers (x_n, x_{n+1}) are strongly correlated. One of the most pragmatic approaches is to choose a starting point x_0 at random out of the parameter space and then discard it together with the first few members of the sequence. This approach is strongly motivated by a clear physical picture: The sequence of random numbers resembles the evolution of the physical system from an arbitrary initial point x_0 toward equilibrium which manifests itself in the condition of detailed balance. Hence, the approach of discarding the first few members of the sequence is referred to as *thermalization*.

The integral of interest, Eq. (14.35) is then approximated with the help of Eq. (14.36), where the random numbers $x_k, x_{k+1}, \ldots, x_{k+N}$ are used, if the thermalization required k steps. There is a remedy which helps to reduce correlations between subsequent random numbers within the sequence which is based on a similar strategy. In particular, the modified sequence

$$x_k, x_{k+\ell}, x_{k+2\ell}, \ldots \,, \tag{14.47}$$

generated by discarding ℓ intermediate random numbers will reduce correlations between the members of this final sequence of random numbers.

Summary

This chapter set the stage for an important numerical tool in Computational Physics: the Monte-Carlo techniques. It started with the conceptual transparent task of how to calculate π using a sequence of uniformly distributed random numbers of the range [0, 1]. This established the so-called hit and miss technique. It moved on to a discussion of Monte-Carlo integration in a more formal way and discussed in detail the error involved by this type of integration as opposed to the error experienced by deterministic methods. The conclusion was, that Monte-Carlo integration was certainly preferable whenever estimates of high dimensional integrals were required and it also had advantages when the integrand was heavily structured. The second part of this chapter dealt with the METROPOLIS algorithm which allowed to generate a sequence of random numbers from some pdf $p(x)$. It was conceptually similar to the rejection method discussed earlier. The mathematical background which is more involved was not discussed within this first contact with the METROPOLIS algorithm. Instead, the emphasis was to demonstrate that this algorithm obeyed detailed balance a property purely based on physics as a condition of thermodynamic equilibrium. It was, furthermore, pointed out that the random numbers generated by this algorithm were highly correlated and some strategies to remedy this problem were discussed.

References

1. Gentle, J.E.: Random Number Generation and Monte Carlo Methods. Statistics and Computing. Springer, Berlin/Heidelberg (2003)
2. Kalos, M.H., Whitlock, P.A.: Monte Carlo Methods, 2nd edn. Wiley, New York (2008)
3. Kroese, D.P., Taimre, T., Botev, Z.I.: Handbook of Monte Carlo Methods. Wiley, New York (2011)
4. Fishman, G.S.: Monte Carlo: Concepts, Algorithms and Applications. Springer Series in Operations Research. Springer, Berlin/Heidelberg (1996)
5. Chow, Y.S., Teicher, H.: Probability Theory. Springer Texts in Statistics, 3rd edn. Springer, Berlin/Heidelberg (1997)
6. Kienke, A.: Probability Theory. Universitext. Springer, Berlin/Heidelberg (2008)
7. Stroock, D.W.: Probability Theory. Cambridge University Press, Cambridge (2011)
8. Schwabl, F.: Statistical Mechanics. Advanced Texts in Physics. Springer, Berlin/Heidelberg (2006)
9. Halley, J.W.: Statistical Mechanics. Cambridge University Press, Cambridge (2006)
10. Pathria, R.K., Beale, P.D.: Statistical Mechanics, 3rd edn. Academic, San Diego (2011)
11. Hardy, R.J., Binek, C.: Thermodynamics and Statistical Mechanics: An Integrated Approach. Wiley, New York (2014)
12. Ballentine, L.E.: Quantum Mechanics. World Scientific, Hackensack (1998)
13. Press, W.H., Teukolsky, S.A., Vetterling, W.T., Flannery, B.P.: Numerical Recipes in C++, 2nd edn. Cambridge University Press, Cambridge (2002)
14. Doucet, A., de Freitas, N., Gordon, N. (eds.): Sequential Monte Carlo Methods in Practice. Information Science and Statistics. Springer, Berlin/Heidelberg (2001)
15. Ripley, B.D.: Stochastic Simulation. Wiley, New York (2006)
16. Landau, D.P., Binder, K.: A Guide to Monte Carlo Simulations in Statistical Physics, 3rd edn. Cambridge University Press, Cambridge (2009)

Chapter 15
The ISING Model

15.1 The Model

Ferromagnetic materials are materials which develop a non-vanishing magnetization M even in the absence of an external magnetic field B. It is an experimental observation, that this magnetization decreases smoothly with increasing temperature, and vanishes above the critical temperature T_C, referred to as CURIE temperature [1]. Above this temperature the magnetization is zero and the material is no longer ferromagnetic but paramagnetic. This typical situation is illustrated in Fig. 15.1 and it is the signature of a *phase transition*. In a theoretical description of this transition the magnetization M serves as an order parameter.[1] At $T = T_C$ the system exhibits a second order phase transition: The magnetization is not differentiable with respect to T; it is, however, continuous.

The microscopic origin of this macroscopic phenomenon is based on the exchange interaction between identical particles, the atoms or molecules forming the material. The exchange interaction is a purely quantum-mechanical effect which is a consequence of the COULOMB interaction in combination with the PAULI exclusion principle.[2] For more detailed information please consult Refs. [2–8].

Given two atoms or molecules with spins S_1 and S_2, where $S_1, S_2 \in \mathbb{R}^3$, the exchange interaction energy is of the form[3]

$$E = JS_1 \cdot S_2 , \tag{15.1}$$

[1]For a short introduction to phase transitions in general please consult Appendix F.

[2]The statement that magnetism is a purely quantum-mechanical phenomenon that cannot explained in classical terms is known as the BOHR-VAN LEEUWEN theorem [3, 4].

[3]In this discussion we regard the spin as a classical quantity. In the quantum mechanic case one has to replace the vectors by vector operators S_i.

B.A. Stickler, E. Schachinger, *Basic Concepts in Computational Physics*,
DOI 10.1007/978-3-319-27265-8_15

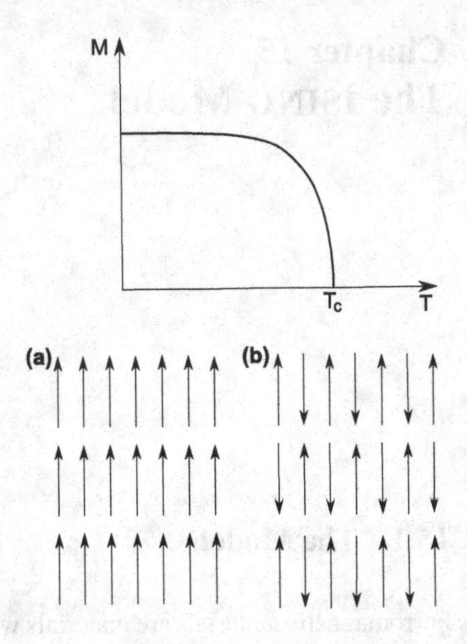

Fig. 15.1 Schematic illustration of the magnetization M as a function of temperature T in a ferromagnetic material

Fig. 15.2 Schematic illustration of the spin-orientation in a **(a)** ferromagnetic ($J < 0$) or **(b)** antiferromagnetic ($J > 0$) two-dimensional crystal

with the exchange constant J. The magnitude of J as well as its sign are determined by overlap integrals which include the COULOMB interaction. If $J < 0$ a parallel orientation of the spins is energetically favorable and ferromagnetism arises if $T < T_C$. On the other hand, if $J > 0$, an antiparallel orientation is established as long as the temperature does not exceed the NÉEL temperature T_N. However, in both cases the system undergoes a phase transition to a paramagnetic state if the temperature T exceeds the CURIE temperature (ferromagnetic case) or the NÉEL temperature (antiferromagnetic case). A schematic illustration of ferro- and antiferromagnetism for a two-dimensional crystal is illustrated in Fig. 15.2. We summarize the different scenarios:

$$\begin{cases} J < 0 & \text{ferromagnetic,} \\ J > 0 & \text{antiferromagnetic,} \\ J = 0 & \text{non-interacting.} \end{cases}$$

We concentrate on a cubic crystal lattice in which the atoms are localized at positions x_ℓ. The spin of atom ℓ will be denoted by $S_\ell \in \mathbb{R}^3$ and the exchange parameter between atom ℓ and atom ℓ' by $J_{\ell\ell'}$. Furthermore, we consider the ferromagnetic case with $J_{\ell\ell'} < 0$. The HAMILTON function [9–11] is of the form

$$H = \frac{1}{2}\sum_{\ell\ell'} J_{\ell\ell'} S_\ell \cdot S_{\ell'} = \frac{1}{2}\sum_{\ell\ell'} J_{\ell-\ell'} S_\ell \cdot S_{\ell'} \,. \tag{15.2}$$

Here $J_{\ell\ell'}$ was replaced by $J_{\ell-\ell'} = J_{\ell'-\ell}$ to account for translational invariance. Moreover, we define that $J_{\ell\ell} = 0$, otherwise we would have to exclude the contributions $\ell = \ell'$ from the above sum. The HAMILTON function (15.2) is genuine to the HEISENBERG model [1]. We note that in this model there is no distinguished direction of spin orientation and, consequently, the HAMILTON function is invariant under a rotation of all spin vectors S_ℓ. The actual spin orientation may be determined by an external magnetic field or by an anisotropy of the crystal lattice. Furthermore, the restriction of the spin orientation to the positive or negative z-direction is the characteristic of the ISING model.

In a quantum mechanical description the HAMILTON operator (Hamiltonian) of the ISING model is defined by

$$H = \frac{1}{2}\sum_{\ell\ell'} J_{\ell-\ell'} S_\ell^z S_{\ell'}^z , \tag{15.3}$$

where S_ℓ^z are the spin operators in z-direction. If spin $1/2$ particles are described by this Hamiltonian, the spin operators S_ℓ^z are replaced by $(\hbar/2)\sigma_\ell^z$ with σ_ℓ^z the PAULI matrix and $\hbar$ the reduced PLANCK's constant. Furthermore, we redefine $J'_{\ell-\ell'} = -(\hbar^2/4)J_{\ell-\ell'}$, $J'_{\ell-\ell'} > 0$, and represent the Hamiltonian in the basis of eigenstates of the operators σ_ℓ^z. These eigenstates have eigenvalues $\sigma_\ell = \pm 1$ which correspond to *spin up* and *spin down* states, respectively. We obtain in this representation

$$H = -\frac{1}{2}\sum_{\ell\ell'} J_{\ell-\ell'}\sigma_\ell\sigma_{\ell'} - h\sum_\ell \sigma_\ell , \tag{15.4}$$

where we dropped the prime on the exchange parameter $J_{\ell-\ell'}$ for the sake of a more compact notation. We added, furthermore, a term which accounts for the possible coupling of the spins to an external magnetic field,[4] where h stands for the reduced field $h = -\mu_B g B/2$.[5]

There are some special cases in which the ISING model can be solved analytically [12, 13]. For instance, one can solve the general case described by Eq. (15.4) with the help of the *mean field* approximation: The contribution h_ℓ acting on site ℓ

$$h_\ell = h + \sum_{\ell'} J_{\ell-\ell'}\sigma_{\ell'} , \tag{15.5}$$

is replaced by its mean value

$$\langle h_\ell \rangle = h + \tilde{J} m , \tag{15.6}$$

[4]We note in passing that the Hamiltonian (15.4) is invariant under a spin flip of all spins if $h = 0$ ($\mathbb{Z}_2$ symmetry). This symmetry is broken if $h \neq 0$, i.e. the spins align with the external field h.

[5]We note that $H \propto \mu \cdot B$ where B is the magnetic field and μ is the magnetic moment. Furthermore, μ can be expressed as $\mu = -\mu_B g S/\hbar = -\mu_B g \sigma/2$, where μ_B is the BOHR magneton, g is the LANDÉ g-factor and σ is the vector of PAULI matrices. The sign is convention.

where $m = \langle\sigma_\ell\rangle$ and $\tilde{J} = \sum_\ell J_\ell$. (The term $\tilde{J}m$ is commonly referred to as the *molecular field*.) With the help of this ansatz it is, for instance, possible to reproduce the experimentally observed CURIE-WEISS – law of ferromagnetic materials: The temperature dependence of the magnetic susceptibility χ for $T > T_C$ can be described by:

$$\chi \propto \frac{1}{T - T_C} . \tag{15.7}$$

Another very interesting special case of the general model (15.4) is the restriction to nearest neighbor (n. n.) interaction with the assumption that the interaction between non-nearest neighbor spins is negligible. One step further goes the approximation that $J_{\ell-\ell'} \equiv J$ for nearest neighbors. Hence, we have

$$J_{\ell-\ell'} = \begin{cases} J & \text{if} \quad \ell, \ell' \quad \text{n. n.} , \\ 0 & \text{otherwise.} \end{cases} \tag{15.8}$$

In this case Eq. (15.4) is rewritten as

$$H = -\frac{J}{2} \sum_{\langle\ell\ell'\rangle} \sigma_\ell \sigma_{\ell'} - h \sum_\ell \sigma_\ell , \tag{15.9}$$

where $\sum_{\langle\ell\ell'\rangle}$ denotes the sum over all nearest neighbors. This model can be solved analytically in one and two dimensions if the system is assumed to be spatially unlimited. The solution in one dimension was published by E. ISING [14]. The solution in two dimensions, which is much more involved, was reported by L. ONSAGER [15].

We briefly discuss ISING's solution in one dimension. The Hamiltonian (15.9) for N-particles aligned in a one-dimensional chain is rewritten as

$$H = -J \sum_{\ell=1}^{N} \sigma_\ell \sigma_{\ell+1} - h \sum_{\ell=1}^{N} \sigma_\ell , \tag{15.10}$$

where we applied periodic boundary conditions, $\sigma_{N+1} = \sigma_1$, and the factor $1/2$ was absorbed into J. Let us now briefly elaborate on the kind of observables we would like to describe within this model. (We note in passing that the following discussion is not restricted to the one-dimensional case.) Given a particular spin configuration $\mathscr{C} = \{\sigma_i\}$, we assume that the probability of finding the system in this configuration is given by the BOLTZMANN distribution $p(\mathscr{C})$[6]:

$$p(\mathscr{C}) = \frac{1}{Z_N} \exp\left[-\frac{E(\mathscr{C})}{k_B T}\right] . \tag{15.11}$$

Here, T is the temperature and k_B is BOLTZMANN's constant. The energy $E(\mathscr{C})$ associated with configuration $\mathscr{C}$ is given by Eq. (15.10). Please note that now, obviously, we have to treat the model in the classical sense, although we consider spin degrees of freedom. The partition function Z_N is given by the sum over all possible configurations $\mathscr{C}$ [3, 4, 16]:

$$Z_N = \sum_{\mathscr{C}} \exp\left[-\frac{E(\mathscr{C})}{k_B T}\right] . \tag{15.12}$$

In general, the task of solving the ISING problem is a problem of how to evaluate the sum (15.12). This is certainly not trivial since, for instance, in the one dimensional case with $N = 100$ grid-points one has $2^N = 2^{100} \approx 1.3 \times 10^{30}$ different configurations $\mathscr{C}$. On the other hand, once Z_N has been determined more information about the properties of the system can be derived [2, 12, 13]. For instance, the expectation value of the energy[7] is given by

$$\langle E \rangle = \sum_{\mathscr{C}} p(\mathscr{C}) E(\mathscr{C}) = k_B T^2 \frac{\partial}{\partial T} \ln Z_N , \tag{15.13}$$

and the expectation value of the magnetization follows from

$$\langle M \rangle = \sum_{\mathscr{C}} p(\mathscr{C}) \mathscr{M}(\mathscr{C}) = k_B T \frac{\partial}{\partial h} \ln Z_N , \tag{15.14}$$

where we defined the magnetization $\mathscr{M}(\mathscr{C})$ of a configuration $\mathscr{C}$ via:

$$\mathscr{M}(\mathscr{C}) = \left(\sum_{\ell} \sigma_{\ell} \right)_{\mathscr{C}} . \tag{15.15}$$

The term $\sum_{\ell} \sigma_{\ell}$ was placed within parenthesis indexed by $\mathscr{C}$ to emphasize its dependence on the particular configuration $\mathscr{C}$. From the observables (15.13) and (15.14) the *fluctuation quantities*, namely, the magnetic susceptibility, χ, and

[6]In particular we assume ergodicity of the system as will be explained in Chap. 16.

[7]$\langle E \rangle$ is also referred to as internal energy U.

the heat capacity, c_h, can be derived. The following relations hold:

$$\chi = \frac{\partial}{\partial h} \langle M \rangle \quad \text{and} \quad c_h = \frac{\partial}{\partial T} \langle E \rangle \, . \tag{15.16}$$

Equation (15.13) is applied to rewrite the expression for the heat capacity:

$$c_h = \sum_{\mathscr{C}} E(\mathscr{C}) \frac{\partial}{\partial T} p(\mathscr{C}) \, . \tag{15.17}$$

Here we made use of the fact that $E(\mathscr{C})$ is independent of temperature T. We evaluate, furthermore, the derivative of $p(\mathscr{C})$ with respect to temperature T:

$$\begin{aligned} \frac{\partial}{\partial T} p(\mathscr{C}) &= \frac{\partial}{\partial T} \left[\frac{\exp\left(-\frac{E(\mathscr{C})}{k_B T}\right)}{Z_N} \right] \\ &= \frac{p(\mathscr{C})}{k_B T^2} \left[E(\mathscr{C}) - \langle E \rangle \right] \, . \end{aligned} \tag{15.18}$$

This is inserted into Eq. (15.17) and results in a final expression for the heat capacity:

$$\begin{aligned} c_h &= \frac{1}{k_B} T^2 \sum_{\mathscr{C}} p(\mathscr{C}) \left[E^2(\mathscr{C}) - E(\mathscr{C}) \langle E \rangle \right] \\ &= \frac{1}{k_B T^2} \left(\langle E^2 \rangle - \langle E \rangle^2 \right) \\ &= \frac{1}{k_B T^2} \operatorname{var}(E) \, . \end{aligned} \tag{15.19}$$

This result justifies why the heat capacity is referred to as a fluctuation quantity.

We determine now, following the same ideas, the magnetic susceptibility using relation (15.14):

$$\chi = \sum_{\mathscr{C}} \mathscr{M}(\mathscr{C}) \frac{\partial}{\partial h} p(\mathscr{C}) \, . \tag{15.20}$$

We note that

$$\frac{\partial}{\partial h} E(\mathscr{C}) = -\mathscr{M}(\mathscr{C}) \, , \tag{15.21}$$

and obtain:

$$\frac{\partial}{\partial h}p(\mathscr{C}) = \frac{\partial}{\partial h}\left[\frac{\exp\left(-\frac{E(\mathscr{C})}{k_B T}\right)}{Z_N}\right]$$
$$= \frac{p(\mathscr{C})}{k_B T}\left[\mathscr{M}(\mathscr{C}) - \langle M\rangle\right] . \tag{15.22}$$

This results in a final expression for the magnetic susceptibility χ which relates it to the variance of the magnetization M:

$$\chi = \frac{1}{k_B T}\sum_{\mathscr{C}} p(\mathscr{C})\left[\mathscr{M}^2(\mathscr{C}) - \mathscr{M}(\mathscr{C})\langle M\rangle\right]$$
$$= \frac{1}{k_B T}\left(\langle M^2\rangle - \langle M\rangle^2\right)$$
$$= \frac{1}{k_B T}\mathrm{var}\,(M) . \tag{15.23}$$

After this excursion, we return to the analytic treatment of the infinite one-dimensional ISING model with nearest neighbor interaction, Eq. (15.10). If it were possible to evaluate the partition function Z_N, the required observables would be directly accessible via the above relations. In most cases this task is not analytically feasible. Nevertheless, in our particular case it appears to be possible because we recognize that we can actually evaluate Eq. (15.12) explicitly by keeping in mind Eq. (15.9):

$$Z_N = \sum_{\mathscr{C}} p(\mathscr{C})$$
$$= \sum_{\mathscr{C}} \exp\left[\frac{1}{k_B T}\left(J\sum_{\ell=1}^{N}\sigma_\ell\sigma_{\ell+1} + h\sum_{\ell=1}^{N}\sigma_\ell\right)\right]$$
$$= \sum_{\mathscr{C}}\prod_{\ell=1}^{N}\exp\left[\frac{J}{k_B T}\sigma_\ell\sigma_{\ell+1} + \frac{h}{2k_B T}(\sigma_\ell + \sigma_{\ell+1})\right] . \tag{15.24}$$

In the last step the sum over σ_ℓ was replaced by an alternative sum

$$\sum_{\ell=1}^{N}\sigma_\ell = \frac{1}{2}\sum_{\ell=1}^{N}(\sigma_\ell + \sigma_{\ell+1}) , \tag{15.25}$$

which is a consequence of the periodic boundary conditions $\sigma_{N+1} = \sigma_1$. Equation (15.24) can be rewritten as

$$Z_N = \operatorname{tr}\left(\mathscr{T}^N\right) , \tag{15.26}$$

where $\operatorname{tr}(\cdot)$ denotes the *trace* operation and we introduced the *transfer* matrix:

$$\mathscr{T}_{\sigma,\sigma'} = \exp\left[\frac{J}{k_B T}\sigma\sigma' + \frac{h}{2k_B T}\left(\sigma + \sigma'\right)\right] . \tag{15.27}$$

Let us briefly clarify this point: The trace operation in the basis of the spin eigenvalues $\sigma = \pm 1$ results in

$$\operatorname{tr}(\mathscr{T}) = \sum_{\sigma} \mathscr{T}_{\sigma,\sigma} = T_{-1,-1} + T_{1,1} . \tag{15.28}$$

Hence, we have

$$\begin{aligned}
\operatorname{tr}\left(\mathscr{T}^N\right) &= \sum_{\sigma} \left(\mathscr{T}^N\right)_{\sigma\sigma} \\
&= \sum_{\sigma} \sum_{\substack{\{\sigma_i\} \\ i=1,\dots,N-1}} \mathscr{T}_{\sigma,\sigma_1} \mathscr{T}_{\sigma_1,\sigma_2} \cdots \mathscr{T}_{\sigma_{N-1},\sigma} \\
&= \sum_{\substack{\{\sigma_i\} \\ i=1,\dots,N}} \mathscr{T}_{\sigma_1,\sigma_2} \mathscr{T}_{\sigma_2,\sigma_3} \cdots \mathscr{T}_{\sigma_N,\sigma_1} .
\end{aligned} \tag{15.29}$$

In the last step we redefined the sum indices and we used the notation $\{\sigma_i\}$ to indicate that the sum runs over all possible values of $\sigma_1, \sigma_2, \dots, \sigma_N$ in order to abbreviate the notation. However, the sum over all possible values of $\sigma_1, \sigma_2, \dots, \sigma_N$ can be replaced by a sum over all configurations $\mathscr{C}$ where one configuration is a specific combination of definite values $\sigma_1, \sigma_2, \dots, \sigma_N$. For these definite values the product of transfer matrices in Eq. (15.29) is equivalent to the product of exponentials in Eq. (15.24) due to our definition of the transfer matrix $\mathscr{T}_{\sigma,\sigma'}$. Hence we demonstrated that expression (15.26) is indeed equivalent to Eq. (15.24).

It follows from definition (15.27) that

$$\mathscr{T} = \begin{pmatrix} \exp\left(\frac{J+h}{k_B T}\right) & \exp\left(-\frac{J}{k_B T}\right) \\ \exp\left(-\frac{J}{k_B T}\right) & \exp\left(\frac{J-h}{k_B T}\right) \end{pmatrix} . \tag{15.30}$$

It is an easy task to determine the eigenvalues of this matrix [17, 18]. The characteristic polynomial

$$\det\begin{pmatrix} \exp\left(\frac{J+h}{k_B T}\right)-\lambda & \exp\left(-\frac{J}{k_B T}\right) \\ \exp\left(-\frac{J}{k_B T}\right) & \exp\left(\frac{J-h}{k_B T}\right)-\lambda \end{pmatrix} \tag{15.31}$$

is of the form

$$\begin{aligned} &\left[\exp\left(\frac{J+h}{k_B T}\right)-\lambda\right]\left[\exp\left(\frac{J-h}{k_B T}\right)-\lambda\right]-\exp\left(-\frac{2J}{k_B T}\right) \\ &= \lambda^2 - 2\lambda\exp\left(\frac{J}{k_B T}\right)\cosh\left(\frac{h}{k_B T}\right)+2\sinh\left(\frac{2J}{k_B T}\right) \\ &\stackrel{!}{=} 0 , \end{aligned} \tag{15.32}$$

which is easily solved. We get for the two eigenvalues $\lambda_{1,2}$

$$\begin{aligned} \lambda_{1,2} = {}& \exp\left(\frac{J}{k_B T}\right)\cosh\left(\frac{h}{k_B T}\right) \\ &\pm\sqrt{\exp\left(\frac{2J}{k_B T}\right)\sinh^2\left(\frac{h}{k_B T}\right)+\exp\left(-\frac{2J}{k_B T}\right)} , \end{aligned} \tag{15.33}$$

and note that $\lambda_1 \geq \lambda_2$ for all temperatures $T \geq 0$.

We now make use of the fact that the trace is invariant under a basis transformation Γ. Hence we can express the transfer matrix in a basis in which it is diagonal and set

$$\mathcal{T}' = \Gamma\mathcal{T}\Gamma^{-1} = \begin{pmatrix} \lambda_1 & 0 \\ 0 & \lambda_2 \end{pmatrix} , \tag{15.34}$$

which immediately results in:

$$Z_N = \lambda_1^N + \lambda_2^N . \tag{15.35}$$

Everything required to calculate the expectation value of energy *per particle* $\langle\varepsilon\rangle$

$$\langle\varepsilon\rangle = \frac{k_B T^2}{N}\frac{\partial}{\partial T}\ln Z_N , \tag{15.36}$$

in the *thermodynamic* limit $N \to \infty$ is now in place and, thus, we can investigate the possibility of a phase transition. First, we consider the limit

$$\lim_{N\to\infty} \frac{1}{N} Z_N = \lim_{N\to\infty} \frac{1}{N} \ln\left(\lambda_1^N + \lambda_2^N\right) = \ln \lambda_1 , \tag{15.37}$$

since $\lambda_1 \geq \lambda_2$ for all $T \geq 0$.[8]

If there is no external field, i.e. $h = 0$, we have

$$\lim_{N\to\infty} \frac{1}{N} Z_N = \ln\left[2\cosh\left(\frac{J}{k_B T}\right)\right] , \tag{15.38}$$

which is a smooth function of T for $T \geq 0$. Consequently, we do not observe a phase transition in the one dimensional ISING model. Even more information about the system can be provided by the spin correlation function $\langle \sigma_\ell \sigma_{\ell'} \rangle$

$$\langle \sigma_\ell \sigma_{\ell'} \rangle = \sum_{\mathscr{C}} p(\mathscr{C}) \sigma_\ell \sigma_{\ell'} . \tag{15.39}$$

A basic, however, tedious calculation shows that in the thermodynamic limit it is described by

$$\langle \sigma_\ell \sigma_{\ell'} \rangle = \left(\frac{\lambda_2}{\lambda_1}\right)^{\ell-\ell'} , \tag{15.40}$$

with the result that the spin correlation decreases with increasing distance $\ell - \ell'$ since $\lambda_2 < \lambda_1$ for $T > 0$.

We move on and briefly sketch the solution of the infinite two-dimensional ISING model according to L. ONSAGER [15]. The HAMILTON function (15.10) changes into:

$$H = -J \sum_{\ell\ell'} \sigma_{\ell,\ell'} \left(\sigma_{\ell+1,\ell'} + \sigma_{\ell-1,\ell'} + \sigma_{\ell,\ell'-1} + \sigma_{\ell,\ell'+1}\right) - h \sum_{\ell,\ell'} \sigma_{\ell,\ell'} . \tag{15.41}$$

---

[8] We transform

$$\lambda_1^N + \lambda_2^N = \lambda_1^N \left[1 + \left(\frac{\lambda_2}{\lambda_1}\right)^N\right] ,$$

and use that

$$\left(\frac{\lambda_2}{\lambda_1}\right)^N \to 0, \qquad \text{as} \qquad N \to \infty .$$

The strategy developed for the one-dimensional case can again be applied: The system is treated as a classical system with spin degrees of freedom. The HAMILTON function (15.41) is inserted into the expression, Eq. (15.12) for the partition function Z_N. With the help of the correct basis Z_N can be described by the trace over a product of transfer matrices. However, in this case the transfer matrix $\mathscr{T}$ is of dimension $2N \times 2N$ rather than 2×2. It is quite obvious that the search for the largest eigenvalue for arbitrary values of N is not a trivial task. Therefore, we limit our discussion to a summary of the most important results for the particular case $h = 0$.

In the two-dimensional case a phase transition is indeed observed: The magnetic susceptibility becomes singular at a particular temperature T_C. This temperature is given as the solution of equation:

$$2\tanh^2\left(\frac{2J}{k_B T_C}\right) = 1 . \tag{15.42}$$

The expectation value of the energy per particle takes on the form

$$\langle \varepsilon \rangle = -J \coth\left(\frac{2J}{k_B T}\right)\left\{1 + \frac{2}{\pi} K_1(\xi)\left[2\tanh^2\left(\frac{2J}{k_B T}\right) - 1\right]\right\}, \tag{15.43}$$

where $K_1(\xi)$ is the complete elliptic integral of the first kind [see Eq. (1.14)] with the argument:

$$\xi = \frac{2\sinh\left(\frac{2J}{k_B T}\right)}{\cosh^2\left(\frac{2J}{k_B T}\right)} . \tag{15.44}$$

The magnetization per particle $\langle m \rangle$ is proved to be determined from

$$\langle m \rangle = \begin{cases} \dfrac{(1+z^2)^{\frac{1}{4}}(1-6z^2+z^4)^{\frac{1}{8}}}{\sqrt{1-z^2}} & \text{for } T < T_C , \\ 0 & \text{for } T > T_C , \end{cases} \tag{15.45}$$

with

$$z = \exp\left(-\frac{2J}{k_B T}\right) .$$

Equation (15.45) clearly describes a phase transition at $T = T_C$.

15.2 Numerics

We study a finite two-dimensional ISING model on a square lattice Ω with grid-points (x_i, y_j), $i, j = 1, 2, \ldots, N$, which will be denoted by (i, j). We write the HAMILTON function in the form

$$H = -J \sum_{\left\langle \substack{ij \\ i'j'} \right\rangle} \sigma_{i,j}\sigma_{i',j'} - h \sum_{ij} \sigma_{i,j} , \tag{15.46}$$

where the $\sigma_{i,j} \in \{-1, 1\}$ are treated as 'classical' spins. We consider nearest neighbor interaction and regard the exchange parameter as independent of the actual positions i, j. The problem is easily motivated: We calculate numerically observables like the expectation value of the energy or of the magnetization which will then be compared with analytic results. Such a procedure provides a rather simple check of the quality of the numerical approach which can then be extended to similar models which cannot any longer be treated analytically. We need numerical methods because summing over all possible configurations in a calculation of the partition function Z_N is simply no longer feasible since, for instance, for $N = 100$ we have $2^{N^2} = 2^{10000} \approx 10^{3000}$ possible configurations which will have to be considered as follows from Eqs. (15.12), (15.13), and (15.14). A more convenient approach would be to approximate the sums with the help of methods we encountered within the context of Monte-Carlo integration in Sect. 14.2. For instance, the estimate of the energy expectation value is given by

$$\langle E \rangle = \frac{1}{M} \sum_{i=1}^{M} E(\mathscr{C}_i) \pm \sqrt{\frac{\operatorname{var}(E)}{M}} . \tag{15.47}$$

Here, $\mathscr{C}_i$, $i = 1, 2, \ldots, M$ are M configurations drawn from the pdf (15.11), the BOLTZMANN distribution. Equation (15.47) is referred to as the *estimator* of the internal energy. We note that we also have to calculate an estimate of the variance of E using a similar approach in order to determine the error induced by this approximation.[9]

Hence, there remains the task to find configurations $\mathscr{C}_i$ which follow the BOLTZMANN distribution (15.11). The inverse transformation method of Sect. 13.2 cannot be applied since $E(\mathscr{C}_i)$ is not invertible. Furthermore, the rejection method

[9] In particular $\operatorname{var}(E) = \langle E^2 \rangle - \langle E \rangle^2$ is to be determined and only the second term is already known. The first term, $\langle E^2 \rangle$, is then estimated with the help of

$$\langle E^2 \rangle = \frac{1}{M} \sum_{i=1}^{M} E_i^2 .$$

is useless since we would need the partition function Z_N to make it work. However, calculating the partition function is a task as difficult as calculating the internal energy (15.13) without any approximations. Therefore, the method of choice will be the METROPOLIS algorithm discussed in Sect. 14.3.

Let $\mathscr{C}$ be a given spin configuration on the two-dimensional square lattice Ω. We modify the spin on one particular grid-point (i,j) and obtain a *trial* spin configuration $\mathscr{C}^t$. According to our discussion in Sect. 14.3, the probability of accepting the new configuration $\mathscr{C}^t$ is then given by

$$\Pr(A|\mathscr{C}^t,\mathscr{C}) = \min\left(\frac{p(\mathscr{C}^t)}{p(\mathscr{C})},1\right) = \min\left\{\exp\left[-\frac{E(\mathscr{C}^t)-E(\mathscr{C})}{k_BT}\right],1\right\}$$
$$= \min\left[\exp\left(-\frac{\Delta E_{ij}}{k_BT}\right),1\right] . \tag{15.48}$$

The spin orientation was changed only on one grid-point (i,j), with $\sigma_{i,j} \to \hat{\sigma}_{i,j} = -\sigma_{i,j}$; thus, the energy difference ΔE_{ij} is easily evaluated using

$$\Delta E_{ij} = 2J\sigma_{i,j}\left(\sigma_{i+1,j}+\sigma_{i-1,j}+\sigma_{i,j-1}+\sigma_{i,j+1}\right) + 2h\sigma_{i,j} . \tag{15.49}$$

with $\sigma_{i,j}$ the original spin orientation.

We focus now on numerical details, some particular to the numerical treatment of the ISING model [19], and some of rather general nature which should be considered whenever a Monte-Carlo simulation is planned.

(1) Lattice Geometry

We regard a two-dimensional $N \times N$ square lattice with periodic boundary conditions[10] in order to reduce finite volume effects. It is of advantage to write a program code which will help to identify the nearest neighbors of some grid-point, since we will need this information in the METROPOLIS run whenever we calculate the energy difference due to a spin flip according to Eq. (15.49). To help with this task a matrix *neighbor*(*site*, *i*) will be generated only once for each choice of the system size N. Here $i = 1,2,3,4$ are the directions to the neighboring grid-points of the grid-point *site*. In a first step the sites of the square lattice are relabeled following

[10]Periodic boundary conditions in two dimensions imply that

$$\sigma_{N+1,j} = \sigma_{1,j} \qquad \text{and} \qquad \sigma_{i,N+1} = \sigma_{i,1} ,$$

for all i, j.

the scheme[11]:

$$\begin{matrix} N(N-1)+1 & \cdots N^2 \\ \vdots & \vdots \\ N+1 & \cdots 2N \\ 1 & 2\cdots N \,. \end{matrix} \tag{15.50}$$

In the next step, the matrix *neighbor* is initialized as an array of size $N^2 \times 4$. Every *site* has four nearest neighbors: *up*, *right*, *down*, and *left*. The corresponding matrix elements for periodic boundary conditions can be evaluated according to the following scheme:

- For *up* we have:
  (a) If $site + N \leq N^2$: $up = site + N$,
  (b) else if $site + N > N^2$: $up = site - N(N-1)$.
- For *right* we have:
  (a) If $\text{mod}(site, N) \neq 0$: $right = site + 1$,
  (b) else if $\text{mod}(site, N) = 0$: $right = site - N + 1$.
- For *down* we have:
  (a) If $site - N \geq 1$: $down = site - N$,
  (b) else if $site - N < 1$: $down = site + N(N-1)$
- For *left* we have:
  (a) If $\text{mod}(site - 1, N) \neq 0$: $left = site - 1$,
  (b) else if $\text{mod}(site - 1, N) = 0$: $left = site + N - 1$.

In a final step, the array elements are rearranged according to

$$neighbor(site, :) = [up, right, down, left] \,, \tag{15.51}$$

where $site = 1, 2, \ldots, N^2$.

(2) Initialization

It has already been discussed in Sect. 14.3 that the quality of random numbers generated with the help of the METROPOLIS algorithm is highly dependent on the choice of initial conditions. This is, in our case, the initial spin configuration $\mathscr{C}_0$.

[11] In the following we will refer to the notation (i), $i = 1, 2, \ldots, N^2$ as the *single-index* notation while the notation (i, j), $i, j = 1, 2, \ldots, N$ will be referred to as the *double-index* notation.

Of course, it would be favorable to start with a configuration which was already drawn from the BOLTZMANN distribution $p(\mathscr{C})$. However, in practice this is not feasible ab initio. But, as will be elucidated in Chap. 16, the METROPOLIS algorithm produces configurations which become independent of the initial state and follow the BOLTZMANN distribution. Hence we can simply start with some arbitrary configuration and discard it together with the first n constituents of the sequence $\mathscr{C}_0, \mathscr{C}_1, \ldots, \mathscr{C}_n$. This method is referred to as *thermalization*.[12] The question arises: can n be determined to ensure that the sequence starting with $\mathscr{C}_{n+1}$ will conform to the pdf $p(\mathscr{C})$?

There are two different ways to approach this problem: (i) The first is to measure auto-correlations between configurations $\mathscr{C}_i$ where it has to be ensured that the set of states is sufficiently large to allow for a significant conclusion. We will discuss auto-correlations in more detail in Chap. 19. (ii) The second approach is to empirically check whether equilibrium has been reached or not. For instance, one could simply plot some selected observables and check when the initial bias vanishes. In this case the observable reaches some saturation value as a function of the number of measurements. A particularly useful method is to start the algorithm with at least two different configurations. As soon as equilibrium has been reached, the observables should approach the same saturation values after a certain (finite) number of measurements. Typical choices are the *cold start* and the *hot start*. Cold start means that the temperature is initially below the critical temperature, i.e. in the ISING model all spins are aligned (ferromagnetic state). Hot start means that the temperature is well above the critical temperature and for the ISING model the spin orientation is chosen at random for any site (paramagnetic state).

(3) Execution of the Code

The METROPOLIS algorithm for the ISING model is executed in the following steps:

1. Choose an initial configuration $\mathscr{C}_0$.
2. We migrate through the lattice sites systematically.[13] Suppose we just reached site (i,j) (we use the double-index notation $i,j = 1,2,\ldots,N$, to improve the readability) and our current configuration is $\mathscr{C}_k$. Then k configurations have been accepted so far. We generate a new configuration $\mathscr{C}^t$ from $\mathscr{C}_k$ by replacing in $\mathscr{C}_k$ the entry $\sigma_{i,j}$ by $-\sigma_{i,j}$.

[12]The number of configurations discarded is referred to as the thermalization length.

[13]A migration through all lattice sites is referred to as a *sweep*.

3. The new configuration is accepted with probability

$$\Pr(A|\mathscr{C}^t, \mathscr{C}_k) = \min\left[\exp\left(-\frac{\Delta E_{ij}}{k_B T}\right), 1\right], \quad (15.52)$$

where ΔE_{ij} is determined from Eq. (15.49). $\mathscr{C}^t$ is accepted if $\Pr(A|\mathscr{C}^t, \mathscr{C}_k)$ is equal to one or if $\Pr(A|\mathscr{C}^t, \mathscr{C}_k) \geq r \in [0, 1]$ otherwise $\mathscr{C}^t$ is rejected. If $\mathscr{C}^t$ was accepted we set $\mathscr{C}_{k+1} = \mathscr{C}^t$.
4. Go to the next lattice site [step 2].

We note that instead of sampling the lattice sites sequentially as suggested in step 2 the lattice sites can also be sampled randomly with the help of

$$i = \text{int}(rN^2) + 1, \quad (15.53)$$

where $r \in [0, 1]$ is a uniformly distributed random number and $\text{int}(\cdot)$ denotes the integer part of a given quantity. Obviously, Eq. (15.53) is only useful in the single-index notation $i = 1, 2, \ldots, N^2$.

(4) Measurement

As soon as thermalization was achieved the procedure to measure interesting observables can be started. Such a procedure consists of collecting the data required and in calculating expectation values as was illustrated in Eq. (15.47) for the case of the expectation value of the energy. A more detailed study of estimator techniques is postponed to Chap. 19. However, there is one crucial point one should be aware of: We already mentioned in our discussion of the Metropolis algorithm in Sect. 14.3 that subsequent configurations $\mathscr{C}_k$ may be strongly correlated. This problem can be circumvented by simply neglecting intermediate configurations. For instance, one may allow a couple of 'empty' sweeps between two measurements.

In the following we discuss some selected results obtained with the numerical approach described above.

15.3 Selected Results

We investigate the two-dimensional Ising model with periodic boundary conditions and we chose $h = 0$ and $J = 0.5$ for all following illustrations.

In a first experiment we plan to check the thermalization process and, thus, measure after every single sampling step and skip thermalization. The observables of interest, the expectation value of the energy per particle, $\langle \varepsilon \rangle$, and the expectation value of the magnetization per particle, $\langle m \rangle$, are illustrated in Fig. 15.3 for 30 sweeps

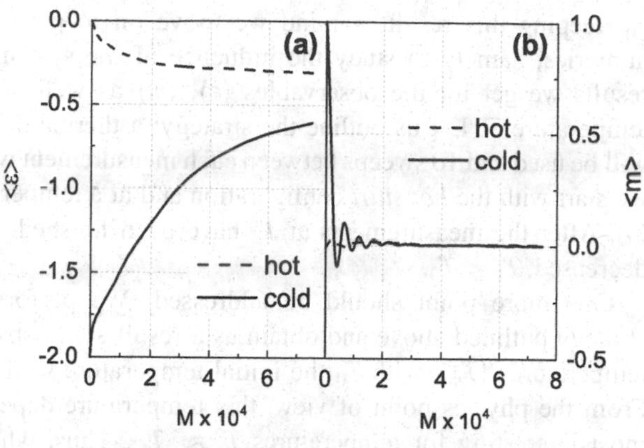

Fig. 15.3 Time evolution of **(a)** the expectation value of the energy per particle $\langle \varepsilon \rangle$ and **(b)** of the expectation value of the magnetization per particle $\langle m \rangle$ vs the number of measurements M. We used a cold start (*solid line*) and a hot start (*dashed line*) to achieve these results

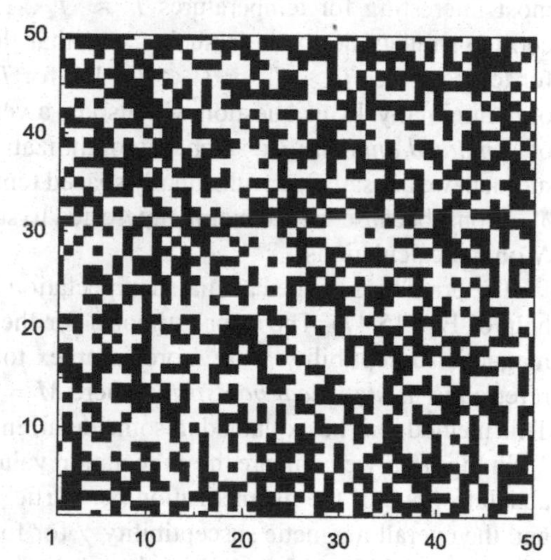

Fig. 15.4 Typical spin configuration for a temperature well above the critical temperature T_C. *Black shaded areas* correspond to spin up sites while the *white areas* are spin down sites

in a system of the size $N = 50$ which corresponds to $m \sim 8 \times 10^4$ measurements. Moreover, we set $k_B T = 3$ which should be well above T_C according to Eq. (15.42). Hence, we expect paramagnetic behavior, i.e. $\langle m \rangle = 0$ in the equilibrium since the acceptance probability is rather large because the spins are randomly orientated. In addition, Fig. 15.4 shows a typical spin configuration for a temperature well above T_C.

According to Fig. 15.3b the expectation value of the magnetization per particle $\langle m \rangle$ approaches indeed zero after a rather short thermalization period independent of the starting procedure. This is certainly not the case for the energy expectation value per particle $\langle \varepsilon \rangle$, Fig. 15.3a, which does not approach saturation even after $M \sim 8 \times 10^4$ measurements for both starting procedures. The consequence is that the thermalization period certainly needs to be longer than only 30 sweeps.

Keeping this result in mind we move on to perform the next check of our numerics, namely to study the influence of the system size N on the numerical results we get for the observables $\langle \varepsilon \rangle$, $\langle m \rangle$ as well as c_h and χ as functions of temperature T. Let us outline the strategy: a thermalization period of 500 sweeps will be used and 10 sweeps between each measurement will be discarded. Moreover, we start with the *hot start* configuration and at a temperature $k_B T_0 = 3$ well above T_C. After the measurements at T_0 have been finished, the temperature is slightly decreased, $T_1 < T_0$.

One more point should be addressed: We perform a simulation using the strategy outlined above and obtain as a result some observable O as a function of temperatures $\{T_n\}$, with T_0 the initial temperature well above T_C and $T_{n+1} < T_n$. From the physics point of view, this temperature dependence will, of course, be most interesting for temperatures $T \approx T_C$. Thus, what we need is an adaptive cooling strategy designed in such a way that the temperature is decreased rapidly for temperatures $T \gg T_C$ or $T \ll T_C$, but for $T \approx T_C$ the temperature is modified only minimally. [This question will also be a very important point in the discussion of *simulated annealing*, a stochastic optimization strategy (see Sect. 20.3).] At the moment we are satisfied with equally spaced temperatures, i.e. $T_{k+1} = T_k - \delta$, where $\delta =$ const because we are mainly interested to study the influence of the system size N on our calculations.

The error bars of the calculated expectation values have been obtained with the help of Eq. (15.47). The error estimates for the heat capacity c_h as well as for the magnetic susceptibility χ are more complex to evaluate. The method employed is referred to as *statistical bootstrap*, where $M = 100$ samples have been generated. This method will be explained in some detail in Chap. 19.

In Fig. 15.5 we compare the expectation value of the energy per particle, $\langle \varepsilon \rangle$, the absolute value of the magnetization per particle $|\langle m \rangle|$, the overall heat capacity c_h and the overall magnetic susceptibility χ for four system sizes $N = 5, 20, 50, 100$. Furthermore, in Fig. 15.6 we show the curves for the system size of $N = 50$ together with corresponding error bars.

We observe that the phase transition becomes sharper with increasing system size. In fact we know, that the phase transition is infinitely sharp as $N \to \infty$ from the analytic solution given by ONSAGER. It is a quite obvious result of this study that the system size N should be greater than 20 to achieve acceptable results.

Furthermore, we presented the absolute value of the magnetization rather than the magnetization itself. The reason is that for $T < T_C$ the ground state is degenerate. In particular, the state with all spins up or all spins down is equally probable since we set the external magnetic field $h = 0$. This is a manifestation of the $\mathbb{Z}_2$ symmetry of the Hamiltonian discussed in Sect. 15.1.

Of particular interest is the region around the critical temperature, referred to as the *critical region*. In this region, the spins are not perfectly aligned and not randomly orientated either. In this region the spins align in so called *magnetic domains*, which are also referred to as WEISS domains [1]. A typical spin configuration which exhibits such domains is presented in Fig. 15.7.

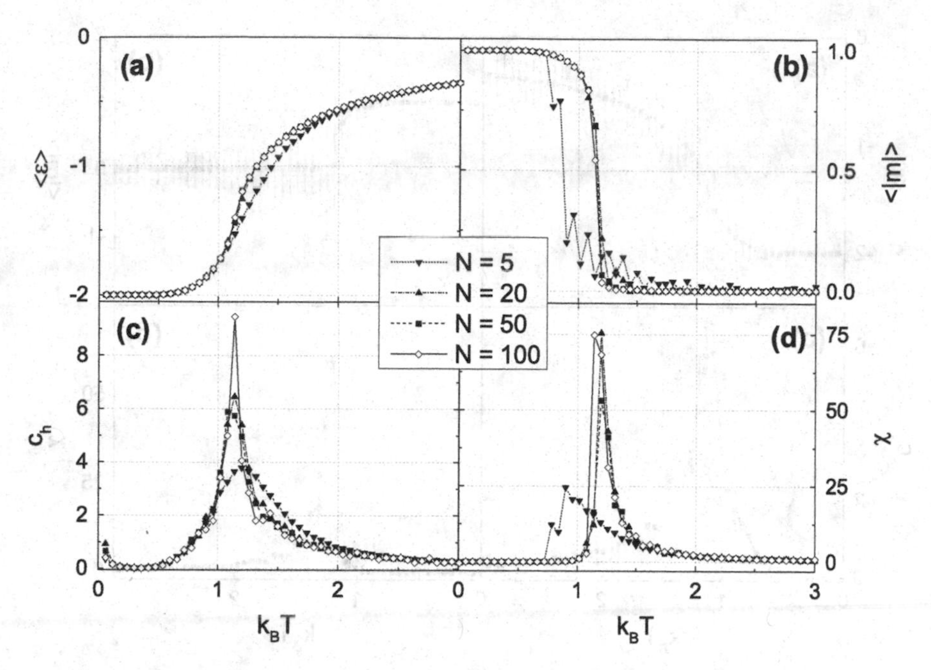

Fig. 15.5 **(a)** The expectation value of the energy per particle $\langle \varepsilon \rangle$, **(b)** the absolute value of the expectation value of the magnetization per particle $|\langle m \rangle|$, **(c)** the heat capacity c_h, and **(d)** the magnetic susceptibility χ vs temperature $k_B T$ for the two-dimensional ISING model. The system sizes are $N = 5, 20, 50, 100$

We conclude this chapter with an interesting note: Fig. 15.6 makes it quite clear that the error of the expectation value of the magnetization and of the energy is biggest for values around the transition temperature. In fact, if we increase the system size the error will become even larger. The reason is quite obvious: The error of our Monte-Carlo integration is proportional to the square root of the variance of the investigated observable. However, since we deal with a second order phase transition, this variance tends to infinity as $N \to \infty$ [4]. There is one cure to the problem: We are dealing here with finite-sized systems, thus, the variance will never actually be infinitely large. Furthermore, according to Eq. (15.47) we can decrease the error by increasing the number of measurements. Hence, if one is confronted with large systems, one has also to perform many measurements in order to reduce the error.[14]

[14] We note from Eq. (15.47) that we have to perform four times as many measurements in order to reduce the error by a factor 2.

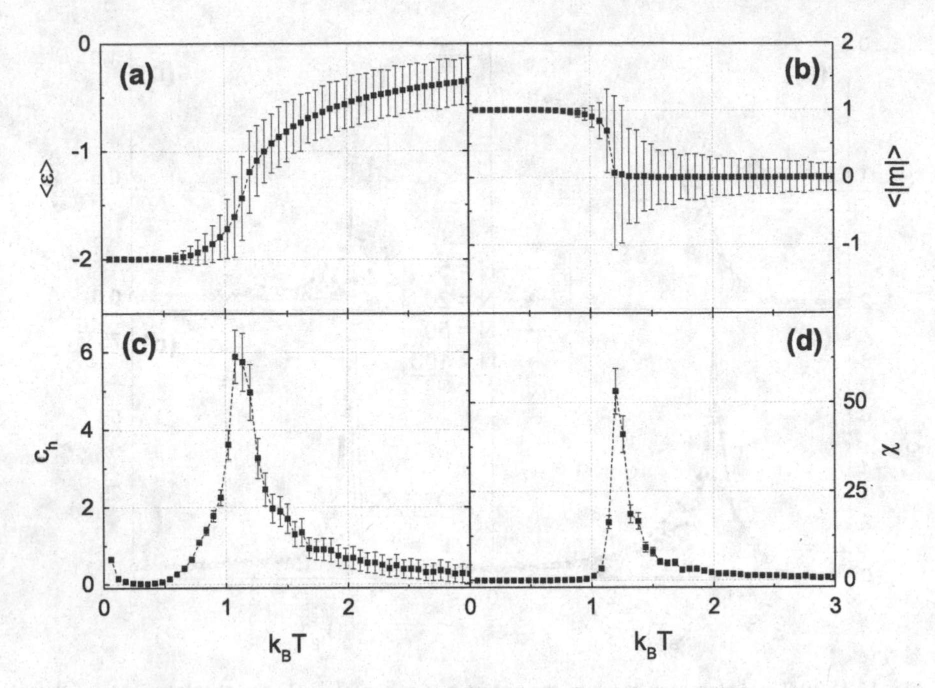

Fig. 15.6 **(a)** The expectation value of the energy per particle $\langle \varepsilon \rangle$, **(b)** the expectation value of the magnetization per particle $|\langle m \rangle|$, **(c)** the heat capacity c_h, and **(d)** the magnetic susceptibility χ with error bars vs temperature $k_B T$ obtained for the two-dimensional ISING model of size $N = 50$

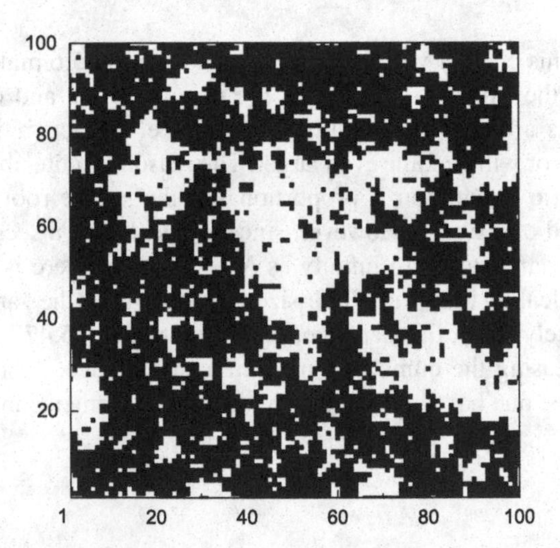

Fig. 15.7 For $T \approx T_C$ the spins organize in WEISS domains. Here we show a typical spin configuration for $N = 100$ and $k_B T = 1.15$. The *black shaded areas* correspond to spin up sites while the *white areas* indicate spin down sites

Summary

The ISING model is a rather simple model which describes effectively a second order phase transition. Such phase transitions are the topic of extensive numerical studies and, therefore, this model served here as a tool to demonstrate how to proceed from the problem analysis to a numerical algorithm which will allow to simulate the physics. The advantage of the ISING model was that under certain simplifications solutions could be derived analytically. In the course of this analysis the important concept of observables was introduced. Observables are certain physical properties of a system which characterize the specific phenomenon of interest. Numerically, observables are certain variables which are to be 'measured' within the course of a simulation. After the extensive analysis of the ISING model the transition to the numerical analysis of the two-dimensional ISING model was a rather easy part. The required modification of spin configurations turned out to be the key element of the simulation and this suggested the application of the METROPOLIS algorithm for sampling. Finally, important problems like initialization of the simulation, thermalization, finite size effects, measurement of observables, and the prevention of correlations between subsequent spin configurations caused by the METROPOLIS algorithm have been discussed on the basis of concrete calculations. The first part of this chapter was motivated by W. S. DORN and D. D. MCCRACKEN [20]:

> Numerical methods are no excuse for poor analysis.

Problems

1. Write a program to simulate the two-dimensional ISING model with periodic boundary conditions with the help of the METROPOLIS algorithm. Follow the strategy outlined in Sect. 15.2 and try to reproduce the results illustrated in Sect. 15.3 for $N = 5, 20, 50$.

   In particular, as a first step write a routine which stores the nearest neighbors of the square lattice in an array. As a second step, write a program which performs a sweep through the lattice geometry. You can either choose the lattice sites systematically or at random. As a third step, set up the main program which calls the sweep routine. Choose some initial configuration and thermalize the system. Measure the expectation value of the energy per particle as well as the absolute value of the expectation value of the magnetization for different temperatures $k_B T$ and determine the respective errors, see Eq. (15.47). Calculate also the overall magnetic susceptibility and the overall heat capacity. The determination of the error is more complicated in this case and can therefore be neglected for the moment.

   Good parameters to start with are $J = 0.5$, $N_{therm} = 500$, $N_{skip} = 10$ and $h = 0.0$.
2. Try also different values of J and $h \neq 0$.

References

1. White, R.M.: Quantum Theory of Magnetism, 3rd edn. Springer Series in Solid-State Sciences. Springer, Berlin/Heidelberg (2007)
2. Landau, L.D., Lifshitz, E.M.: Course of Theoretical Physics. Statistical Physics, vol. 5. Pergamon Press, London (1963)
3. Halley, J.W.: Statistical Mechanics. Cambridge University Press, Cambridge (2006)
4. Pathria, R.K., Beale, P.D.: Statistical Mechanics, 3rd edn. Academic, San Diego (2011)
5. Baym, G.: Lectures on Quantum Mechanics. Lecture Notes and Supplements in Physics. The Benjamin/Cummings, London/Amsterdam (1969)
6. Cohen-Tannoudji, C., Diu, B., Laloë, F.: Quantum Mechanics, vol. I. Wiley, New York (1977)
7. Sakurai, J.J.: Modern Quantum Mechanics. Addison-Wesley, Menlo Park (1985)
8. Ballentine, L.E.: Quantum Mechanics. World Scientific, Hackensack (1998)
9. Arnol'd, V.I.: Mathematical Methods of Classical Mechanics, 2nd edn. Graduate Texts in Mathematics, vol. 60. Springer, Berlin/Heidelberg (1989)
10. Goldstein, H., Poole, C., Safko, J.: Classical Mechanics, 3rd edn. Addison-Wesley, Menlo Park (2013)
11. Scheck, F.: Mechanics, 5th edn. Springer, Berlin/Heidelberg (2010)
12. Mandl, F.: Statistical Physics, 2nd edn. Wiley, New York (1988)
13. Schwabl, F.: Statistical Mechanics. Advanced Texts in Physics. Springer, Berlin/Heidelberg (2006)
14. Ising, E.: Beitrag zur Theorie des Ferromagnetismus. Z. Phys. **31**, 253 (1925)
15. Onsager, L.: Crystal statistics, I. A two-dimensional model with an order-disorder transition. Phys. Rev. **65**, 117–149 (1944). doi:10.1103/PhysRev.65.117
16. Hardy, R.J., Binek, C.: Thermodynamics and Statistical Mechanics: An Integrated Approach. Wiley, New York (2014)
17. Kwak, J.H., Hong, S.: Linear Algebra. Springer, Berlin/Heidelberg (2004)
18. Strang, G.: Introduction to Linear Algebra, 4th edn. Cambridge University Press, Cambridge (2009)
19. Stauffer, D., Hehl, F.W., Ito, N., Winkelmann, V., Zabolitzky, J.G.: Computer Simulation and Computer Algebra, pp. 79–84. Springer, Berlin/Heidelberg (1993)
20. Dorn, W.S., McCracken, D.D.: Numerical Methods with Fortran IV Case Studies. Wiley, New York (1972)

Chapter 16
Some Basics of Stochastic Processes

16.1 Introduction

This chapter is devoted to an introduction to some basic concepts of stochastic processes. This introduction serves two purposes: First of all, it allows for a more systematic treatment of non-deterministic methods in Computational Physics which is certainly necessary if we really aim at an understanding of these methods. The second reason can be found in the elementary importance of stochastics in modern theoretical physics and chemistry in general. Hence, many of the concepts elaborated within this chapter will be of profound importance in subsequent chapters. For instance, we present a discussion of diffusion theory in Chap. 17 as a motivating example.

The reader not familiar with the basics of probability theory [1–4] is highly encouraged to at least consult Appendix E before proceeding. In particular, we are going to apply the notation introduced in Appendix E throughout this chapter.

The basics of stochastic processes will be discussed within five sections including this introduction. In Sect. 16.2 we introduce basic definitions associated with stochastic processes in general. Here we discuss concepts which will serve as a basis for an understanding of the methods presented within the subsequent sections. Section 16.3 deals with a special class of stochastic processes, the so called MARKOV processes. As we shall see, these processes are of fundamental importance for statistical physics and for computational methods. Moreover, in Sect. 16.4 we consider so called MARKOV-chains which are discrete MARKOV processes defined on a discrete time span. This will serve as the basis of a very important method in computational physics, the so called MARKOV-Chain Monte Carlo technique. We already encountered a simple example of this method in Sect. 14.3 and in Chap. 15, the METROPOLIS algorithm. Finally, in Sect. 16.5 continuous-time MARKOV-chains will be discussed, in particular, discrete MARKOV processes on a continuous time span. These processes are very important, for instance, in diffusion theory as will be demonstrated in Chap. 17.

B.A. Stickler, E. Schachinger, *Basic Concepts in Computational Physics*,
DOI 10.1007/978-3-319-27265-8_16

A discussion of detailed balance will also be included in the section on MARKOV processes, Sect. 16.3, although detailed balance follows from physical arguments. Detailed balance has already been introduced in our discussion of the METROPOLIS algorithm, Sect. 14.3.

16.2 Stochastic Processes

The following discussion is primarily restricted to one-dimensional processes.

A stochastic process is a time dependent process depending on randomness [5–7]. From a mathematical point of view, a stochastic process $Y_X(t)$ is a random variable Y which is a function of another random variable X and time $t \geq 0$:

$$Y_X(t) = f(X, t) . \tag{16.1}$$

Here we apply the notation of Appendix E and denote random variables by capital letters, such as X, and their realization by lower case characters, such as x. Consequently, the realization of a stochastic process is described by

$$Y_x(t) = f(x, t) . \tag{16.2}$$

The set of all possible realizations of $Y_X(t)$ spans the *state space* of the stochastic process. We note that it is in principle not necessary to define t as the time in a classical sense. It suffices to denote $t \in T$, where T is a totally ordered set such as, for instance, $T = \mathbb{N}$ the natural numbers. The set T is referred to as the *time span*. We distinguish four different scenarios:

- discrete state space, discrete time span,
- continuous state space, discrete time span,
- discrete state space, continuous time span,
- continuous state space, continuous time span.

Stochastic processes on a continuous time span are referred to as continuous-time stochastic processes.

Suppose the random variable X follows the pdf $p_X(x)$. It is then an easy task to calculate averages $\langle Y(t) \rangle$ of the stochastic process $Y_X(t)$ via

$$\langle Y(t) \rangle = \int \mathrm{d}x \, Y_x(t) p_X(x) . \tag{16.3}$$

This concept is easily extended to multiple times $t_1, t_2, \ldots, t_n$ by

$$\langle Y(t_1) Y(t_2) \cdots Y(t_n) \rangle = \int \mathrm{d}x \, Y_x(t_1) Y_x(t_2) \cdots Y_x(t_n) p_X(x) , \tag{16.4}$$

which defines the *moments* of the stochastic process [1, 5, 8]. Similar to the concept of the correlation coefficient (see Appendix, Sect. E.10) we define the so called *auto-correlation function* $\kappa(t_1, t_2)$:

$$\begin{aligned}\kappa(t_1, t_2) &= \frac{\langle [Y(t_1) - \langle Y(t_1)\rangle][Y(t_2) - \langle Y(t_2)\rangle]\rangle}{\sqrt{\left\langle [Y(t_1) - \langle Y(t_1)\rangle]^2\right\rangle\left\langle [Y(t_2) - \langle Y(t_2)\rangle]^2\right\rangle}} \\ &= \frac{\langle Y(t_1)Y(t_2)\rangle - \langle Y(t_1)\rangle\langle Y(t_2)\rangle}{\sqrt{\text{var}[Y(t_1)]\text{var}[Y(t_2)]}} \\ &= \frac{\gamma[Y(t_1), Y(t_2)]}{\sqrt{\text{var}[Y(t_1)]\text{var}[Y(t_2)]}} . \end{aligned} \tag{16.5}$$

The function $\gamma[Y(t_1), Y(t_2)]$ is referred to as the *auto-covariance function* and is defined as

$$\gamma[Y(t_1), Y(t_2)] = \text{cov}[Y(t_1), Y(t_2)] . \tag{16.6}$$

We proceed by defining the pdf of a stochastic process $Y_X(t)$. The pdf $p_1(y, t)$, which describes the probability that the stochastic process $Y_X(t)$ takes on its representation y at time t, is given by (see Sect. 14.2)

$$p_1(y, t) = \int \mathrm{d}x\, \delta[y - Y_x(t)]p_X(x) . \tag{16.7}$$

We define, in analogy, the pdf $p_n(y_1, t_1, y_2, t_2, \ldots, y_n, t_n)$ which describes the probability that the stochastic process takes on the realization y_1 at time t_1, y_2 at time t_2, ..., and y_n at time t_n for arbitrary n:

$$\begin{aligned} p_n(y_1, t_1, y_2, t_2, \ldots, y_n, t_n) = \int \mathrm{d}x\, \delta[y_1 - Y_x(t_1)]\delta[y_2 - Y_x(t_2)] \cdots \\ \times \delta[y_n - Y_x(t_n)]p_X(x) . \end{aligned} \tag{16.8}$$

This is referred to as the *hierarchy of pdfs*. We note the following important properties of the pdf $p_n(y_1, t_1, y_2, t_2, \ldots, y_n, t_n)$ [8]:

- $p_n(y_1, t_1, y_2, t_2, \ldots, y_n, t_n) \geq 0$, (16.9)
- $p_n(\ldots, y_k, t_k \ldots, y_\ell, t_\ell, \ldots) = p_n(\ldots, y_\ell, t_\ell \ldots, y_k, t_k, \ldots)$, (16.10)
- $\int \mathrm{d}y_n\, p_n(y_1, t_1, \ldots, y_n, t_n) = p_{n-1}(y_1, t_1, \ldots, y_{n-1}, t_{n-1})$, (16.11)
- $\int \mathrm{d}y p_1(y, t) = 1$. (16.12)

The moments defined in Eq. (16.4) can also be expressed with the help of the pdfs p_n by

$$\langle Y(t_1)Y(t_2)\cdots Y(t_n)\rangle = \int \mathrm{d}y_1 \ldots \mathrm{d}y_n y_1 \cdots y_n p_n(y_1, t_1, \ldots, y_n, t_n) . \tag{16.13}$$

Conditional pdfs $p_{\ell|k}$ can also be introduced. They describe the probability that we have y_{k+1} at $t_{k+1}, \ldots, y_{k+\ell}$ at $t_{k+\ell}$ if there existed y_1 at $t_1, \ldots, y_k$ at t_k via

$$p_{\ell|k}(y_{k+1}, t_{k+1}, \ldots, y_{k+\ell}, t_{k+\ell}|y_1, t_1, \ldots, y_k, t_k) = \frac{p_{k+\ell}(y_1, t_1, \ldots, y_{k+\ell}, t_{k+\ell})}{p_k(y_1 t_1, \ldots, y_k, t_k)} . \tag{16.14}$$

It follows that

$$\int \mathrm{d}y_2 \, p_{1|1}(y_2, t_2|y_1, t_1) = 1 . \tag{16.15}$$

Let us give some further definitions [8]:

- A stochastic process is referred to as a *stationary process* if the moments defined in Eq. (16.4) are invariant under a time-shift Δt:

$$\langle Y(t_1)Y(t_2)\cdots Y(t_n)\rangle = \langle Y(t_1 + \Delta t)Y(t_2 + \Delta t)\cdots Y(t_n + \Delta t)\rangle . \tag{16.16}$$

  In particular, one has $\langle Y(t)\rangle = \text{const}$ and the auto-covariance depends only on the time difference $|t_1 - t_2|$:

$$\gamma(t_1, t_2) = \mathrm{cov}[Y(t_1), Y(t_2)] = \mathrm{cov}[Y(0), Y(|t_1 - t_2|)] \equiv \gamma(t_1 - t_2) . \tag{16.17}$$

  It is understood that $\gamma(t) = \gamma(-t)$. Moreover, we have

$$p_n(y_1, t_1 + \Delta t, \ldots, y_n, t_n + \Delta t) = p_n(y_1, t_1, \ldots, y_n, t_n) , \tag{16.18}$$

  and in particular, $p_1(y, t) = p_1(y)$.
- A *time-homogeneous* process is a stochastic process whose conditional pdfs are stationary

$$p_{1|1}(y_2, t_2|y_1, t_2 - \tau) = p_{1|1}(y_2, s_2|y_1, s_2 - \tau) , \tag{16.19}$$

  for all t_2, τ, s_2. The pdf $p_{1|1}$ is referred to as *transition probability*.
- A *process of stationary increments* is a stochastic process $Y_X(t)$ for which the difference $Y_X(t_2) - Y_X(t_1)$ is stationary for all $t_2 - t_1$, with $t_2 > t_1 \geq 0$. This means, in particular, that the pdf of this process depends only on the time difference $t_2 - t_1$. The quantities $Y_X(t_2) - Y_X(t_1)$ are referred to as *increments*.

- A *process of independent increments* is a stochastic process $Y_X(t)$ for which the differences

$$Y_X(t_2) - Y_X(t_1), Y_X(t_3) - Y_X(t_2), \ldots, Y_X(t_n) - Y_X(t_{n-1}),$$

are independent for all $t_n > t_{n-1} > \ldots > t_2 > t_1$.
- A LÉVY *process* is a continuous-time stochastic process with stationary independent increments which starts with $Y_X(0) = 0$.
- A *Gaussian process* is a stochastic process $Y_X(t)$ for which all finite linear combinations of $Y_X(t)$, $t \in T$ follow a normal distribution (see Appendix, Sect. E.7). We shall come back to this kind of process in Chap. 17.
- A WIENER *process* is a continuous-time stochastic process with independent increments which starts with $Y_X(0) = 0$ and for which the increments $Y_X(t_2) - Y_X(t_1)$ follow a normal distribution with mean 0 and variance $t_2 - t_1$. The WIENER process is a special case of a LÉVY process. One of the main applications of the WIENER process is to study *Brownian motion* or diffusion. This process will be discussed in more detail in Sect. 16.3 and in Chap. 17.
- The *random walk* is the discrete analogy to the WIENER process [9–11]. This means in particular that if the step size of the random walk goes to zero, the WIENER process is reestablished. This point will be elucidated in Chap. 17.

After stating the most important definitions, we proceed to the next section in which the attention is on a special class of stochastic processes, the so called MARKOV processes.

16.3 MARKOV Processes

A MARKOV process is a stochastic process $Y_X(t)$ for which the conditional pdf $p_{1|n-1}$ satisfies for arbitrary n and $t_1 < t_2 < \cdots < t_n$ the relation

$$p_{1|n-1}(y_n, t_n | y_1, t_1, \ldots, y_{n-1}, t_{n-1}) = p_{1|1}(y_n, t_n | y_{n-1}, t_{n-1}) . \tag{16.20}$$

Hence, a MARKOV process is a process in which any state y_n, t_n is uniquely defined by its precursor state y_{n-1}, t_{n-1} and is independent of the entire rest of the past [12]. MARKOV processes are of particular importance in natural sciences because of their rather simple structure. This will become clear throughout the rest of this book.

We note in passing that a process with independent increments is *always* a MARKOV process because

$$Y_X(t_{n+1}) = Y_X(t_n) + [Y_X(t_{n+1}) - Y_X(t_n)] , \tag{16.21}$$

is satisfied. Since the increment $Y_X(t_{n+1}) - Y_X(t_n)$ is independent of all previous increments which gave rise to $Y_X(t_n)$ *by definition*, $Y_X(t_{n+1})$ depends only on $Y_X(t_n)$, which is exactly the MARKOV property (16.20).

The quantity $p_{1|1}(y_n, t_n | y_{n-1}, t_{n-1})$ which appears in Eq. (16.20) is referred to as *transition probability*. Given the transition probability $p_{1|1}$ together with the pdf p_1, one can construct the whole hierarchy of pdfs (16.8) of the MARKOV process by calculating successively [8]:

$$\begin{aligned} p_2(y_1, t_1, y_2, t_2) &= p_{1|1}(y_2, t_2 | y_1, t_1) p_1(y_1, t_1) , \\ p_3(y_1, t_1, y_2, t_2, y_3, t_3) &= p_{1|2}(y_3, t_3 | y_1, t_1, y_2, t_2) p_2(y_1, t_1, y_2, t_2) , \\ &= p_{1|1}(y_3, t_3 | y_2, t_2) p_{1|1}(y_2, t_2 | y_1, t_1) p_1(y_1, t_1) , \\ \vdots &\quad \vdots \end{aligned} \tag{16.22}$$

Here we employed definition (16.14) and in the second step of the second equation we employed for $p_{1|2}$ the MARKOV property (16.20).

The fact that the whole hierarchy of pdfs can be constructed by repeating the steps illustrated in Eq. (16.22) reveals the rather simple structure of MARKOV processes. However, Eq. (16.22) contains another useful information. We regard the pdf p_3 of (16.22) for three successive times $t_1 < t_2 < t_3$. First we integrate the left-hand side with respect to y_2 which yields with the help of property (16.10):

$$\int dy_2 \, p_3(y_1, t_1, y_2, t_2, y_3, t_3) = p_2(y_1, t_1, y_3, t_3) . \tag{16.23}$$

Hence, we have

$$p_2(y_1, t_1, y_3, t_3) = p_1(y_1, t_1) \int dy_2 \, p_{1|1}(y_3, t_3 | y_2, t_2) p_{1|1}(y_2, t_2 | y_1, t_1) , \tag{16.24}$$

or after dividing both sides by $p_1(y_1, t_1)$ and by keeping in mind Eq. (16.14) we arrive at:

$$p_{1|1}(y_3, t_3 | y_1, t_1) = \int dy_2 \, p_{1|1}(y_3, t_3 | y_2, t_2) p_{1|1}(y_2, t_2 | y_1, t_1) . \tag{16.25}$$

This equation is known as the CHAPMAN-KOLMOGOROV equation [8, 13]. The interpretation of this equation is straight-forward: the transition probability from (y_1, t_1) to (y_3, t_3) is equivalent to the transition probability from (y_1, t_1) to (y_2, t_2) multiplied by the transition probability from (y_2, t_2) to (y_3, t_3) when summed over all intermediate positions y_2. This is illustrated in Fig. 16.1.

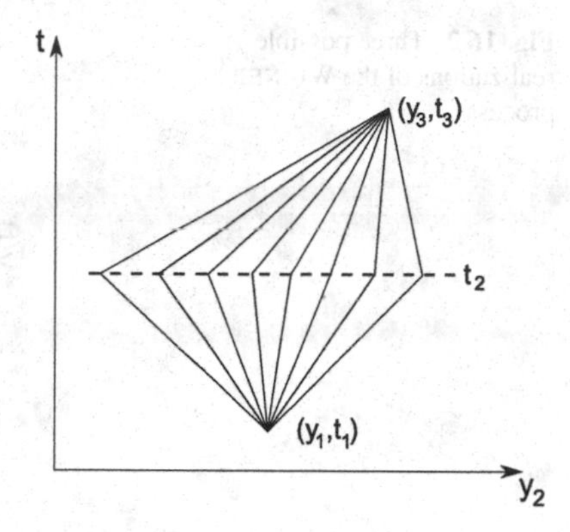

Fig. 16.1 Illustration of the CHAPMANN-KOLMOGOROV equation

We state a very important theorem: Any two non-negative functions $p_{1|1}$ and p_1 uniquely define a MARKOV process if the CHAPMAN-KOLMOGOROV equation (16.25) is obeyed and if

$$p_1(y_2, t_2) = \int dy_1 p_{1|1}(y_2, t_2|y_1, t_1) p_1(y_1, t_1) , \tag{16.26}$$

which follows immediately from the first equation in Eqs. (16.22) in combination with property (16.10).

As a first example we consider one of the most important MARKOV processes in physics, the WIENER process [10]. Its importance stems from its application to the description of Brownian motion, the random motion of dust particles on a fluid surface. (In Chap. 17 we take a closer look at diffusion phenomena.) The transition probability of the WIENER process is of the form[1]

$$p_{1|1}(y_2, t_2|y_1, t_1) = \frac{1}{\sqrt{2\pi(t_2 - t_1)}} \exp\left[-\frac{(y_2 - y_1)^2}{2(t_2 - t_1)}\right] . \tag{16.27}$$

The initial condition is given by

$$p_1(y_1, t_1 = 0) = \delta(y_1) . \tag{16.28}$$

A straight-forward calculation proves that (16.27) indeed obeys the CHAPMAN-KOLMOGOROV equation (16.25). Moreover, we deduce from Eqs. (16.28) together with (16.26) that

$$p_1(y, t) = \frac{1}{\sqrt{2\pi t}} \exp\left(-\frac{y^2}{2t}\right) . \tag{16.29}$$

[1]This form is equivalent to the above definition of the WIENER process, in particular to the requirement of normally distributed increments with variance $t_2 - t_1$.

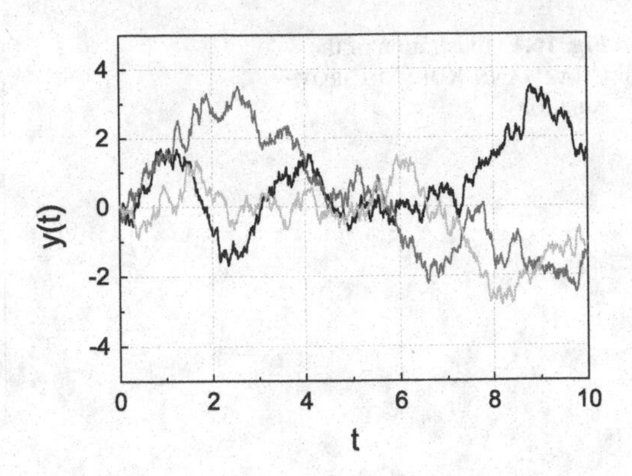

Fig. 16.2 Three possible realizations of the WIENER process

The WIENER process is easily realized on the computer. We regard the one-dimensional case and start at the origin $Y_X(0) = 0$. As per definition the increments $Y_X(t + \mathrm{d}t) - Y_X(t)$ follow a normal distribution $\mathcal{N}(\mathrm{d}y|0, \mathrm{d}t)$ of mean zero and variance $\mathrm{d}t$. Hence, we start with $y_0 = 0$ at time $t_0 = 0$, sample the displacement $\mathrm{d}y$ within a time-step $\mathrm{d}t$ from $\mathcal{N}(\mathrm{d}y|0, \mathrm{d}t)$ and calculate the next position at time $t_1 = t_0 + \mathrm{d}t$ which is given by: $y_1 = y_0 + \mathrm{d}y$.[2] This process is repeated until a certain time limit has been reached. Figure 16.2 presents the result of three such calculations which have been started using different seeds.

Let us mention a second very important MARKOV process, the POISSON process. The POISSON process is particularly interesting for problems involving *waiting times*, such as the decay of some radioactive nucleus. However, we shall also come across the POISSON process within the context of diffusion in Chap. 17. The transition probability of the POISSON process is defined as

$$p_{1|1}(n_2, t_2|n_1, t_1) = \frac{(t_2 - t_1)^{n_2 - n_1}}{(n_2 - n_1)!} \exp\left[-(t_2 - t_1)\right] . \tag{16.30}$$

Here it is understood that $n_1, n_2 \in \mathbb{N}$ and $n_2 > n_1$. Hence, the POISSON process counts the number of occurrences n_2 of a certain event until the time t_2 is reached under the premise that n_1 events have already occurred at time t_1. The POISSON process is initialized by the pdf

$$p_1(n_1, t_1 = 0) = \delta_{n_1 0} , \tag{16.31}$$

here δ_{ij} is the KRONECKER-δ. Hence we have according to Eq. (16.26)

$$p_1(n, t) = \sum_{n_1} p_{1|1}(n, t|n_1, t_1 = 0) p_1(n_1, t_1 = 0) = \frac{t^n}{n!} \exp(-t) , \tag{16.32}$$

[2]Alternatively, we may sample $\mathrm{d}y$ from a normal distribution with variance 1 and multiply it by $\sqrt{\mathrm{d}t}$. This follows from a simple transformation of variables.

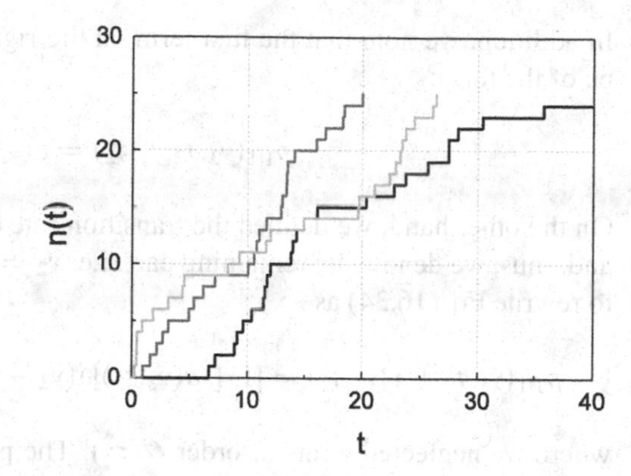

Fig. 16.3 Three possible realizations of a POISSON process

which is a POISSON distribution (see Appendix, Sect. E.4 and, for instance, Ref. [14]). Let us briefly consider the time between two events. Suppose we had n_1 events at time t_1. Then, we calculate the probability that at time $t_2 = t_1 + \tau$ we still counted $n_2 = n_1$ events, thus, nothing happened. We have

$$p_{1|1}(n_1, t_1 + \tau | n_1, t_1) = \exp(-\tau) \tag{16.33}$$

and the waiting times are independent and follow an exponential distribution. We may simulate the POISSON process by starting at $t_1 = 0$ with $n_1 = 0$ and by increasing $n_2, n_3, \ldots$ by one, i.e. $n_{i+1} = n_i + 1$ after successive waiting times $\tau_1, \tau_2, \ldots$ which we sample from the exponential distribution (see Sect. 13.2) until a final count N has been reached. The result of such a procedure is illustrated in Fig. 16.3 where $n(t)$ vs t has been plotted for three runs started by different seeds.

Finally, we remark that for a time-homogeneous MARKOV process the transition probability $p_{1|1}(y_2, t_2|y_1, t_1)$ depends by definition on the time difference $t_2 - t_1 \equiv \tau$ rather than explicitly on the two times t_1 and t_2 and is usually denoted by $T_\tau(y_2, y_1)$.

We turn now our attention to another very important general concept of MARKOV processes, the *master equation* [8]. This equation is in fact the differential form of the CHAPMAN-KOLMOGOROV equation. We regard the CHAPMAN-KOLMOGOROV equation (16.25) for three successive times $t_1 < t_2 < t_3 = t_2 + \tau$ where τ is assumed to be small, i.e. $\tau \ll 1$. We expand the conditional pdf $p_{1|1}$ in a TAYLOR series with respect to τ:

$$p_{1|1}(y_3, t_2 + \tau|y_2, t_2) = p_{1|1}(y_3, t_2|y_2, t_2) + \tau \frac{\partial}{\partial \tau} p_{1|1}(y_3, t_2 + \tau|y_2, t_2)\Big|_{\tau=0} + \mathcal{O}(\tau^2) \,. \tag{16.34}$$

In order to transform this equation into a more transparent form, we introduce the transition rate $w(y_3|y_2, t_2)$ from y_2 to y_3, with $y_2 \neq y_3$:

$$w(y_3|y_2, t_2) = \frac{\partial}{\partial \tau} p_{1|1}(y_3, t_2 + \tau|y_2, t_2)\Big|_{\tau=0} \,. \tag{16.35}$$

In addition, we note that the first term on the right-hand side of Eq. (16.34) has to be of the form:

$$p_{1|1}(y_3, t_2|y_2, t_2) = \delta(y_3 - y_2) . \tag{16.36}$$

On the other hand, we defined the transition rate (16.35) only for elements $y_2 \neq y_3$ and, thus, we denote the remaining part (i.e. $y_2 = y_3$) by $a(y_2, t_2)$. All this allows us to rewrite Eq. (16.34) as

$$p_{1|1}(y_3, t_2 + \tau|y_2, t_2) = [1 + a(y_2, t_2)]\delta(y_3 - y_2) + \tau w(y_3|y_2, t_2) , \tag{16.37}$$

where we neglected terms of order $\mathcal{O}(\tau^2)$. The pdf $p_{1|1}(y_3, t_2 + \tau|y_2, t_2)$ is subject to the normalization (16.15) and this provides us with the required condition to determinc $a(y_2, t_2)$:

$$a(y_2, t_2) = -\tau \int dy_3 w(y_3|y_2, t_2) . \tag{16.38}$$

Hence, the term $1 + a(y_2, t_2)$ describes the probability that no event occurs within the time interval $[t_2, t_2 + \tau]$. The expansion (16.37) is inserted into the CHAPMAN-KOLMOGOROV equation (16.25) with the result:

$$\begin{aligned} \frac{p_{1|1}(y_3, t_2 + \tau|y_1, t_1) - p_{1|1}(y_3, t_2|y_1, t_1)}{\tau} = \int dy_2 \big[&w(y_3|y_2, t_2)p_{1|1}(y_2, t_2|y_1, t_1) \\ &- w(y_2|y_3, t_2)p_{1|1}(y_3, t_2|y_1, t_1)\big] . \end{aligned} \tag{16.39}$$

Finally, we arrive, in the limit $\tau \to 0$, at the master equation:

$$\begin{aligned} \frac{\partial}{\partial t} p_{1|1}(y, t|y', t') = \int dy'' \big[&w(y|y'', t)p_{1|1}(y'', t|y', t') \\ &- w(y''|y, t)p_{1|1}(y, t|y', t')\big] . \end{aligned} \tag{16.40}$$

We multiply this equation by $p_1(y', t')$ and integrate over y'. This results in the master equation for the pdf $p_1(y, t)$

$$\frac{\partial}{\partial t} p_1(y, t) = \int dy' \left[w(y|y', t)p_1(y', t) - w(y'|y, t)p_1(y, t)\right] , \tag{16.41}$$

where we made use of the property (16.26).

Let us briefly discuss this result: In its derivation we assumed the state space to be continuous. However, if a master equation for a discrete state space is required the integral is to be replaced by a sum over the discrete states. On the other hand, the physical interpretation of such an equation is straight-forward: The time evolution of the quantity $p_1(y, t)$ is governed by the sum over all transitions into state y (first term) minus all transitions out of state y. We remark that master equations occur commonly in physical applications; for instance, the collision integral of the BOLTZMANN equation is of a similar form. The transitions rates $w(y|y', t)$ can be determined explicitly in many applications in physics.[3]

Furthermore, if the system is in a stationary state then it is described by a stationary distribution $p_1(y, t) = p_1(y)$ and we obtain from (16.41) the relation

$$\int \mathrm{d}y' w(y|y', t) p_1(y') = \int \mathrm{d}y' w(y'|y, t) p_1(y) , \tag{16.42}$$

which is referred to as *global balance*. The much stronger condition

$$w(y|y', t) p_1(y') = w(y'|y, t) p_1(y) , \tag{16.43}$$

is referred to as *detailed balance* and will be discussed next.

The task now is to prove that the equilibrium distribution function $p_e(X)$ of a classical physical system will, under certain restrictions, indeed fulfill detailed balance. (This proof was given by N.G. VAN KAMPEN [5].) The next steps of the proof become more transparent if a vector $x = (q_k, p_k)^T \in \mathbb{R}^{6N}$ is introduced which represents the phase space trajectory of the N particles constituting the system under investigation. This trajectory is determined by HAMILTON's equations of motion [16–18]:

$$\dot{q}_k = \frac{\partial}{\partial p_k} H(x) , \qquad \dot{p}_k = -\frac{\partial}{\partial q_k} H(x) . \tag{16.44}$$

Furthermore, $Y_X(t)$ denotes a stochastic process which describes some observable of the physical system. We require that $Y_X(t)$ is invariant under time reversal. We also assume the equilibrium distribution function $p_e(x)$ to be invariant under time reversal, which in most cases is equivalent to the requirement that the HAMILTON function $H(x)$ is invariant under time reversal. The operation of time reversal will be indicated by bared variables:

$$\bar{t} = -t , \quad \bar{x} = (q_k, -p_k)^T . \tag{16.45}$$

[3] As an example we quote FERMI's golden rule [15], where the transition rate $w_{nn'}$ from unperturbed states n to n' is of the form

$$w_{nn'} = \frac{2\pi}{\hbar} |H'_{nn'}| \rho(E_n) ,$$

where $H'_{nn'}$ are the matrix elements of the perturbation Hamiltonian H' and $\rho(E_n)$ denotes the density of states of the unperturbed system.

Hence, the above assumptions result in

$$\overline{Y_x(t)} = Y_{\overline{x}}(\overline{t}) = Y_{\overline{x}}(-t) = Y_x(t) , \tag{16.46}$$

and

$$\overline{p_e(x)} = p_e(\overline{x}) = p_e(x) . \tag{16.47}$$

In particular, we deduce from Eq. (16.46) that

$$Y_{\overline{x}}(0) = Y_x(0) , \tag{16.48}$$

and

$$Y_{\overline{x}}(t) = Y_x(-t) . \tag{16.49}$$

We calculate the pdf p_2 from

$$p_2(y_1, 0, y_2, t) = \int \mathrm{d}x \delta[y_1 - Y_x(0)] \delta[y_2 - Y_x(t)] p_e(x) . \tag{16.50}$$

However, since we integrate over the whole phase space we recognize that the volume is invariant under a change $\mathrm{d}x \to \mathrm{d}\overline{x}$. Thus, we can change the variable of integration from x to $\overline{x}$ and this results in:

$$\begin{aligned} p_2(y_1, 0, y_2, t) &= \int \mathrm{d}\overline{x} \delta[y_1 - Y_{\overline{x}}(0)] \delta[y_2 - Y_{\overline{x}}(t)] p_e(\overline{x}) \\ &= \int \mathrm{d}x \delta[y_1 - Y_x(0)] \delta[y_2 - Y_x(-t)] p_e(x) \\ &= p_2(y_2, -t, y_1, 0) \\ &= p_2(y_2, 0, y_1, t) . \end{aligned} \tag{16.51}$$

We obtain immediately:

$$p_{1|1}(y_2, t|y_1, 0) p_e(y_1) = p_{1|1}(y_1, t|y_2, 0) p_e(y_2) . \tag{16.52}$$

Differentiation of this equation with respect to t together with definition (16.35) yields for small values of t

$$w(y_2|y_1) p_e(y_1) = w(y_1|y_2) p_e(y_2) , \tag{16.53}$$

which is the condition of detailed balance, Eq. (16.43), for stationary distributions.

It should be noted at this point that detailed balance in physical systems is strongly connected to the entropy growth (the H-theorem by BOLTZMANN [19]).

Here, detailed balance was based on the condition that the stochastic process $Y_X(t)$ was invariant under time reversal and that the equilibrium distribution $p_e(x)$ had the same property. This has in most cases the consequence that the HAMILTON function is also invariant under time reversal transformations. However, a detailed discussion of these properties is far beyond the scope of this book.

We continue our presentation with so called MARKOV-chains which are a special class of MARKOV processes. MARKOV-chains will prove to be very important for the understanding of MARKOV-chain Monte Carlo techniques, such as the METROPOLIS algorithm.

16.4 MARKOV-Chains

A MARKOV-chain is a time-homogeneous MARKOV process defined on a discrete time span and in a discrete state space [20–22]. Hence, we express the time instances by integers $T = \mathbb{N}$, $t_n = n$ where $n \in \mathbb{N}$ and possible outcomes are indexed by integers $Y_X(t_n) \in \{m\}$ where $m \in \mathbb{N}$. As a first consequence of the discreteness of the state space we replace all pdfs $p(\cdot)$ by probabilities $\Pr(\cdot)$. Hence the MARKOV property (16.20) reads

$$\Pr(Y_{n+1} = y | Y_n = y_n, \ldots, Y_1 = y_1) = \Pr(Y_{n+1} = y | Y_n = y_n) , \tag{16.54}$$

where we applied the notation $Y_n \equiv Y_X(t_n)$ and $y_n \in \{m\}$ is one particular realization out of the discrete state space. Since we assume the transition probabilities to be independent of the actual time, we can define a *transition matrix* $P = \{p_{ij}\}$ via

$$p_{ij} = \Pr(Y_{n+1} = j | Y_n = i) . \tag{16.55}$$

Consequently, we write

$$\Pr(Y_n = i_n, Y_{n-1} = i_{n-1}, \ldots, Y_0 = i_0) = \Pr(Y_0 = i_0) p_{i_0 i_1} p_{i_1 i_2} \cdots p_{i_{n-1} i_n} . \tag{16.56}$$

We note that the transition matrix is a *stochastic matrix*, a matrix with only non-negative elements such that the sum of each row is equal to one. Furthermore, one can prove that the product of two stochastic matrices results, again, in a stochastic matrix.

We define the *state vector at time n*, $\pi^{(n)} = \{\pi_i^{(n)}\}$ as

$$\pi_i^{(n)} = \Pr(Y_n = i) . \tag{16.57}$$

From the marginalization rule (Appendix, Sect. E.6) follows for the particular case $n = 1$

$$\Pr(Y_1 = i) = \sum_k \Pr(Y_0 = k) \Pr(Y_1 = i | Y_0 = k) , \tag{16.58}$$

or with the help of the definitions (16.55) and (16.57):

$$\pi_i^{(1)} = \sum_k p_{ki} \pi_k^{(0)} , \quad \forall i. \tag{16.59}$$

Hence, we get for $n = 1$

$$\pi^{(1)} = \pi^{(0)} P , \tag{16.60}$$

and for $n = 2$

$$\pi^{(2)} = \pi^{(1)} P = \pi^{(0)} P^2 . \tag{16.61}$$

Obviously,

$$\pi^{(n)} = \pi^{(0)} P^n , \tag{16.62}$$

follows for arbitrary n. Hence the probability matrix for an n step transition $P^{(n)}$ is given by $P^{(n)} = P^n$. We immediately deduce that the CHAPMAN-KOLMOGOROV equation for MARKOV-chains is fulfilled since

$$P^{(n)} P^{(m)} = P^n P^m = P^{n+m} = P^{(n+m)} , \tag{16.63}$$

for two integers n and m.

Let us cite some further definitions in order to classify MARKOV-chains [5, 8, 20–22]:

- The notation $i \to j$ means *state i leads to state j* and is true whenever there is a path of length n, $i_0 = i, i_1, \ldots, i_n = j$ such that all $p_{i_k i_{k+1}} > 0$ for $k = 0, 1, \ldots, n-1$. This is equivalent to $(P^n)_{ij} > 0$.
- The notation $i \leftrightarrow j$ means *state i communicates with state j*. This relation is true whenever $i \to j$ and $j \to i$.
- A class of states is given if (i) all states within one class communicate with each other and (ii) two states of different classes never communicate with each other. These classes are referred to as the *irreducible* classes of the MARKOV-chain.
- An *irreducible* MARKOV-*chain* is a MARKOV-chain in which the whole state space forms an irreducible class, i.e. *all* states communicate with each other.
- A *closed set of states* is a set of states which never leads to states which are outside of this set.

- An *absorbing state* is a state which does not lead to any other states: It forms itself a closed set. An absorbing state can be reached from the outside but there is no escape from it.
- A state is referred to as *transient* if the probability of returning to the state is less than one.
- A state is referred to as *recurrent* if the probability of returning to the state is equal to one.
- Furthermore, we call a state *positive recurrent* if the expectation value of the first return time is less than infinity and *null recurrent* if it is infinity. We may formulate this in a more mathematical language: The time of first return to state i is defined via

$$T_i = \inf (n \geq 1 : X_n = i | X_0 = i) \ . \tag{16.64}$$

The probability that we return to state i for the first time after n steps is defined as

$$f_{ii}^n = \Pr(T_i = n) \ . \tag{16.65}$$

Hence, a state is referred to as recurrent if

$$F_i = \sum_n f_{ii}^n = 1 \ , \tag{16.66}$$

positive recurrent if

$$\langle T_i \rangle = \sum_n n f_{ii}^n < \infty \ , \tag{16.67}$$

and null recurrent if

$$\langle T_i \rangle = \sum_n n f_{ii}^n = \infty \ . \tag{16.68}$$

We note that we also have $\langle T_i \rangle = \infty$ if state i is transient. Furthermore, one can show that a state is only recurrent if

$$\sum_n p_{ii}^n = \infty \ . \tag{16.69}$$

- A state is referred to as *periodic* if the return time of the state can only be a multiple of some integer $d > 1$.
- A state is referred to as *aperiodic* if $d = 1$.
- We call a state *ergodic* if it is positive recurrent and aperiodic.
- A MARKOV-chain is called *ergodic* if all its states are *ergodic*.

We give some useful theorems in the context of the above definitions: First of all, it can be proved that if a MARKOV-chain is irreducible it follows that either *all* states are transient, or *all* states are null recurrent, or *all* states are positive recurrent.

Furthermore, a theorem by KOLMOGOROV states that if a MARKOV-chain is irreducible and aperiodic then the limit

$$\pi_j = \lim_{n\to\infty} \pi_j^{(n)} = \frac{1}{\langle T_j \rangle}, \tag{16.70}$$

exists. It follows from the above discussion that if all states j are transient or null recurrent we have

$$\pi_j = 0, \tag{16.71}$$

and if all states j are positive recurrent, we have

$$\pi_j \neq 0. \tag{16.72}$$

In this case the state vector $\pi = \{\pi_j\}$ is referred to as the *stationary distribution* or *equilibrium distribution*. We note that in this context the term equilibrium does not mean that nothing changes, but that the system *forgets* its own past. In particular, as soon as the system reaches the stationary distribution, it is independent of the initial state $\pi^{(0)}$.

We concentrate now on equilibrium distributions. It follows from Eq. (16.62) that π satisfies:

$$\pi = \pi P. \tag{16.73}$$

Thus, π is the *left-eigenvector* to the transition probability matrix P with eigenvalue 1. We note that Eq. (16.73) states a homogeneous eigenvalue problem: The solution is only determined up to a constant multiplicator (see Sect. 8.3). However, it is clear that the vector π satisfies

$$\sum_j \pi_j = 1. \tag{16.74}$$

One can prove that the unique solution of the eigenvalue problem (16.73) together with the normalization condition (16.74) for n states can be written as

$$\pi = e \cdot (P - E - I)^{-1}, \tag{16.75}$$

where e is an n-element row vector containing only ones, E is a $n \times n$ matrix containing only ones and I is the $n \times n$ identity.

Let us briefly elaborate on this point: if it is possible to construct a MARKOV-chain which possesses a unique stationary distribution, we know that it will definitely reach this distribution independent of the choice of initial conditions. The existence as well as the form of the stationary distribution is clearly determined by the transition probabilities p_{ij}. A sufficient condition for a unique stationary distribution to exist is the requirement of reversibility. A MARKOV-chain is referred to as *reversible* if

$$p_{ij}\pi_i = p_{ji}\pi_j , \quad \forall i,j, \tag{16.76}$$

i.e. if the transition probabilities ensure *detailed balance* for the stationary distribution π.

Now we are in a position to understand better why detailed balance was such an important concept of the METROPOLIS algorithm discussed in Sect. 14.3: Invoking the detailed balance condition ensures that for all possible initial states the MARKOV-chain converges toward the equilibrium distribution for which detailed balance is fulfilled. Of course, the convergence time will highly depend on the choice of the initial state as well as on the choice of the transition matrix. Hence, we can generate random numbers with the help of such a MARKOV-chain and after a *thermalization* period these numbers will follow the required pdf. Methods based on this concept are commonly referred to as MARKOV-*chain Monte Carlo sampling methods* [23–26].

We give a brief example, the *spread of a rumor*. Let Z_1 and Z_2 be two distinct versions of a report. If a person receives report Z_1 it will pass this report on as Z_1 with probability $(1-p)$ or as Z_2 with probability p. An alternative is that the person receives Z_2 and passes it on as Z_2 with probability $(1-q)$ or modifies it to Z_1 with probability q. We summarize

- $\Pr(Z_1 \to Z_1) = (1-p) = p_{11}$,
- $\Pr(Z_1 \to Z_2) = p = p_{12}$,
- $\Pr(Z_2 \to Z_1) = q = p_{21}$,
- $\Pr(Z_2 \to Z_2) = (1-q) = p_{22}$.

The transition matrix is of the form

$$P = \begin{pmatrix} 1-p & p \\ q & 1-q \end{pmatrix} . \tag{16.77}$$

We note that the two states communicate with each other $Z_1 \leftrightarrow Z_2$, hence the MARKOV-chain is irreducible. Furthermore, since the process can reach either state Z_1 or Z_2 within a single time step, it is clearly aperiodic. Let us briefly investigate the probabilities of first recurrence f_{ii}^n after n steps. Due to the theorem by KOLMOGOROV it is sufficient to investigate the state Z_1 since the MARKOV-chain is irreducible and it follows that also Z_2 has the same recurrence properties. We note the following possible paths for a first return to state Z_1:

- $1: \quad \Pr(Z_1 \to Z_1) = (1-p) = f_{11}^1$,
- $2: \quad \Pr(Z_1 \to Z_2 \to Z_1) = pq = f_{11}^2$,

- $3: \quad \Pr(Z_1 \to Z_2 \to Z_2 \to Z_1) = p(1-q)q = f_{11}^3$,
- $n: \quad \Pr(Z_1 \to Z_2 \to \cdots \to Z_2 \to Z_1) = p(1-q)^{n-2}q = f_{11}^n$.

The probability of returning to Z_1 is, see Eq. (16.66),

$$\begin{aligned}
F_1 &= \sum_{n=1}^{\infty} f_{11}^n \\
&= (1-p) + pq\sum_{n=0}^{\infty}(1-q)^n \\
&= (1-p) + pq\frac{1}{1-(1-q)} \\
&= 1 ,
\end{aligned} \tag{16.78}$$

where we employed that $0 < (1-q) < 1$ as well as the convergence of the geometric series. Hence state Z_1 is recurrent and, therefore, also state Z_2. We calculate the expectation value of the first return time $\langle T_1 \rangle$:

$$\begin{aligned}
\langle T_1 \rangle &= \sum_{n=1}^{\infty} n f_{11}^n \\
&= (1-p) + pq\sum_{n=0}^{\infty}(n+2)(1-q)^n \\
&= (1-p) + 2pq\underbrace{\sum_{n=0}^{\infty}(1-q)^n}_{=\frac{1}{q}} + pq\sum_{n=0}^{\infty} n(1-q)^n \\
&= 1 + p - pq(1-q)\frac{\mathrm{d}}{\mathrm{d}q}\underbrace{\sum_{n=0}^{\infty}(1-q)^n}_{=\frac{1}{q}} \\
&= 1 + p + \frac{p}{q}(1-q) \\
&= \frac{p+q}{q} .
\end{aligned} \tag{16.79}$$

Hence, the states Z_1 and Z_2 are positive recurrent as long as $p \neq 0$ and $q \neq 0$. This means that an equilibrium distribution exists and it can be obtained from Eq. (16.70).

We have

$$\pi_1 = \frac{1}{\langle T_1 \rangle} = \frac{q}{p+q} . \tag{16.80}$$

Due to the normalization condition (16.74) we obtain

$$\pi_2 = 1 - \pi_1 = \frac{p}{p+q} , \tag{16.81}$$

and, therefore,

$$\langle T_2 \rangle = \frac{p+q}{p} . \tag{16.82}$$

Since all states are positive recurrent and aperiodic, the above MARKOV-chain is ergodic. Finally, we remark that this example also fulfills detailed balance since

$$\pi_1 p_{12} = \frac{qp}{p+q} = \pi_2 p_{21} . \tag{16.83}$$

Let us briefly interpret this example: Suppose the original, true version Z_1 of a report is '*Mr. X is going to resign*' while Z_2 is just the opposite: '*Mr. X is* not *going to resign*'. The property of irreducibility of the MARKOV-chain reflects the fact that there is no version of the report which cannot be reached or modified. Moreover, we just demonstrated that the process is positive recurrent: Even if the probability p that Z_1 was modified to Z_2 is very small and the probability q that Z_2 was modified to Z_1 is very high, the report will infinitely often return to version Z_2 with probability one. This means that the public will be told infinitely often that Mr. X is *not* going to resign with probability one. The equilibrium probabilities π_1 and π_2 display the asymptotic probability of versions one and two of the report, respectively. However, as has already been emphasized, this does not mean that the report cannot be modified in equilibrium, it simply displays the fact that the probabilities reached a steady state. Finally, we note an interesting effect in passing: Suppose that the probabilities that any of the two versions is modified is very small but equal, i.e. $p = q \ll 1$. Then the equilibrium distribution is

$$\pi_1 = \pi_2 = \frac{1}{2} , \tag{16.84}$$

and the public will believe Z_1 and Z_2 with the same probability after some time independently of the initial version and also independent of the actual decision of Mr. X. Detailed balance expresses the property that the probability of receiving Z_2 and passing it on as Z_1 is the same as the probability of receiving Z_2 and passing it on as Z_1.

We close this section with a final remark: It is an easy task to generalize the ideas of MARKOV-chains to continuous state spaces since we already introduced the required tools in Sect. 16.3. Let $\pi(x)$ denote the stationary distribution density and $p(x|y)$ the accompanying transition rate pdf. Then relation (16.73) transforms into

$$\pi(x) = \int \mathrm{d}y \pi(y) p(x|y) , \tag{16.85}$$

together with

$$\int \mathrm{d}x \pi(x) = 1 , \tag{16.86}$$

the usual normalization of pdfs. In this case, the condition of detailed balance is given by

$$\pi(x) p(y|x) = \pi(y) p(x|y) , \tag{16.87}$$

which is equivalent to Eq. (16.52).

16.5 Continuous-Time MARKOV-Chains

A generalization of the results of the previous sections to a continuous time span is straight-forward. We define the continuous-time MARKOV-chain as a time-homogeneous MARKOV process on a discrete state space but with a continuous time span, $t \geq 0$. Thus

$$\Pr[X(t+s) = n | X(s) = m] = p_{nm}(t) , \tag{16.88}$$

is *independent* of $s \geq 0$. In this case the transition matrix $P(t) = \{p_{ij}(t)\}$ is an explicit function of time t. Its elements $p_{nm}(t)$ have the following four properties:

(a) All matrix elements $p_{nm}(t)$ of the transition matrix P are positive:

$$p_{nm}(t) \geq 0 , \qquad \forall t > 0 . \tag{16.89}$$

(b) The usual normalization of the rows of the transition matrix P is valid:

$$\sum_m p_{nm}(t) = 1 , \qquad \forall n \text{ and } t > 0 . \tag{16.90}$$

(c) As for every MARKOV process, the transition matrix of the continuous time MARKOV-chain obeys the CHAPMAN-KOLMOGOROV equation:

$$\sum_k p_{nk}(t)p_{km}(t') = p_{nm}(t+t') , \tag{16.91}$$

which can alternatively be expressed as

$$P(t+t') = P(t)P(t') . \tag{16.92}$$

(d) We assume $p_{nm}(t)$ to be a continuous function of t and that:

$$\lim_{t\to 0} p_{nm}(t) = \begin{cases} 1 & \text{for } n = m , \\ 0 & \text{for } n \neq m . \end{cases} \tag{16.93}$$

It follows from this equation that the matrix elements $p_{nm}(t)$ can be written as

$$p_{nm}(t) = \begin{cases} 1 + q_{nn}\,t + \mathcal{O}(t^2) & \text{for } n = m , \\ q_{nm}\,t + \mathcal{O}(t^2) & \text{for } n \neq m , \end{cases} \tag{16.94}$$

where we introduced with $\{q_{nm}\} = Q$ the *transition rate matrix*. The transition rate matrix Q obeys:

(a) All off-diagonal elements q_{nm}, $n \neq m$, are non-negative since

$$q_{nm} = \lim_{t\to 0} \frac{p_{nm}(t)}{t} \geq 0 \qquad \text{for } n \neq m . \tag{16.95}$$

(b) All diagonal elements q_{nn} are non-positive since

$$q_{nn} = -\lim_{t\to 0} \frac{1 - p_{nn}(t)}{t} \leq 0 . \tag{16.96}$$

(c) Differentiating Eq. (16.90) with respect to t yields that the sum over all elements in a row is equal to zero. Therefore, we conclude:

$$q_{nn} = -\sum_{\substack{m \\ n\neq m}} q_{nm} . \tag{16.97}$$

Moreover, differentiating the CHAPMAN-KOLMOGOROV equation with respect to t or t' gives the KOLMOGOROV forward – or KOLMOGOROV backward equations

$$\dot{P}(t) = P(t)Q \qquad \text{and} \qquad \dot{P}(t) = QP(t) , \tag{16.98}$$

respectively. We obtain $P(t) = \exp(Qt)$ where the exponential function of a matrix has to be interpreted as

$$\exp(Qt) = \sum_{k=0}^{\infty} \frac{t^k}{k!} Q^k , \tag{16.99}$$

where $Q^0 = I$ is the identity matrix.

We define s as the time of the first jump of our process for the particular case $X(0) = n$

$$s = \min [t|X(t) \neq X(0)] . \tag{16.100}$$

It can be shown that $P_n(s > t)$, the probability that the jump occurs at *some* time $s > t$, is given by

$$P_n(s > t) = \exp(q_{nn}t) , \tag{16.101}$$

where we note that $q_{nn} \leq 0$. Moreover,

$$P_n[X(s) = m] = -\frac{q_{nm}}{q_{nn}} , \tag{16.102}$$

and the process starts again at time s and in state m. This means that in a continuous-time MARKOV-chain the waiting times between two consecutive jumps are *exponentially distributed*. One of the simplest examples of a continuous time MARKOV-chain is the POISSON process, discussed in Sect. 16.3.

Summary

This chapter introduced the concept of stochastic processes $Y_X(t)$ as 'time' dependent processes depending on randomness. Y was a random variable which depended on another random variable X and t, the time. All realizations of $Y_X(t)$ spanned the state space. Each stochastic process was coupled to a pdf which described the probability that the process took on the realization y at time t. In the course of this introduction a series of general properties which classify such processes were defined. This was followed by the discussion of a particular class of stochastic processes, the MARKOV processes. They had the remarkable property that a future realization of the process solely depended on its current realization and not on the history how this current realization had been reached (MARKOV property). A huge class of processes in physics and related sciences is Markovian in nature. The next refinement in our discussion was the introduction of MARKOV-chains. These were processes defined on a discrete time span and in a discrete state space.

This allowed to replace the pdfs by probabilities. Again, various specific properties of MARKOV-chains opened the possibility of a distinctive classification. A very important observation was that under certain conditions a MARKOV-chain reached a stationary or equilibrium distribution and that it definitely arrived at this distribution independent of the choice of initial conditions. Moreover, detailed balance was obeyed by this equilibrium condition. This observation was the backbone of MARKOV-chain Monte Carlo sampling methods, in particular of the METROPOLIS algorithm. Finally, continuous-time MARKOV-chains were discussed.

Problems

1. Write a program to simulate the WIENER process in one dimension. Follow the method explained in Sect. 16.2 and perform the following analysis:
   a. Illustrate graphically some typical sample paths.
   b. Calculate the mean $\langle x(t) \rangle$ and the variance $\text{var}[x(t)]$ by restarting the process several times with different seeds and plot the result.
   c. Measure the position x of the particle at a particular time t for several runs (with different seeds) and illustrate the result $p(x, t)$ graphically.
2. Realize numerically a POISSON process according to the instructions given in Sect. 16.2. Again, plot some typical sample paths. Moreover, calculate the mean waiting time $\langle \tau \rangle$ as well as the variance $\text{var}(\tau)$ numerically as well as analytically.

References

1. Chow, Y.S., Teicher, H.: Probability Theory, 3rd edn. Springer Texts in Statistics. Springer, Berlin/Heidelberg (1997)
2. Kienke, A.: Probability Theory. Universitext. Springer, Berlin/Heidelberg (2008)
3. Stroock, D.W.: Probability Theory. Cambridge University Press, Cambridge (2011)
4. von der Linden, W., Dose, V., von Toussaint, U.: Bayesian Probability Theory. Cambridge University Press, Cambridge (2014)
5. van Kampen, N.G.: Stochastic Processes in Physics and Chemistry. Elsevier, Amsterdam (2008)
6. Gardiner, C.: Stochastic Methods. Springer Series in Synergetics. Springer, Berlin/Heidelberg (2009)
7. Chiasson, J.: Introduction to Probability Theory and Stochastic Processes. Wiley, New York (2013)
8. Breuer, H.P., Petruccione, F.: Open Quantum Systems, chap. 1. Clarendon Press, Oxford (2010)
9. Montroll, E.W., Schlesinger, M.F.: The wonderful world of random walks. In: Lebowitz, J.L., Montroll, E.W. (eds.) Non-equilibrium Phenomena II. Studies in Statistical Mechanics, vol. 11. North-Holland, Amsterdam (1984)
10. Rudnik, J., Gaspari, G.: Elements of the Random Walk. Cambridge University Press, Cambridge (2010)

11. Ibe, O.C.: Elements of Random Walk and Diffusion Processes. Wiley, New York (2013)
12. Marcus, M.B., Rosen, J.: Markov Processes, Gaussian Processes, and Local Times. Cambridge University Press, Cambridge (2011)
13. Papoulis, A., Pillai, S.: Probability, Random Variables and Stochastic Processes. McGraw Hill, New York (2001)
14. Knuth, D.: The Art of Computer Programming, vol. II, 3rd edn. Addison Wesley, Menlo Park (1998)
15. Ballentine, L.E.: Quantum Mechanics. World Scientific, Hackensack (1998)
16. Arnol'd, V.I.: Mathematical Methods of Classical Mechanics, 2nd edn. Graduate Texts in Mathematics, vol. 60. Springer, Berlin/Heidelberg (1989)
17. Scheck, F.: Mechanics, 5th edn. Springer, Berlin/Heidelberg (2010)
18. Goldstein, H., Poole, C., Safko, J.: Classical Mechanics, 3rd edn. Addison-Wesley, Menlo Park (2013)
19. Pathria, R.K., Beale, P.D.: Statistical Mechanics, 3rd edn. Academic, San Diego (2011)
20. Norris, J.R.: Markov Chains. Cambridge Series in Statistical and Probabilistic. Cambridge University Press, Cambridge (1998)
21. Modica, G., Poggiolini, L.: A First Course in Probability and Markov Chains. Wiley, New York (2012)
22. Graham, C.: Markov Chains: Analytic and Monte Carlo Computations. Wiley, New York (2014)
23. Fishman, G.S.: Monte Carlo: Concepts, Algorithms and Applications. Springer Series in Operations Research. Springer, Berlin/Heidelberg (1996)
24. Doucet, A., de Freitas, N., Gordon, N. (eds.): Sequential Monte Carlo Methods in Practice. Information Science and Statistics. Springer, Berlin/Heidelberg (2001)
25. Kalos, M.H., Whitlock, P.A.: Monte Carlo Methods, 2nd edn. Wiley, New York (2008)
26. Kroese, D.P., Taimre, T., Botev, Z.I.: Handbook of Monte Carlo Methods. Wiley, New York (2011)

Chapter 17
The Random Walk and Diffusion Theory

17.1 Introduction

Diffusion is one of the most widely spread processes in science. Its occurrence ranges from random motion of dust particles on fluid surfaces, historically known as Brownian motion, to the motion of particles in numerous physical systems [1, 2], the spreading of malaria by migration of mosquitoes [3], or even to the description of fluctuations in stock markets [4].

For instance, let us regard N neutral, identical, classical particles which solely interact through collisions, for instance an H_2-gas in a box, where $N = N_A \approx 6.022 \times 10^{23}$. We are interested in the dynamics of one particle under the influence of all others and under no influence by an external force; we expect that diffusion will be the dominating process. From the microscopic point of view such a situation can be described with the help of N coupled NEWTON's equations of motion. (See Chap. 7.) Anyhow, such a task will not be feasible due to the size of the system – the magnitude of N. However, a statistical description can be obtained from BOLTZMANN's equation [5]

$$\frac{\mathrm{d}}{\mathrm{d}t} f(r, \eta, t) = \frac{\partial}{\partial t} f(r, \eta, t)\bigg|_{\text{coll.}} , \tag{17.1}$$

where $f(r, \eta, t)$ is the phase space distribution function. Hence, $f(r, \eta, t)\mathrm{d}r\mathrm{d}\eta$ is the number of particles of momentum η within the phase-space volume $\mathrm{d}r\mathrm{d}\eta$ which is centered around position r at time t. We have, in particular:

$$\frac{\partial}{\partial t} f(r, \eta, t) + \frac{\eta}{m} \cdot \frac{\partial}{\partial r} f(r, \eta, t) + F \cdot \frac{\partial}{\partial \eta} f(r, \eta, t) = C[f](r, \eta, t) . \tag{17.2}$$

Here $C[f](r, \eta, t)$ is the collision integral and F describes an external force. In cases where collisions result solely from two-body interactions between particles that

B.A. Stickler, E. Schachinger, *Basic Concepts in Computational Physics*,
DOI 10.1007/978-3-319-27265-8_17

are assumed to be uncorrelated prior to the collision,[1] the collision integral can be described by

$$C[f](r,\eta,t) = \int d\xi_1 \int d\xi_2 \int d\xi_3 g(\xi_1,\xi_2,\xi_3,\eta)\,[f(r,\xi_1,t)f(r,\xi_2,t) \\ -f(r,\eta,t)f(r,\xi_3,t)] \,, \tag{17.3}$$

where $g(\xi_1,\xi_2,\xi_3,\eta)$ accounts for the probability that a collision between two particles of initial moments ξ_1 and ξ_2 and final momenta ξ_3 and η occurs. This function depends on the particular type of particles under investigation and has, in general, to be determined from a microscopic theory.[2] We now define the particle density $\rho(r,t)$ as a function of space r and time t via

$$\rho(r,t) = \int d\eta f(r,\eta,t) \,. \tag{17.4}$$

A complicated mathematical analysis of Eq. (17.1) results in a diffusion equation of the well-known form

$$\frac{\partial}{\partial t}\rho(r,t) = D\frac{\partial^2}{\partial r^2}\rho(r,t) \,, \tag{17.5}$$

if collisions dominate the dynamics (*diffusion limit*). Here $D =$ const is the diffusion coefficient of dimension length$^2\times$ time^{-1}. Note that

$$\int dr \rho(r,t) = N \,, \tag{17.6}$$

is the number of particles within our system.[3] Thus, in our example we can interpret diffusion as the average evolution of the integrated phase space distribution function governed by collisions between particles. Such an interpretation will certainly not hold in the case of fluctuations in stock markets or in the case of the spreading of malaria because typically mosquitoes do not *collide* with humans.

It is the aim of the first part of this chapter to present a purely stochastic approach to diffusion, the so called *random walk* model [7, 8]. This stochastic

[1]This assumption is known as the approximation of *molecular chaos*. In fact it represents the MARKOV approximation to the dynamics of a many particle system.

[2]For instance, one can employ FERMI's golden rule [6] to obtain this function on a quantum mechanical level. We already came across an expression of the form (17.3) on the right hand side of the master equation, see Sect. 16.3, Eq. (16.42). However, the collision integral of the BOLTZMANN equation is non-linear.

[3]The function $\rho(r,t)$ is referred to as a *physical distribution function* due to the normalization condition (17.6). This is in contrast to distribution functions we encountered so far within this book, which are normalized to unity.

description will prove to have several precious advantages: (i) We will be able to identify criteria for the validity of the diffusion model even for systems lacking a straight-forward physical interpretation. (ii) The stochastic formulation will give us the opportunity to perform diffusion 'experiments' on the computer without much computational effort as the methods employed are based on algorithms discussed in previous chapters. (iii) Within this framework it will be an easy task to generalize the approach to stochastic models of anomalous diffusion [9]: The *fractal time random walk* and LÉVY *flight* models [10]. These models play an increasingly important role in modern statistical physics.

17.2 The Random Walk

The random walk is one of the classical examples of MARKOV-chains [11–13]. In this section we discuss some of the basic properties of random walks in one dimension. For convenience, we are going to use the familiar picture of one diffusing particle.

Basics

The random walk [8] is defined as the motion of a single particle which moves at the time instances

$$0, \Delta t, 2\Delta t, \ldots, n\Delta t, \ldots \,, \tag{17.7}$$

between grid-points

$$\ldots, -n\Delta x, \ldots, -\Delta x, 0, \Delta x, \ldots, n\Delta x, \ldots \,. \tag{17.8}$$

For a more transparent notation the lattice point $n\Delta x$, with $n \in \mathbb{Z}$, will be denoted by x_n and the instance $k\Delta t$, with $k \in \mathbb{N}$, will be denoted by t_k. This notation follows the conventions of Chap. 2. The initial position is given by

$$\Pr[X(t_0 = 0) = x_i] = \delta_{i0} \,, \tag{17.9}$$

and the transition rates p_{ij} from position i to position j within a single time step Δt are defined as

$$\Pr[X(t_{n+1}) = x_i | X(t_n) = x_j] = p\delta_{ji-1} + q\delta_{ji+1} + r\delta_{ij} \,. \tag{17.10}$$

Here p denotes the probability that the particle jumps to the neighboring grid-point on the right-hand side, q stands for the probability that the particle jumps to the neighboring grid-point on the left-hand side, and r denotes the probability of staying at the same-grid point within this time step. Naturally, we have

$$p + q + r = 1 \,. \tag{17.11}$$

Consequently, we have a MARKOV-chain with time instances t_n and a state space spanned by the positions x_k. Moreover, we note that the stochastic process is clearly irreducible since all states communicate with each other (see Sect. 16.4). Hence, it follows that either all states are recurrent or all states are transient. Furthermore, in the case that $r \neq 0$ the MARKOV-chain is aperiodic, otherwise the chain is periodic with periodicity $d = 2$ because it takes at least two steps to return to the starting position.

We concentrate first on the *classical random walk* that is a one-dimensional random walk with $\Delta t = \Delta x = 1$, $r = 0$, and $p + q = 1$. This ensures that the probability of remaining in the actual position within one time step is equal to zero. If, furthermore, $p = q = 1/2$ the random walk is referred to as *unbiased* and for $p \neq q$ we call it *biased*. We write the position $X(t_n) = x_n$ at time $t_n = n$ as

$$x_n = \sum_{i=1}^{n} \xi_i \,, \tag{17.12}$$

where $\xi_i \in \{-1, 1\}$ and $\Pr(\xi_i = +1) = p$, $\Pr(\xi_i = -1) = q$. Let us assume that within these n steps the particle moved m times to the right and, consequently, $n - m$ times to the left. The actual position x_n after n steps can then be determined from

$$x_n = m - (n - m) = 2m - n \equiv k \,, \tag{17.13}$$

where we used that $x_0 = 0$. It is interesting to calculate the probability $\Pr(x_n = k)$ to find the particle after n time steps at some particular position k. This is simply the sum over all paths along which the particle moved $m = (n + k)/2$ times to the right and $n - m = (n - k)/2$ times to the left multiplied by the probability for m steps to the right and $n - m$ steps to the left. In total, this yields $\binom{n}{m} = \binom{n}{(n+k)/2}$ different contributions and we have

$$\begin{aligned} \Pr(x_n = k) &= \binom{n}{m} p^m q^{n-m} \\ &= \binom{n}{(n+k)/2} p^{\frac{n+k}{2}} q^{\frac{n-k}{2}} \,. \end{aligned} \tag{17.14}$$

In particular, we find for the unbiased random walk:

$$\Pr(x_n = k) = \binom{n}{(n+k)/2} \left(\frac{1}{2}\right)^n \,. \tag{17.15}$$

Due to the aperiodicity of the classical random walk, k can only take on the values $k = -n, -n+2, \dots, n-2, n$. Consequently $n \pm k$ has to be even. For all other values of k we have $\Pr(x_n = k) = 0$. Furthermore,

$$\sum_{\substack{k=-n \\ n\pm k\,\text{even}}}^{n} \Pr(x_n = k) = \sum_{m=0}^{n} \binom{n}{m} p^m q^{n-m} = (p+q)^n = 1 , \tag{17.16}$$

and the probability of finding the particle at time n within $[-n, n]$ is equal to one. A simple algorithm to simulate the one-dimensional biased random walk consists of the following steps:

1. Define values x_0, p, and $q = 1 - p$.
2. Draw a uniformly distributed random number $r \in [0, 1]$.
3. If $r < p$ set $x_{n+1} = x_n + 1$, otherwise set $x_{n+1} = x_n - 1$.
4. Return to step 2.

In Fig. 17.1 we present three different realizations of an unbiased one-dimensional random walk for (a) $N = 50$, (b) $N = 100$, and (c) $N = 1000$ consecutive steps.

Comparison between Figs. 17.1 and 16.2 already suggests a connection between the random walk and the WIENER process and we shall come back to this point in the course of this chapter.

Moments

Let us briefly elaborate on the moments of the random walk (see Appendix, Sect. E.2). The first moment or expectation value $\langle x_n \rangle$ is given by

$$\langle x_n \rangle = \sum_{\substack{k=-n \\ n\pm k\,\text{even}}}^{n} k \binom{n}{(n+k)/2} p^{(n+k)/2} q^{(n-k)/2} = \sum_{m=0}^{n} (2m-n) \binom{n}{m} p^m q^{n-m} = (2\langle m \rangle - n) = n(2p-1) . \tag{17.17}$$

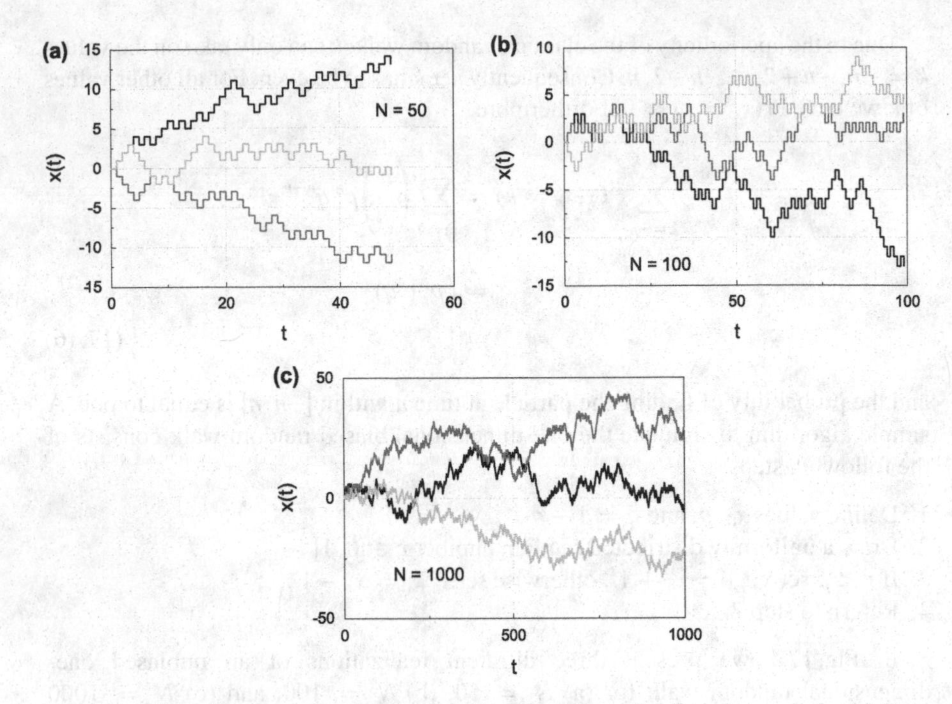

Fig. 17.1 Three different realizations of an unbiased one-dimensional random walk for **(a)** $N = 50$, **(b)** $N = 100$, and **(c)** $N = 1000$ time steps and different seeds

We now introduce a *bias* v such that

$$p = \frac{1}{2}(1 + v) \quad \text{and} \quad q = \frac{1}{2}(1 - v) , \tag{17.18}$$

and obtain

$$\langle x_n \rangle = nv . \tag{17.19}$$

We calculate the second moment $\langle x_n^2 \rangle$ using the above method and get:

$$\langle x_n^2 \rangle = n(1 - v^2) + n^2 v^2 . \tag{17.20}$$

The variance var (x_n) follows immediately:

$$\text{var}\,(x_n) = \langle x_n^2 \rangle - \langle x_n \rangle^2 = n(1 - v^2) . \tag{17.21}$$

We note the following: The expectation value $\langle x_n \rangle$ moves according to Eq. (17.19) with a uniform velocity defined by the bias $v = p - q$. In particular, for the unbiased random walk $v = 0$ and, thus, $\langle x_n \rangle = 0$ for all n. Furthermore, we observe that var (x_n) increases linearly with time n – a property we already noted for the WIENER process in Sect. 16.3 – and it is maximal for $v = 0$. For $v = \pm 1$, which describes a pure drift motion in the positive or negative x direction, the variance is equal to zero.

Recurrence

Let us briefly investigate the recurrence behavior of the random walk. We are interested in the probability $f_{00}^{(2\ell)}$ of a first return to the origin $x_0 = 0$ after 2ℓ steps. We already know that $f_{00}^{(2\ell)} \propto p^\ell q^\ell$ from our previous analysis. In the very first time step the particle moves either to $x_1 = 1$ or to $x_1 = -1$ and, consequently, within the following $2\ell - 2$ steps it must not cross or touch the line $x_k = 0$ and the particle has to terminate at position $x_{2\ell-1} = x_1$. Therefore, the walker performs $\ell - 1$ steps to the left and $\ell - 1$ steps to the right within these $2\ell - 2$ steps. The total number of possible paths N from x_1 to $x_{2\ell-1} = x_1$ is, thus, given by

$$N = \binom{2\ell - 2}{\ell - 1} . \tag{17.22}$$

Moreover, N may also be written as the sum of N_c paths which cross or touch the line $x_k = 0$ and N_{nc} paths which do not cross or touch the line $x_k = 0$, i.e.

$$N = N_c + N_{nc} . \tag{17.23}$$

Obviously, we are only interested in the paths which do not cross or touch the line $x_k = 0$. We employ the reflection principle to solve this problem. In general, the number of paths which go from $x_1 = i > 0$ to $x_{k+1} = j > 0$ within k-steps and cross the line $x_\ell = 0$ is equal to the total number of paths which go from $x_1 = -i$ to $x_{k+1} = j$, as is schematically illustrated in Fig. 17.2.

Let us regard the case $x_1 = 1$: The walker moved in the first step to the right. Hence, from the reflection principle we obtain that the number of paths from x_1 to $x_{2\ell-2} = x_1$ in $2\ell - 2$ steps which cross or touch the line $x_k = 0$ is given by the total

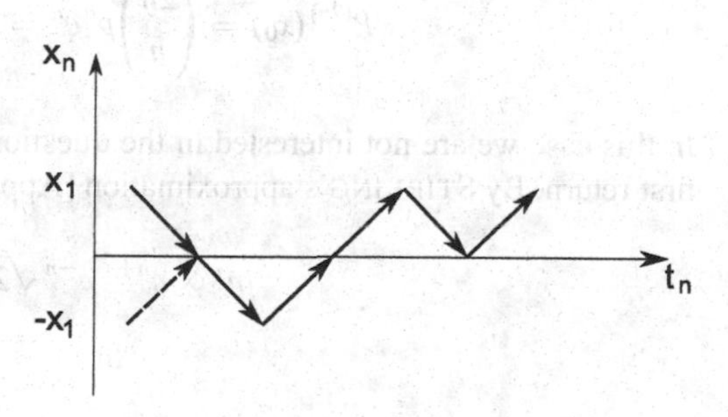

Fig. 17.2 Illustration of the reflection principle

number of paths from $-x_1$ to $x_{2\ell-2} = x_1$. Thus, N_c is determined by:

$$N_c = \binom{2\ell-2}{\ell} . \tag{17.24}$$

We note that in this picture, the walker moves ℓ steps to the right and $\ell - 2$ steps to the left. Hence, we obtain that the number of paths which do not cross or touch the line $x_k = 0$ is given by

$$2N_{nc} = 2(N - N_c) = \frac{1}{2\ell - 1}\binom{2\ell}{\ell} . \tag{17.25}$$

The prefactor 2 accounts for the fact that the walker can move in its first step either to $x_1 = -1$ or to $x_1 = 1$. Thus, the probability for the first return of the particle after 2ℓ steps is described by:

$$f_{00}^{(2\ell)} = \frac{1}{2\ell - 1}\binom{2\ell}{\ell} p^\ell q^\ell . \tag{17.26}$$

We calculate the recurrence probability according to Eq. (16.66) and this results in

$$\sum_{\ell=0}^{\infty} f_{00}^{(2\ell)} = \begin{cases} 1 & \text{for } p = q = \frac{1}{2} , \\ 2p & \text{for } p < q , \\ 2q & \text{for } p > q , \end{cases} \tag{17.27}$$

with the consequence that the one-dimensional random walk is only recurrent in the unbiased case $v = 0$.

Another possibility to demonstrate the recurrence of the unbiased one-dimensional random walk is provided by Eq. (16.69). The probability that a walker returns to $x_0 = 0$ after $2n$ steps is given by

$$P^{(2n)}(x_0) = \binom{2n}{n} p^n q^n = \frac{(2n)!}{n!n!}(pq)^n . \tag{17.28}$$

In this case we are not interested in the question whether or not it is the particle's first return. By STIRLING's approximation [Appendix, Eq. (E.20)]

$$n! \propto n^{n+\frac{1}{2}} e^{-n} \sqrt{2\pi} , \tag{17.29}$$

and we obtain for $P^{(2n)}(x_0)$:

$$P^{(2n)}(x_0) \propto \frac{(4pq)^n}{\sqrt{n\pi}} \,. \tag{17.30}$$

We assume $p \leq 1/2$ and since $pq = p(1-p) \leq 1/4$ one gets

$$\sum_{n=0}^{\infty} P^{(2n)}(x_0) \to \infty \quad \text{only for} \quad p = q = \frac{1}{2} \,. \tag{17.31}$$

The same argument holds for $p > 1/2$ since we can write $pq = (1-q)q \leq 1/4$. According to Eq. (16.69) this means that the process is recurrent only for $p = q$, in accordance with our previous result (17.27), and transient otherwise. We note that this agrees also with the more physical picture of an external force inducing a bias or *drift* velocity $v \neq 0$.

It should be noted that the unbiased random walk in two dimensions is also recurrent while it can be proved to be transient in higher dimensions. For instance, the recurrence probability is approximately 0.34[4] in 3D.

17.3 The WIENER Process and Brownian Motion

It is the purpose of this section to demonstrate that the WIENER process is the scaling limit of the random walk. Moreover, we discuss briefly the LANGEVIN equation and derive the diffusion equation.

As a starting point we consider the one-dimensional unbiased random walk on an equally spaced grid according to Eq. (17.8) and time instances given by Eq. (17.7). We denote the stochastic process by $X_n = X(t_n)$ and it is described by

$$X_n = \sum_{i=1}^{n} \xi_i \Delta x \,, \tag{17.32}$$

where $\xi \in \{-1, 1\}$ together with $X_0 = 0$. Since we regard the unbiased case $\Pr(\xi_i = \pm 1) = 1/2$, $\langle \xi_i \rangle = 0$, and $\mathrm{var}\,(\xi_i) = 1$. This is equivalent to

$$\langle X_n \rangle = 0 \quad \text{and} \quad \mathrm{var}\,(X_n) = n\Delta x^2 \,, \tag{17.33}$$

[4]This is one of PÓLYA's random walk constants [14–16].

as we already demonstrated in the previous section, Eq. (17.21). The variance $\operatorname{var}(X_n)$ can be reformulated as

$$\operatorname{var}(X_n) = t_n \frac{\Delta x^2}{\Delta t}, \tag{17.34}$$

using the definition $t_n \equiv n\Delta t$. The simultaneous limit $\Delta t, \Delta x \to 0$ is now performed in such a way that

$$\lim_{\substack{\Delta t \to 0 \\ \Delta x \to 0}} \frac{\Delta x^2}{\Delta t} = D = \text{const}, \tag{17.35}$$

with D the diffusion coefficient. This limit is known as the *continuous limit* and it will be denoted by the operator $\mathscr{L}$. Hence, in the continuous limit Eq. (17.34) results in

$$\mathscr{L}\left[\operatorname{var}(X_n)\right] = Dt, \tag{17.36}$$

where we renamed $t_n \equiv t$. We also note that the limit $\Delta t \to 0$ for constant t is equivalent to $n \to \infty$ and we obtain in accordance with the central limit theorem (see Appendix, Sect. E.8):

$$\mathscr{L}(X_n) \to W_t \sim \mathscr{N}(0, Dt). \tag{17.37}$$

Here $\mathscr{N}(0, Dt)$ denotes the normal distribution of mean zero and variance Dt, Appendix Eq. (E.43). Furthermore, the symbol W_t was introduced to represent the WIENER process and the symbol '$\sim$' stands, within this context, for the notion *follows the distribution*. If W_t describes a WIENER process it is necessary to prove that W_t has independent increments $W_{t_2} - W_{t_1}$ which follow, according to Sect. 16.3, a normal distribution with mean zero and a variance proportional to $t_2 - t_1$. This is demonstrated quite easily: We learn from our discussion of the random walk that

$$X_n - X_m = \sum_{i=1}^{n} \xi_i - \sum_{i=1}^{m} \xi_i = \sum_{i=m+1}^{n} \xi_i, \tag{17.38}$$

and, therefore, $X_n - X_m$ and $X_m - X_k$ are clearly independent for $n > m > k$ and it follows that also $W_t - W_s$ and $W_s - W_u$ are independent. Furthermore, we have

$$X_n - X_m \stackrel{d}{=} X_{n-m}, \tag{17.39}$$

where the symbol '$\stackrel{d}{=}$' stands for the notion *to follow the same distribution* or *to be distributionally equivalent*. Therefore, in the limit $\mathscr{L}$ for $t > s$

$$W_t - W_s \stackrel{d}{=} W_{t-s} \sim \mathscr{N}\left[0, D(t-s)\right] , \tag{17.40}$$

which completes the proof. We note that the particular case $D = 1$ is commonly referred to as the *standard* WIENER *process*. We remark that in many cases the terms WIENER process and Brownian motion are used as synonyms for a stochastic process satisfying the above properties. However, strictly speaking, the stochastic process is the WIENER process while Brownian motion is the physical phenomenon which can be described by the WIENER process.

If we suppose that $p \neq q$ then

$$\mathscr{L}\left(\langle X_n \rangle\right) = \nu t , \tag{17.41}$$

with the drift constant ν, describes a WIENER process with a drift term

$$\mathscr{L}(X_n) \to \tilde{W}_t = \nu t + W_t . \tag{17.42}$$

This process behaves like W_t with the only difference that it fluctuates around mean νt instead of mean zero. Note that for $\nu > 0$ the mean $\langle \tilde{W}_t \rangle$ increases, while for $\nu < 0$ it decreases with time t.

Another interesting property of the WIENER process is its *self-similarity*. In particular, we have the property that for $\alpha > 0$

$$W_t \stackrel{d}{=} \alpha^{-\frac{1}{2}} W_{\alpha t} , \tag{17.43}$$

with the consequence that it is completely sufficient to study the properties of the WIENER process for $t \in [0, 1]$ to know its properties for arbitrary time intervals. Relation (17.43) follows from the fact that $W_t \sim \mathscr{N}(0, Dt)$.

Furthermore, *white noise*, $\eta(t)$, is defined as the formal derivative of the WIENER process W_t with respect to time. We give its most important properties without going into details[5]:

$$\langle \eta(t) \rangle = 0, \quad \text{and} \quad \langle \eta(t)\eta(s) \rangle = \delta(t-s) . \tag{17.44}$$

[5]In fact, it can be shown that W_t is non-differentiable with probability one. This is the reason why it is defined as the *formal* derivative of W_t. Let $\varphi(t)$ be a test function and $f(t)$ an arbitrary function which does not need to be differentiable with respect to t. Then the formal derivative $\dot{f}(t)$ is defined by

$$\int_0^\infty \mathrm{d}t \dot{f}(t)\varphi(t) = -\int_0^\infty \mathrm{d}t f(t)\dot{\varphi}(t) .$$

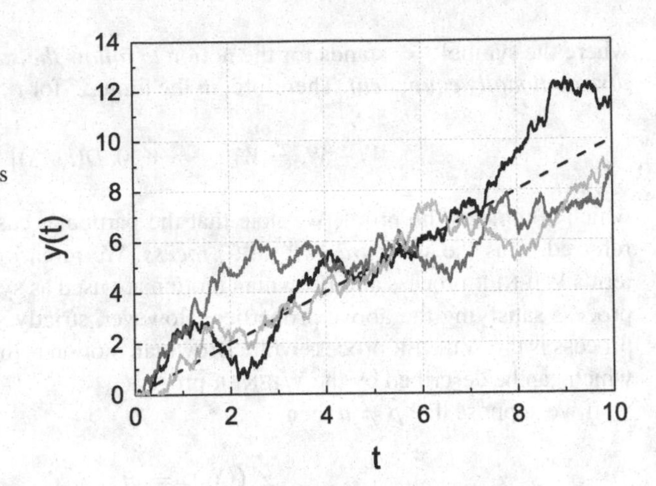

Fig. 17.3 Three different realizations of the standard WIENER process with drift $\nu = 1$ according to Eq. (17.42). The expectation value $\langle x \rangle = \nu t$ of the process is presented as a *dashed line*

White noise is referred to as *Gaussian white noise* if $\eta(t)$ follows a normal distribution.

Figure 17.3 presents three different realizations of the standard WIENER process with drift according to Eq. (17.42). The curves in this figure were generated using the procedure outlined in Sect. 16.3 in connection with Fig. 16.2.

Let us derive the diffusion equation from the random walk model. The probability $\Pr(x, t)$ of finding the particle at time t at position x is expressed by

$$\begin{aligned}
\Pr(x, t) &= \Pr(x, t - \Delta t)r + \Pr(x - \Delta x, t - \Delta t)p \\
&\quad + \Pr(x + \Delta x, t - \Delta t)q \\
&= \Pr(x, t - \Delta t)(1 - p - q) + \Pr(x - \Delta x, t - \Delta t)p \\
&\quad + \Pr(x + \Delta x, t - \Delta t)q\,,
\end{aligned} \tag{17.45}$$

where we made use of relation (17.11). The interpretation of this equation is straight-forward: The probability to find the particle at the position-time point (x, t) is the sum of three terms. The first term describes the probability that the particle arrived already at position x in the previous time step $t - \Delta t$ and that it will stay there during the next time step. The remaining two terms describe the probability that the particle arrived at position $x - \Delta x$ ($x + \Delta x$) in the previous time step and that it will move one step to the right (left) in the next time step. Each particular term is now expanded into a TAYLOR series up to order $\mathcal{O}(\Delta x^2)$ and $\mathcal{O}(\Delta t)$, respectively. This requires the transition from a discrete to a continuous state space and, consequently, the probabilities $\Pr(\cdot)$ are replaced by pdfs $p(\cdot)$. We get

$$\begin{aligned}
p(x, t) &= (1 - p - q)\left[p(x, t) - \Delta t \frac{\partial p(x, t)}{\partial t}\right] \\
&\quad + p\Big[p(x, t) - \Delta t \frac{\partial p(x, t)}{\partial t} - \Delta x \frac{\partial p(x, t)}{\partial x}
\end{aligned}$$

$$+\frac{1}{2}\Delta x^2 \frac{\partial^2 p(x,t)}{\partial x^2}\bigg]$$

$$+q\bigg[p(x,t) - \Delta t\frac{\partial p(x,t)}{\partial t} + \Delta x\frac{\partial p(x,t)}{\partial x}$$

$$+\frac{1}{2}\Delta x^2 \frac{\partial^2 p(x,t)}{\partial x^2}\bigg] , \tag{17.46}$$

and furthermore:

$$\frac{\partial p(x,t)}{\partial t} = -\frac{(p-q)\Delta x}{\Delta t}\frac{\partial p(x,t)}{\partial x} + \frac{(p+q)\Delta x^2}{2\Delta t}\frac{\partial^2 p(x,t)}{\partial x^2} . \tag{17.47}$$

We draw the continuous limit and define the drift constant

$$\nu = \mathscr{L}\left[(p-q)\frac{\Delta x}{\Delta t}\right] = \lim_{\substack{\Delta t\to 0\\ \Delta x\to 0}} \frac{(q-p)}{\Delta t}\Delta x , \tag{17.48}$$

the diffusion constant

$$D = \mathscr{L}\left[(p+q)\frac{\Delta x^2}{2\Delta t}\right] = \lim_{\substack{\Delta t\to 0\\ \Delta x\to 0}} \frac{(p+q)}{2\Delta t}\Delta x^2 , \tag{17.49}$$

and arrive at the one-dimensional diffusion equation with drift term:

$$\frac{\partial p(x,t)}{\partial t} = \nu\frac{\partial p(x,t)}{\partial x} + D\frac{\partial^2 p(x,t)}{\partial x^2} . \tag{17.50}$$

This equation is referred to as a FOKKER-PLANCK equation [17]. In the specific case $p = q$ the drift term disappears and we obtain, as expected, the classical diffusion equation

$$\frac{\partial}{\partial t}p(x,t) = D\frac{\partial^2}{\partial x^2}p(x,t) , \tag{17.51}$$

which we solved already numerically in Chaps. 9 and 11. It follows from this discussion that the position of a diffusing particle can be described as a stochastic process where, in the continuous limit, the jump-lengths follow a normal distribution. In addition, we know from our discussion of continuous-time MARKOV-chains in Sect. 16.5, that the waiting times between two successive jumps will certainly follow an exponential distribution. These insights will serve as a starting point in the discussion of general diffusion models in Sect. 17.4. Moreover, we note that the anisotropy of the jump-length distribution is a model for the presence of an external field which manifests itself in a drift term.

An alternative approach to the formal description of Brownian motion goes back to LANGEVIN. He considered the classical equation of motion of a particle in a fluid which reads

$$\dot{v} = -\beta v \,, \tag{17.52}$$

where β denotes the friction coefficient and we set the particle's mass m equal to one. LANGEVIN argued that this equation may only be valid for the average motion of the particle which corresponds to the long time behavior of the motion of massive particles. However, if the particle is not heavy at all its trajectory can be highly affected by collisions with solvent's molecules. He supposed that a reasonable generalization of Eq. (17.52) should be of the form [18]

$$\dot{v} = -\beta v + F(t) \,, \tag{17.53}$$

where $F(t)$ is a *random force*. In particular, $F(t)$ is a stochastic process which satisfies

$$\langle F(t) \rangle = 0 \quad \text{and} \quad \langle F(t)F(s) \rangle = A\delta(t - t') \,, \tag{17.54}$$

where A is a constant and we obtain

$$F(t) \stackrel{d}{=} \sqrt{A}\eta(t) \,. \tag{17.55}$$

Equation (17.53) is referred to as the LANGEVIN equation and it is *the* prototype *stochastic differential equation*. Based on the definition of white noise $\eta(t)$ the LANGEVIN equation can be rewritten:

$$\mathrm{d}v = -\beta v \mathrm{d}t + \sqrt{A}\mathrm{d}W_t \,. \tag{17.56}$$

The solution of the LANGEVIN equation describes a stochastic process referred to as the ORNSTEIN-UHLENBECK *process* [19]. This process is essentially the only stochastic process which is stationary, Gaussian and Markovian. Its master equation is a FOKKER-PLANCK equation of the form [17]

$$\frac{\partial}{\partial t} p(v,t) = \beta \frac{\partial}{\partial v} v p(v,t) + \frac{A}{2} \frac{\partial^2}{\partial v^2} p(v,t) \,, \tag{17.57}$$

where $p(v,t)$ is the pdf of the ORNSTEIN-UHLENBECK process. If the initial velocity v_0 is given then the pdf $p(v,t)$ can be proved to be

$$p(v,t) = \frac{\sqrt{\beta}}{\sqrt{\pi A \left(1 - e^{-2\beta t}\right)}} \exp\left[-\frac{\beta \left(v - v_0 e^{-\beta t}\right)^2}{A \left(1 - e^{-2\beta t}\right)} \right] . \tag{17.58}$$

It is possible to solve the LANGEVIN equation (17.53) analytically with the result:

$$v(t) = v_0 \exp(-\beta t) + \sqrt{A} \int_0^t \mathrm{d}t' \eta(t') \exp\left[-\beta(t-t')\right] . \quad (17.59)$$

We write in particular

$$v(t_{n+1}) = v(t_n) \exp(-\beta \Delta t) + Z_n , \quad (17.60)$$

with Z_n defined as:

$$Z_n = \sqrt{A} \int_0^{\Delta t} \mathrm{d}t' \eta(t_n + t') \exp\left[-\beta(\Delta t - t')\right] . \quad (17.61)$$

Since $\eta(t)$ was assumed to be Gaussian white noise, Z_n can be proved to be described by

$$Z_n \sim \mathcal{N}\left\{0, \frac{A}{2\beta}\left[1 - \exp(-2\beta\Delta t)\right]\right\} , \quad (17.62)$$

which offers a very convenient way to simulate the ORNSTEIN-UHLENBECK process. This particular formulation of Brownian motion allows to model this process by sampling changes in the velocity Z_n from the normal distribution with mean zero and the variance given in Eq. (17.62). The walker's position $x(t)$ can then be obtained by approximating the velocity $v = \dot{x}$ with the help of finite difference derivatives, as described in Chap. 2. In conclusion we remark that although the LANGEVIN equation was introduced in a heuristic manner, it represents a very useful tool due to its rather simple interpretation.

Figure 17.4 presents three different realizations of the ORNSTEIN-UHLENBECK process based on three different initial velocities v_0. The corresponding random trajectories $x(t)$ of the Brownian particle are illustrated in Fig. 17.5.

17.4 Generalized Diffusion Models

We formulate now a very general approach to diffusive behavior which is based on continuous random variables. We start with the introduction of the pdf $\Lambda(x, t)$. Its purpose is to describe the event that a particle arrives at time t at position x. It can be expressed as [20, 21]

$$\Lambda(x, t) = \int \mathrm{d}x \int_0^t \mathrm{d}t' \psi(x, t; x', t') \Lambda(x', t') , \quad (17.63)$$

where $\psi(x, t; x', t')$ is the *jump pdf*. We offer the following interpretation: $\psi(x, t; x', t')$ describes the probability for an event that a particle which arrived

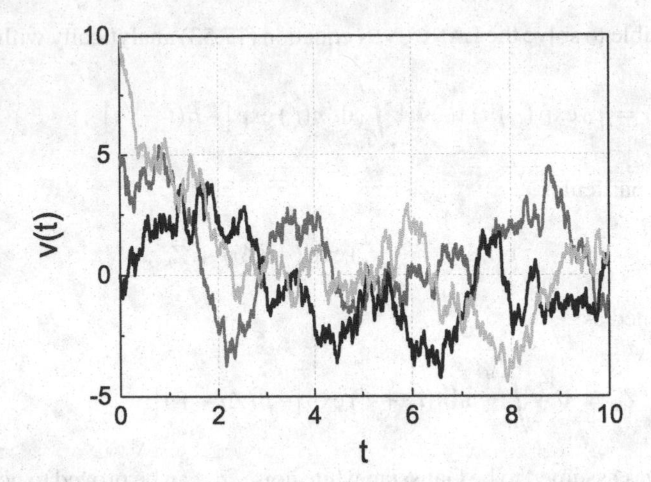

Fig. 17.4 Three different realizations of the ORNSTEIN-UHLENBECK process $v(t)$ vs t. For this simulation we chose $\beta = 1$, $A = 5$, $\mathrm{d}t = 10^{-2}$ and $N = 10^3$ time steps. Furthermore, we chose three different initial velocities, i.e. $v_0 = 0$ (*black*), $v_0 = 5$ (*gray*) and $v_0 = 10$ (*light gray*)

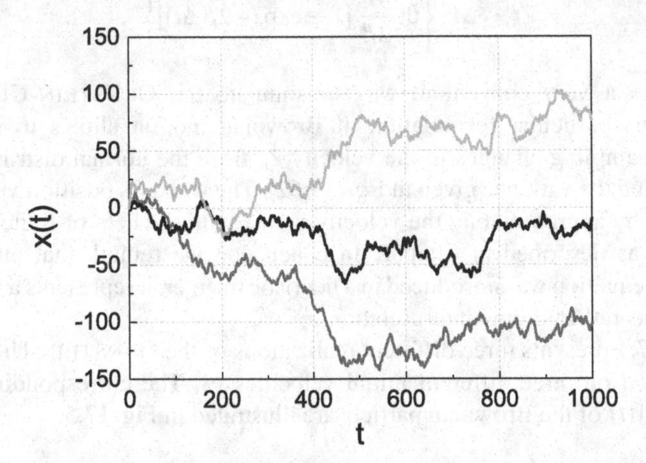

Fig. 17.5 Random trajectories $x(t)$ vs t of the Brownian particle which correspond to the velocities $v(t)$ illustrated in Fig. 17.4 with initial position $x_0 = 0$. Note that we used for this figure $N = 10^5$ time steps

at time t' at position x' – with pdf $\Lambda(x', t')$ – waited at position x' until the time t was reached and then jumped within an infinitesimal time interval from position x' to x. If we regard a space and time homogeneous process then $\psi(x, t; x', t')$ is replaced by $\psi(x - x', t - t')$. This allows the introduction of a jump length pdf $p(x)$ and of a waiting time pdf $q(t)$. They are related to the jump pdf by

$$p(x) = \int_0^\infty \mathrm{d}t' \, \psi(x, t') \quad \text{and} \quad q(t) = \int_{-\infty}^\infty \mathrm{d}x' \, \psi(x', t) \, . \tag{17.64}$$

If the jump length pdf and the waiting time pdf are conditionally independent one can simply write $\psi(x,t) = p(x)q(t)$. The probability $\varphi(x,t)$ of finding a particle at position x at time t is, furthermore, given by

$$\varphi(x,t) = \int_0^t dt' \Lambda(x,t')\Psi(t-t') , \tag{17.65}$$

where $\Psi(t)$ is the probability, that a particle stayed at least for a time interval t at the same position, i.e.

$$\Psi(t) = 1 - \int_0^t dt' q(t-t') . \tag{17.66}$$

Finally, the jump length variance σ^2 and the characteristic waiting time τ are given by

$$\sigma^2 = \int_{-\infty}^{\infty} dx' x'^2 p(x') \quad \text{and} \quad \tau = \int_0^{\infty} dt' t' q(t') . \tag{17.67}$$

We conclude from our discussion of the WIENER process that for Brownian motion the jump length pdf is a Gaussian and the waiting time pdf is an exponential distribution:

$$p(x) = \frac{1}{\sqrt{2\pi\sigma^2}} \exp\left(-\frac{x^2}{2\sigma^2}\right) \quad \text{and} \quad q(t) = \frac{1}{\tau} \exp\left(-\frac{t}{\tau}\right) . \tag{17.68}$$

The characteristic function [Appendix Eq. (E.54)] of the waiting time pdf $q(t)$ is given by

$$\hat{q}(s) = \int_0^{\infty} dt\, q(t) e^{-st} = \frac{1}{s\tau + 1} , \tag{17.69}$$

and we find for the jump length pdf $p(x)$:

$$\hat{p}(k) = \int dx e^{-ikx} p(x) = \exp\left(-\sigma^2 k^2/2\right) . \tag{17.70}$$

For $x, t \to \infty$, i.e. $k, s \to 0$, the characteristic functions $\hat{q}(s)$ and $\hat{p}(k)$ develop the asymptotic behavior

$$\lim_{s \to 0} \frac{1}{1 + s\tau} \approx 1 - s\tau + \mathcal{O}(s^2) , \tag{17.71}$$

and

$$\lim_{k \to 0} \exp\left(-\sigma^2 k^2/2\right) \approx 1 - \sigma^2 k^2/2 + \mathscr{O}(k^4) . \tag{17.72}$$

In fact, it can be shown that any pair of jump length and waiting time pdfs lead in first order to the same asymptotic behavior, namely $\mathscr{O}(\tau)$ and $\mathscr{O}(\sigma^2)$, as long as the moments τ and σ^2 exist.

However, there is a variety of processes which cannot be accounted for within the basic framework of Brownian motion. Such processes are described within the concept of anomalous diffusion [9, 20]. Examples are, for instance, the foraging behavior of spider monkeys, particle trajectories in a rotating flow, diffusion of proteins across cell membranes, diffusion of tracers in polymer-like breakable micelles, the traveling behavior of humans, charge carrier transport in disordered organic molecules, etc.

We concentrate now on two particular models of anomalous diffusion. The first model can, from a qualitative point of view, be characterized as a diffusion process which consists of small clustering jumps which are intersected by very long *flights*. Such behavior is, for instance, encountered in the context of human travel behavior, Fig. 17.6 [22], charge carrier transport in disordered solids, etc. The incorporation of these long jumps on a stochastic level is referred to as LÉVY *flight* [10]. The second model, which is referred to as the *fractal time random walk* incorporates

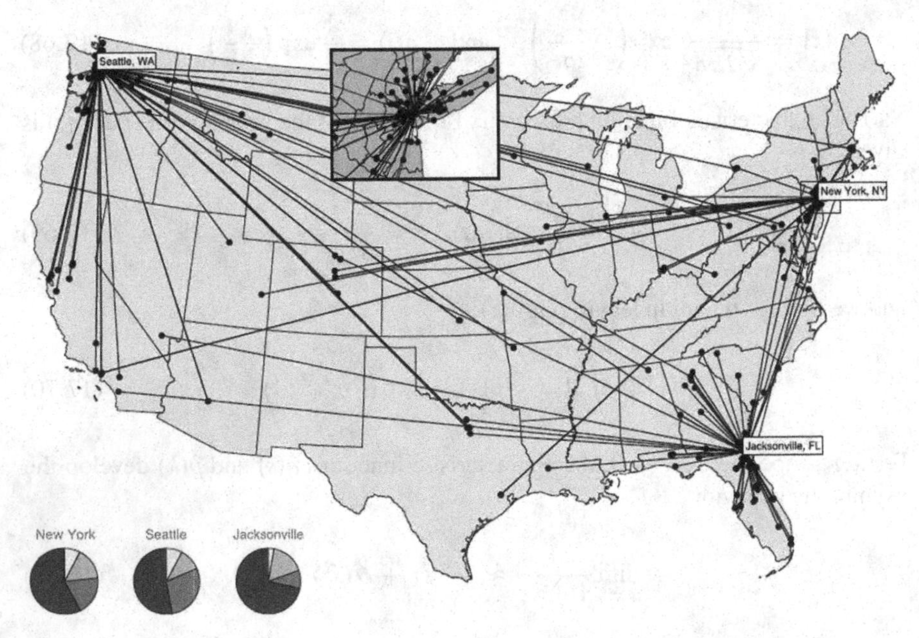

Fig. 17.6 Traveling behavior of humans (Adapted from [22]. Copyright © 2006, Rights Managed by Nature Publishing Group)

anomalously long waiting times between two successive jumps. In particular, these long waiting times account for non-Markovian effects which could be due to, for instance, trapping processes of charge carriers in disordered solids. It has to be emphasized at this point that the resulting diffusion models are still linear in the pdf $\varphi(x, t)$. The inclusion of non-linear effects will not be discussed here, however, can be achieved within the framework of *non-extensive thermodynamics* [23].

Let us start with LÉVY flights. In this case one modifies the asymptotic behavior of the characteristic function of the jump length pdf according to

$$\hat{p}(k) \propto 1 - (\sigma |k|)^{\alpha} , \tag{17.73}$$

where $\alpha \in (0, 2]$. We recognize that this is the asymptotic behavior $|k| \to 0$ of the characteristic function of a symmetric LÉVY α-stable distribution [19] following Appendix Eq. (E.69). In the limit $\alpha \to 2$ normal, Gaussian behavior is recovered. According to Appendix Eq. (E.70) the characteristic function (17.73) corresponds to a jump length pdf:

$$p(x) \propto |x|^{-\alpha-1} \qquad \text{for} \qquad |x| \to \infty . \tag{17.74}$$

It is commonly referred to as a *fat-tailed jump length pdf* because of its asymptotic behavior.

A LÉVY flight is, in principle, a random walk where the length of the jumps at discrete time instances t_n follow the pdf (17.74). In the continuous time limit, the waiting times are distributed exponentially as was illustrated in Sect. 16.5. It has to be noted that in such a case the jump length variance diverges, i.e. $\Sigma^2 \to \infty$. Consequently, LÉVY α-stable distributions are *not* subject to the central limit theorem (see Appendix, Sect. E.8). In particular, the distance from the origin after some finite time t follows a LÉVY α-stable distribution. Moreover, we note that if $0 < \alpha < 1$ even the mean jump length $\langle x \rangle$ diverges. A detailed mathematical analysis proves, that Lévy flights result in a diffusion equation of the form

$$\frac{\partial}{\partial t} p(x, t) = D_{\alpha} \mathscr{D}^{\alpha}_{|x|} p(x, t) , \tag{17.75}$$

where D_{α} is the fractional diffusion coefficient of dimension length$^{\alpha}\times$ time^{-1} and $\mathscr{D}^{\alpha}_{|x|}$ is the symmetric RIESZ fractional derivative operator of order $\alpha \in (1, 2)$[6]:

$$\mathscr{D}^{\alpha}_{|x|} f(x) = \frac{1}{2\Gamma(2-\alpha)\cos\left(\frac{\alpha\pi}{2}\right)} \int \mathrm{d}x' \frac{f''(x')}{|x - x'|^{\alpha-1}} , \tag{17.76}$$

where $f''(x)$ is the second spatial derivative of f.

[6] A short introduction to fractional derivatives and integrals can be found in Appendix G.

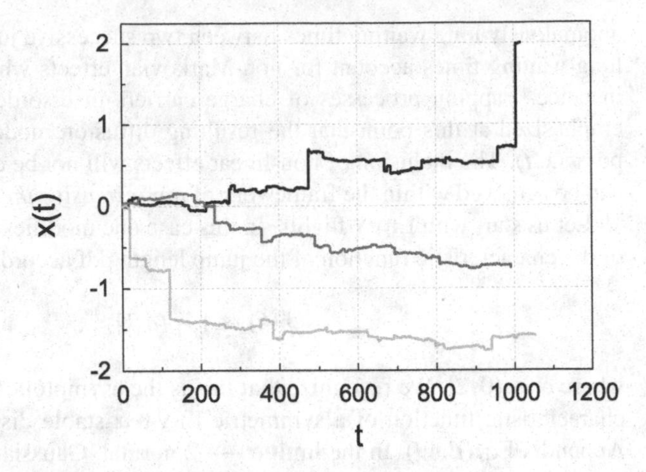

Fig. 17.7 Three different realizations of the one dimensional Lévy flight. The parameters are $\ell = 0.001$, $\alpha = 1.3$ and we performed $N = 1000$ time steps

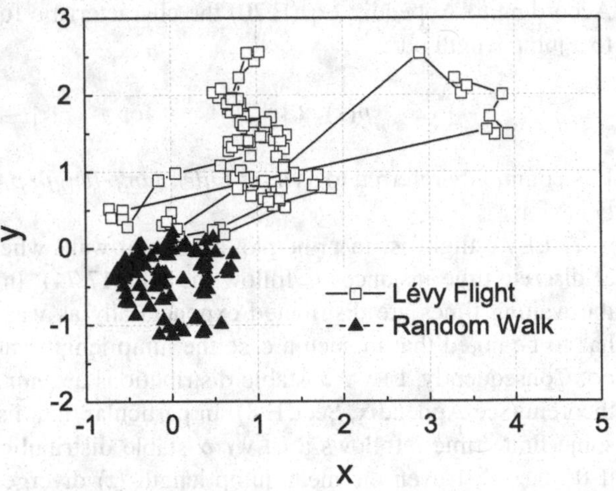

Fig. 17.8 Comparison between the two-dimensional WIENER process (*solid up-triangles*) and the two-dimensional LÉVY flight (*open squares*) for $\alpha = 1.3$. The minimal flight length of the LÉVY flight as well as the jump length variance of the WIENER process were set $\ell = \Sigma^2 = 0.1$ and we performed $N = 100$ time steps

Figure 17.7 illustrates a one-dimensional LÉVY flight and Fig. 17.8 presents a comparison between a two-dimensional LÉVY flight and a two-dimensional WIENER process. These figures were generated by sampling an exponential distribution with mean $\langle t \rangle = 1$ for the waiting times. On the other hand a jump length pdf

$$p(x) = \alpha \ell^{\alpha} \frac{\Theta(x - \ell)}{x^{\alpha+1}}, \quad x > 0. \tag{17.77}$$

was sampled for the jump length of the LÉVY flight. Here α is referred to as the LÉVY index, $\Theta(\cdot)$ denotes the HEAVISIDE Θ function and $\ell > 0$ is the *minimal flight length*. We introduced this particular form of the pdf because it can rather easily be sampled with the help of the inverse transformation method – Sect. 13.2 – and it obeys the asymptotic behavior, Eq. (17.74). Moreover, it can be proved that it gives the correct behavior in the limit $\ell \to 0$. Finally, the direction of the jump has to be sampled in an additional step. Figure 17.8 is particularly instructive because the different physics described by these two models becomes immediately apparent.

Let us turn our attention to the second scenario, the fractal time random walk. In this case the asymptotic behavior of the waiting time pdf is modified according to

$$\hat{q}(s) \propto 1 - (Ts)^{\beta} , \tag{17.78}$$

where $\beta \in (0, 1]$ and for $\beta \to 1$ regular behavior, an exponentially distributed waiting time, is recovered. A pdf of such a form is commonly referred to as a *fat-tailed waiting time pdf*. After an inverse LAPLACE transform we obtain

$$q(t) \propto t^{-\beta-1} \quad \text{for} \quad t \to \infty . \tag{17.79}$$

We note that in this case the mean waiting time $T = \langle t \rangle$ diverges for $\beta < 1$. This clearly indicates a non-Markovian time evolution since we demonstrated in Sect. 16.5 that every Markovian discrete time process converges in the continuous time limit to a process with exponentially distributed waiting times. Again, the ansatz

$$q(t) = \beta \tau^{\beta} \frac{\Theta(t-\tau)}{t^{\beta+1}} , \tag{17.80}$$

is employed, where $\tau > 0$ is the *minimal waiting time*. The process is essentially a random walk with waiting times distributed according to Eq. (17.80), i.e. the jump length Δx is constant. In the continuous space limit $\Delta x \to 0$ the jump lengths follow a Gaussian, as in the case of a regular random walk. A detailed analysis proves that in the limit $\tau \to 0$ the corresponding diffusion equation is given by

$${}^{C}D_t^{\beta} p(x,t) = D_{\beta} \frac{\partial^2}{\partial x^2} p(x,t) , \tag{17.81}$$

where the diffusion constant D_{β} is of dimension length$^2\times$ time$^{-\beta}$. Here, ${}^{C}D_t^{\beta}$ is the CAPUTO fractional time derivative of order $\beta \in (0, 1)$ (see Appendix G). It is of the form

$${}^{C}D_t^{\beta} f(t) = \frac{1}{\Gamma(1-\beta)} \int_0^t \mathrm{d}t' \frac{\dot{f}(t')}{(t-t')^{\beta}} . \tag{17.82}$$

It follows from the properties of fractional derivatives that an alternative form of Eq. (17.81) can be found, namely

$$\frac{\partial}{\partial t} p(x,t) = D_{\beta} \frac{\partial^2}{\partial x^2} D_t^{\beta} p(x,t) , \tag{17.83}$$

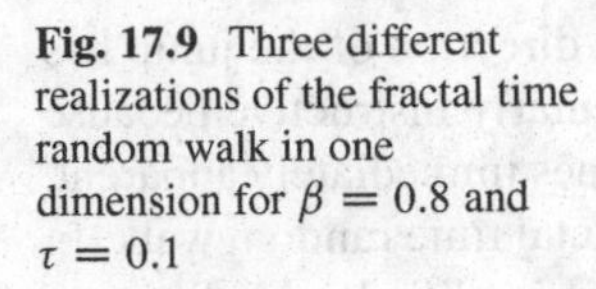
Fig. 17.9 Three different realizations of the fractal time random walk in one dimension for $\beta = 0.8$ and $\tau = 0.1$

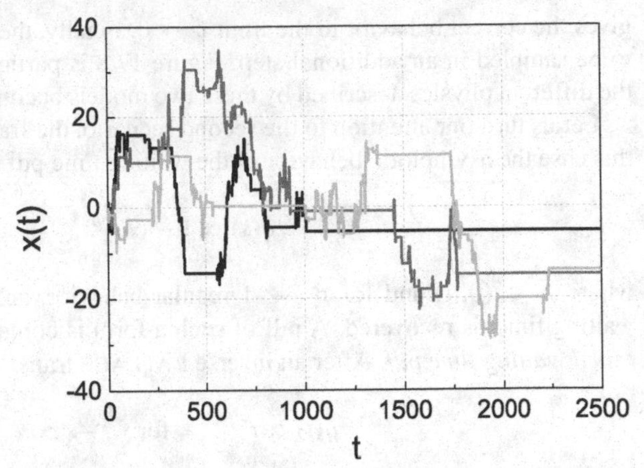

where D_t^β is the RIEMANN-LIOUVILLE fractional derivative of order β (see Appendix G).

Figure 17.9 presents three different realizations of the fractal time random walk. The waiting times were sampled from the pdf (17.80) with the help of the inverse transformation method – Sect. 13.2 – and the jump lengths were sampled from a normal distribution with jump length variance $\Sigma^2 = 1$.

It is a straight-forward task to combine fractal time random walks and LÉVY flights to so called *fractal time* LÉVY *flights*. The resulting diffusion equation can be written as

$$ {}^C D_t^\beta p(x,t) = D_{\alpha\beta} \mathscr{D}_{|x|}^\alpha p(x,t) \,, \tag{17.84} $$

where the diffusion constant $D_{\alpha\beta}$ has units length$^\alpha \times$ time$^{-\beta}$ and ${}^C D_t^\beta$ and $\mathscr{D}_{|x|}^\alpha$ are the fractional CAPUTO and RIESZ derivatives, respectively.

Figure 17.10 illustrates three different realizations of such a diffusion process. The waiting times were sampled from the pdf (17.80) where we set $\tau = 0.1$ and $\beta = 0.8$. The jump lengths were sampled from the pdf (17.77) with $\alpha = 1.3$ and $\ell = 0.01$. Finally, the direction of the jump was sampled in an additional step.

We close this chapter with a short discussion: The description of diffusion processes with the help of stochastics proofed to be one of the most powerful methods in modern theoretical physics. Within this chapter we discussed several different paths toward a description of Brownian motion, namely the random walk, the WIENER process, and the LANGEVIN equation, as well as models which describe phenomena beyond Brownian motion. It has to be emphasized that the field of anomalous diffusion in general is still developing rapidly, however, its importance for the description of various phenomena in science is already impressive. We refer the interested reader to the excellent review articles by R. METZLER and J. KLAFTER on anomalous diffusion [20, 21].

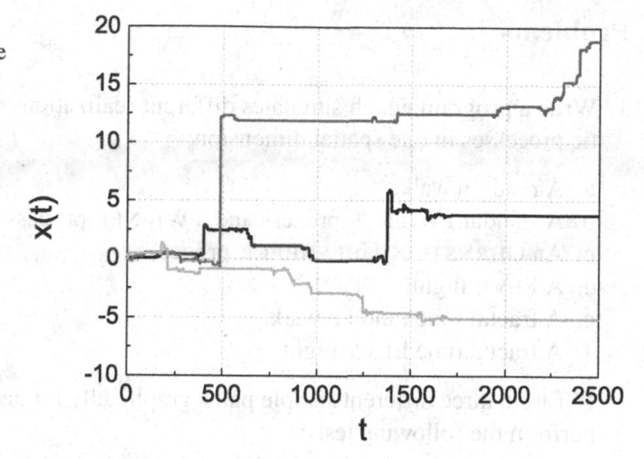

Fig. 17.10 Three possible realizations of the fractal time LÉVY flight in one dimension. The parameters are $\tau = 0.1$, $\beta = 0.8$, $\ell = 0.01$ and $\alpha = 1.3$

Summary

The random walk, a classical example of MARKOV-chains, was used to open the door to the realm of diffusion theory. Random walks have been used for a long time to simulate Brownian motion and related problems. From a theoretical point of view random walks were described by the scaling limit of the WIENER process. The biased WIENER process was then used to demonstrate that the FOKKER-PLANCK equation followed in the limit of a continuous state space, as the classical diffusion equation followed from the unbiased WIENER process in the same limit. Brownian motion was also the basis for the rather heuristic introduction of the stochastic differential equation by LANGEVIN. A direct consequence of this equation was the ORNSTEIN-UHLENBECK process with its master equation, the FOKKER-PLANCK equation. It was the only stationary, Gaussian, and Markovian process in this class of stochastic diffusion processes. An extension of these processes was then possible by the introduction of a jump pdf which in turn allowed to define a jump length pdf and a waiting time pdf. These two pdfs resulted in a more general description of diffusion processes in a space and time homogeneous environment. Furthermore, the observation that many diffusive processes (not only in physics) cannot be understood within the framework of 'classical' Brownian motion resulted in the introduction of LÉVY flights. This was particularly motivated by the need for a process whose jump-length variance diverges which enabled, for instance the simulation of human travel behavior. In the very last step the fractal time random walk was introduced. It was characterized by a specific form of the waiting time pdf which made it possible to describe on a stochastic level anomalously long waiting times between two consecutive jumps. Such behavior can, for instance, be observed by trapping phenomena in solids. The combination of both extensions resulted in the fractal time LÉVY flight.

Problems

1. Write a program which simulates different realizations of the following stochastic processes in one spatial dimension:

   a. A random walk.
   b. A standard WIENER process and a WIENER process with drift.
   c. An OHRNSTEIN-UHLENBECK process.
   d. A LÉVY flight.
   e. A fractal time random walk.
   f. A fractal time LÉVY flight.

   Illustrate three different sample paths graphically for each process. Furthermore, perform the following tests:

   a. Calculate the expectation value $\langle x_n \rangle$ and the variance var (x_n) of the random walk numerically by restarting the process several times with different seeds.
   b. In a similar fashion, calculate numerically $\langle W_t \rangle$ and var (W_t).
   c. Try different parameters α, β for LÉVY flights and fractal time random walks.

2. Write a program which simulates the WIENER process in two dimensions. This can be achieved by drawing the jump length from a normal distribution and sampling the *jump angle*, i.e. the direction, in an additional step. Augment this program with LÉVY flight jump lengths pdfs.

References

1. Glicksman, M.E.: Diffusion in Solids. Wiley, New York (1999)
2. Cussler, E.L.: Diffusion. Cambridge University Press, Cambridge (2009)
3. Dubitzky, W., Wolkenhauer, O., Cho, K.H., Yokota, H. (eds.): Encyclopedia of Systems Biology, p. 1596. Springer, Berlin/Heidelberg (2013)
4. Tapiero, C.S.: Risk and Financial Management: Mathematical and Computational Methods. Wiley, New York (2004)
5. Harris, S.: An Introduction to the Theory of the Boltzmann Equation. Dover, Mineola (2004)
6. Ballentine, L.E.: Quantum Mechanics. World Scientific, Hackensack (1998)
7. Rudnik, J., Gaspari, G.: Elements of the Random Walk. Cambridge University Press, Cambridge (2010)
8. Ibe, O.C.: Elements of Random Walk and Diffusion Processes. Wiley, New York (2013)
9. Pekalski, A., Sznajd-Weron, K. (eds.): Anomalous Diffusion. Lecture Notes in Physics. Springer, Berlin/Heidelberg (1999)
10. Shlesinger, M.F., Zaslavsky, G.M., Frisch, U. (eds.): Lévy Flights and Related Topics in Physics. Lecture Notes in Physics. Springer, Berlin/Heidelberg (1995)
11. Norris, J.R.: Markov Chains. Cambridge Series in Statistical and Probabilistic. Cambridge University Press, Cambridge (1998)
12. Modica, G., Poggiolini, L.: A First Course in Probability and Markov Chains. Wiley, New York (2012)
13. Graham, C.: Markov Chains: Analytic and Monte Carlo Computations. Wiley, New York (2014)

14. Pólya, G.: Über eine Aufgabe der Wahrscheinlichkeitsrechnung betreffend die Irrfahrt im Straßennetz. Math. Ann. **84**, 149–160 (1921)
15. Finch, S.R. (ed.): Mathematical Constants, pp. 322–331. Cambridge University Press, Cambridge (2003)
16. Weisstein, E.W.: Pólya's random walk constants. http://mathworld.wolfram.com/PolyasRandomWalkConstants.html
17. Risken, H., Frank, T.: The Fokker-Planck Equation. Springer Series in Synergetics. Springer, Berlin/Heidelberg (1996)
18. Coffey, W.T., Kalmykov, Y.P.: The Langevin Equation, 3rd edn. World Scientific Series in Contemporary Chemical Physics: Volume 27. World Scientific, Hackensack (2012)
19. Breuer, H.P., Petruccione, F.: Open Quantum Systems, chap. 1. Clarendon Press, Oxford (2010)
20. Metzler, R., Klafter, J.: The random walk's guide to anomalous diffusion: a fractional dynamics approach. Phys. Rep. **339**, 1–77 (2000)
21. Metzler, R., Klafter, J.: The restaurant at the end of random walk: recent developments in the description of anomalous transport by fractional dynamics. J. Phys. A **37**, R161 (2004)
22. Brockmann, D., Hufnagl, L., Geisl, T.: The scaling laws of human travel. Nature (London) **439**, 462–465 (2006). doi:10.1038/nature04292
23. Tsallis, C.: Introduction to Nonextensive Statistical Mechanics. Springer, Berlin/Heidelberg (2009)

Chapter 18
MARKOV-Chain Monte Carlo and the POTTS Model

18.1 Introduction

This chapter discusses in more detail the concept of MARKOV-chain Monte Carlo techniques [1–4]. We already came across the METROPOLIS algorithm in Sect. 14.3, where the condition of *detailed balance* proved to be the crucial point of the method. The reason for imposing such a condition was explained in all required detail during our discussion of MARKOV-chains within Sect. 16.4. The ISING model, analyzed in Chap. 15, served as a first illustration of the applicability of MARKOV-chain Monte Carlo methods in physics.

Let us briefly summarize what we learned so far: We discussed several methods to sample pseudo random numbers from a given distribution in Chap. 13. The two most important methods, the inverse transformation method and the rejection method, were based on an exact knowledge of the analytic form of the distribution function which the random numbers were supposed to follow. However, when simulating the physics of the ISING model it was required to draw random configurations from the equilibrium distribution of the system and, unfortunately, the exact analytic form of this distribution was unknown. On the other hand, in the discussion of MARKOV-chains we came across the condition of detailed balance. Invoking this condition ensured that the constructed MARKOV-chain converged toward a stationary distribution, independent of the initial conditions. Consequently, we can also sample random numbers by constructing a MARKOV-chain with a stationary distribution which is equal to the distribution from which we would like to obtain our random numbers. In such a case the distribution function has to be known, at least in principle. However, the formulation of the METROPOLIS algorithm allowed for an unknown normalization constant of the distribution function which, in turn, makes this method such a powerful tool in computational physics.

Here we plan to discuss MARKOV-chain Monte Carlo techniques in greater detail. We start with the introduction of the concept of *importance sampling*, review the METROPOLIS algorithm, and discuss the straight forward generalization to the

B.A. Stickler, E. Schachinger, *Basic Concepts in Computational Physics*,
DOI 10.1007/978-3-319-27265-8_18

METROPOLIS-HASTINGS algorithm. Finally, the applicability of the METROPOLIS-HASTINGS algorithm will be demonstrated by simulating the physics of the q-states POTTS model [5] which is closely related to the ISING model. This chapter is closed with a brief presentation of some of the more advanced techniques within this context.

18.2 MARKOV-Chain Monte Carlo Methods

Before turning our focus toward the MARKOV-chain Monte Carlo methods we shall briefly discuss *importance sampling*. Let $p(x)$ be a certain pdf from which we would like to draw a sequence of random numbers $\{x_i\}$, $i \geq 1$. Furthermore, let $f(x)$ be some arbitrary function and we would like to estimate its expectation value $\langle f \rangle_p$ which is determined by the integral

$$\langle f \rangle_p = \int \mathrm{d}x f(x) p(x) \; . \tag{18.1}$$

But $\langle f \rangle_p$ can also be regarded as the expectation value $\langle a \rangle_u$ of the function $a(x) := f(x)p(x)$, with $u(x)$ the pdf of the uniform distribution. Hence, we may evaluate $\langle a \rangle_u$ by drawing uniformly distributed random numbers on a given interval $[a, b] \subset \mathbb{R}$ and by estimating the expectation value by its arithmetic mean as discussed in Sect. 14.2. This approach is the easiest version of a method referred to as *simple sampling*. On the other hand, we might approximate $\langle f \rangle_p$ by sampling x_i according to $p(x)$ and by employing the central limit theorem (see Appendix, Sect. E.8) as was demonstrated in Sect. 14.2. The basic idea of *importance sampling*, however, is to improve this approach by sampling from a different distribution $q(x)$ which is in most cases chosen in such a way that the expectation value $\langle f \rangle_p$ is easier to evaluate.

Let $g(x)$ be some function with $g(x) > 0$ for all x. Then

$$\langle f \rangle_p = \int \mathrm{d}x f(x) p(x) = \int \mathrm{d}x \frac{f(x)}{g(x)} p(x) g(x) = c \left\langle \frac{f}{g} \right\rangle_q \; , \tag{18.2}$$

where we defined the function $q(x) = p(x)g(x)/c$ and c is chosen in such a way that $\int \mathrm{d}x q(x) = 1$. We note that $g(x)$ can be *any* positive function. Hence, such an approach might be interesting in two different scenarios: (i) if it is easier to sample from the distribution $q(x)$ rather than from $p(x)$ and, (ii) if such a sampling results in a variance reduction which is equivalent to a decrease in error, and less random numbers are to be sampled to obtain comparable results.

Let us briefly elaborate on this point: we have

$$\mathrm{var}\left(\frac{f}{g}\right)_q = \left\langle \left(\frac{f}{g} - \left\langle \frac{f}{g} \right\rangle_q \right)^2 \right\rangle_q \; , \tag{18.3}$$

and for the particular choice $g(x) \equiv f(x)$ we obtain that

$$\text{var}\left(\frac{f}{g}\right)_q = 0\,, \tag{18.4}$$

and the error of our Monte Carlo integration vanishes. However, the ideal case of $g(x) \equiv f(x)$ is unrealistic because we obtain for the normalization constant c

$$c = \int \mathrm{d}x\, p(x) g(x) = \langle g \rangle_p \equiv \langle f \rangle_p\,, \tag{18.5}$$

which is exactly the integral we want to evaluate. Nevertheless, the function $g(x)$ can be adapted to improve the result. We choose $g(x)$ in such a way that the integral $\langle g \rangle_p$ is easily evaluated and that $g(x)$ follows $f(x)$ as closely as possible; in other words, the quotient $f(x)/g(x)$ becomes as constant as possible. This means that we no longer sample from $p(x)$ within a given interval but only from points which are of importance for the particular function $f(x)$. Such an approach is referred to as *importance sampling* [6–9].

The attentive reader might have observed that, on a first glance, importance sampling has nothing to do with MARKOV-chain Monte Carlo methods in general. Nevertheless, it can be demonstrated that MARKOV-chain Monte Carlo methods correspond indeed to importance sampling.

To prove this, we remember that MARKOV-chain Monte Carlo techniques are based on the generation of a sequence of configurations $S^{(n)}$:

$$S^{(1)} \to S^{(2)} \to \ldots \to S^{(n)} \to \ldots\,. \tag{18.6}$$

Each individual configuration $S^{(n)}$ is generated from the previous configuration $S^{(n-1)}$ at random with a certain transition probability $P(S^{(n-1)} \to S^{(n)})$. These transition probabilities obey

$$P(S \to S') \geq 0 \quad \text{and} \quad \sum_{S'} P(S \to S') = 1\,, \tag{18.7}$$

and this property ensures that the sequence (18.6) is a MARKOV-chain. In Sect. 16.4 we observed that the condition of detailed balance for a stationary distribution $P(S)$

$$P(S)P(S \to S') = P(S')P(S' \to S) \tag{18.8}$$

guarantees convergence of the MARKOV-chain toward the stationary distribution. Hence, the remaining task is to find transition probabilities which fulfill detailed balance. In a typical situation, the transition probabilities can be written as

$$P(S \to S') = P_p(S \to S')P_a(S \to S')\,, \tag{18.9}$$

where $P_p(S \to S')$ is the probability that a configuration S' is proposed and $P_a(S \to S')$ is the probability that the proposed configuration is accepted. In many cases one simplifies the situation by assuming that

$$P_p(S \to S') = P_p(S' \to S) , \tag{18.10}$$

and, thus, the condition of detailed balance changes into

$$P(S)P_a(S \to S') = P(S')P_a(S' \to S) . \tag{18.11}$$

The METROPOLIS algorithm uses *one* possible choice for the acceptance probability, namely

$$P_a(S \to S') = \min\left[1, \frac{P(S')}{P(S)}\right] . \tag{18.12}$$

It was demonstrated in Sect. 14.3 that Eq. (18.12) indeed fulfills Eq. (18.11). The execution of the algorithm has already been illustrated in Chap. 15 in its application to the numerics of the ISING model.

A rather straight forward generalization of the METROPOLIS algorithm (18.12) is found when an asymmetric proposal probability $P_p(S \to S')$ is considered. It is easily demonstrated that the choice

$$P_a(S \to S') = \min\left[1, \frac{P(S')}{P(S)} \frac{P_p(S' \to S)}{P_p(S \to S')}\right] , \tag{18.13}$$

also fulfills detailed balance (18.8). The choice (18.13) is referred to as the METROPOLIS-HASTINGS algorithm [10].[1]

By exploiting the MARKOV property in order to sample configurations according to the BOLTZMANN distribution we perform *importance sampling* as illustrated above. An alternative approach would be to select different configurations according to a uniform distribution which obviously increases the numerical cost of the method by magnitudes. Hence, sampling with the help of a MARKOV-chain yields a variance reduction in comparison to the crude approach of simple sampling. Furthermore, the algorithm can be optimized by a clever choice of $P_p(S \to S')$ which does not need to be symmetric. Clearly, this choice will have to depend on the particular problem at hand.

We shall briefly discuss two alternative approaches to MARKOV-chain Monte Carlo sampling, namely GIBBS *sampling* [11] and *slice sampling* [12]: Suppose we want to sample a sequence of m-dimensional variables $x^{(n)} = (x_1^{(n)}, x_2^{(n)}, \ldots, x_m^{(n)})^T$ from a multivariate distribution function $p(x) = p(x_1, x_2, \ldots, x_m)$. In such a case

[1]Please note that it is common in the literature to refer even to Eq. (18.12) as a METROPOLIS-HASTINGS algorithm, despite the fact that here $P_p(S' \to S) = P_p(S \to S')$.

GIBBS sampling is particularly interesting if the joint distribution function is well-known and simple to sample. The acceptance probability for a particular component of the vector $x^{(n)}$ is set to:

$$P_a(x_j^{(n+1)}|x^{(n)}) = p(x_j^{(n+1)}|x_1^{(n+1)}, \ldots, x_{j-1}^{(n+1)}, x_{j+1}^{(n)}, \ldots, x_m^{(n)}) . \tag{18.14}$$

This is possible as we have

$$\begin{aligned} p(x_j|x_1, \ldots, x_{j-1}, x_{j+1}, \ldots, x_m) &= \frac{p(x_1, \ldots, x_m)}{p(x_1, \ldots, x_{j-1}, x_{j+1}, \ldots, x_m)} \\ &\propto p(x_1, \ldots, x_m) , \end{aligned} \tag{18.15}$$

because the denominator of the left hand side of Eq. (18.15) is independent of x_j. It can, therefore, be treated as a normalization constant when x_j is sampled.

Let us briefly discuss *slice sampling*: For reasons of simplicity we shall regard the uni-variate case where $p(x)$ denotes the pdf from which we would like to sample. We apply the following algorithm:

1. Choose some initial value x_0.
2. Sample a uniformly distributed random variable y_0 from the interval $[0, p(x_0)]$.
3. Sample the next random variable x_1 uniformly from the *slice* $\mathscr{S}[p^{-1}(y_0)]$.
4. Sample a uniformly distributed random number $y_1 \in [0, p(x_1)]$.
5. Sample the next random variable x_2 uniformly from the slice $\mathscr{S}[p^{-1}(y_1)]$.
6. ...

⋮

The final sequence $\{x_j\}$ is constructed by ignoring the y_n-values (even steps). This procedure is illustrated schematically in Fig. 18.1.

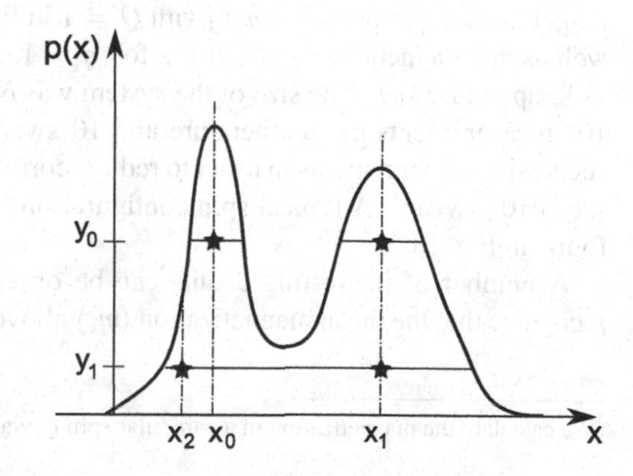

Fig. 18.1 Schematic illustration of *slice sampling* for the uni-variate case described by the pdf $p(x)$. The relevant steps of the algorithm are indicated by *solid asterisks*

18.3 The POTTS Model

We studied in Chap. 15 the two-dimensional ISING model as an example for the METROPOLIS algorithm. Here we expand on this discussion and investigate the q-states POTTS model [5] which can be regarded as a generalization of the ISING model. The model is characterized by the HAMILTON function

$$H = -\sum_{ij} J_{ij}\delta_{\sigma_i\sigma_j} , \tag{18.16}$$

where the notation used for the ISING model applies. In particular, we shall regard the case $J_{ij} = J$ for i,j nearest neighbors and $J_{ij} = 0$ otherwise. In contrast to the ISING model, the spin realizations σ_i on grid-point i can take integer values $\sigma_i = 1, 2, \ldots, q$. For $q = 2$, the POTTS model is equivalent to the ISING model which can be easily proved by rewriting the HAMILTON function as

$$H = \frac{J}{2}\sum_{\langle ij\rangle} 2\left(\frac{1}{2} - \delta_{\sigma_i\sigma_j}\right) - J\sum_i \frac{1}{2} . \tag{18.17}$$

Here $\langle ij\rangle$ denotes sum over nearest neighbors. We observe that $2\left(\frac{1}{2} - \delta_{\sigma_i\sigma_j}\right)$ is equal to -1 for $\sigma_i = \sigma_j$ and $+1$ for $\sigma_i \neq \sigma_j$. Moreover, the constant energy shift in Eq. (18.17) can be neglected and we recover the ISING model of Chap. 15.

The method to calculate the observables of interest, like $\langle E\rangle$, $\langle M\rangle$, c_h and χ, and the basic algorithm can be adopted from Sect. 15.2 as is.[2] There is one important difference. It occurs in step 3 of the algorithm: Instead of setting $\sigma_{ij} = -\sigma_{ij}$ we sample the new value of σ_{ij} uniformly distributed from $1, 2, \ldots, q$ under exclusion of the old value of σ_{ij}.

Figures 18.2, 18.3, 18.4, and 18.5 display the mean energy per particle, the mean magnetization per particle $\langle m_1\rangle$ [with $Q = 1$ in Eq. (18.18)], the heat capacity c_h as well as the magnetic susceptibility χ for $q = 1, 2, \ldots, 8$ and $J = 0.5$ [Eq. (18.16)] vs temperature k_BT. The size of the system was $N \times N$ with $N = 40$. We performed 10^4 measurements per temperature and 10 sweeps where discarded between two successive measurements in order to reduce correlations. The equilibration time was set to 10^3 sweeps. A typical spin configuration for $q = 4$ and $k_BT = 0.47$ can be found in Fig. 18.6.

A number of interesting details can be observed in Fig. 18.3. First of all, we recognize that the mean magnetization $\langle m_1\rangle$ above the critical temperature decreases

---

[2]We calculate the magnetization in a particular spin Q via

$$\mathscr{M}_Q(\mathscr{C}) = \left(\sum_i \delta_{\sigma_i,Q}\right)_{\mathscr{C}} . \tag{18.18}$$

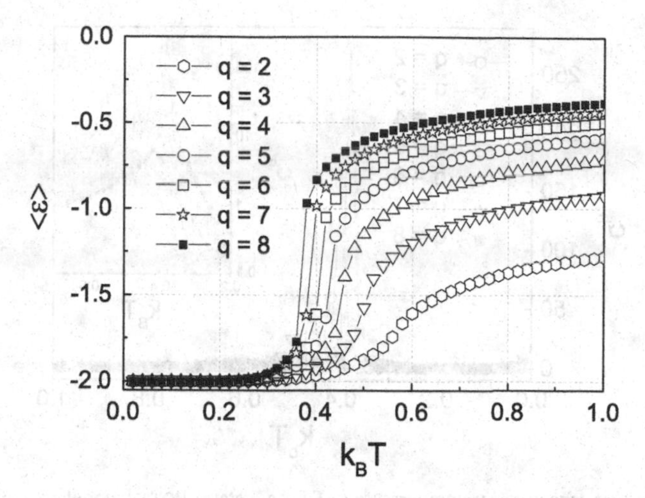

Fig. 18.2 The mean free energy per particle $\langle \varepsilon \rangle$ vs temperature $k_B T$ for a q-states POTTS model on a 40×40 square lattice, with $q = 1, 2, \ldots, 8$ and $J = 0.5$. 10^4 measurements have been performed

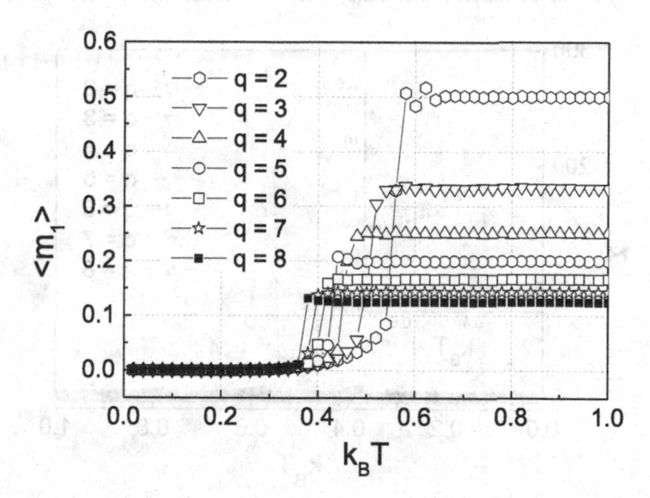

Fig. 18.3 The mean magnetization per particle $\langle m_1 \rangle$ vs temperature $k_B T$ for a q-states POTTS model on a 40×40 square lattice, for $q = 1, 2, \ldots, 8$ and $J = 0.5$. 10^4 measurements have been performed

with increasing values of q. The reason is that the mean magnetization $\langle m_1 \rangle$ represents for $T \gg T_c$ the probability of finding a particular spin in state $\sigma_i = 1$. This is equivalent to $1/q$ for a uniform distribution and therefore decreases with increasing q. Please note that the expectation value of the magnetization $\langle m_Q \rangle$ is restricted to take the values from $\{0, 1\}$ for $T \ll T_C$ due to the modified definition of $\mathcal{M}_Q(\mathscr{C})$. This is in contrast to the ISING model where $\langle m \rangle \in \{-1, 1\}$ for $T < T_c$. Hence, we obtain for $T \ll T_C$ $\langle m_1 \rangle = 0$ with probability $(q-1)/q$ and $\langle m_1 \rangle = 1$ with probability $1/q$. However, the particular definition (18.18) of $\mathcal{M}_Q(\mathscr{C})$ is not

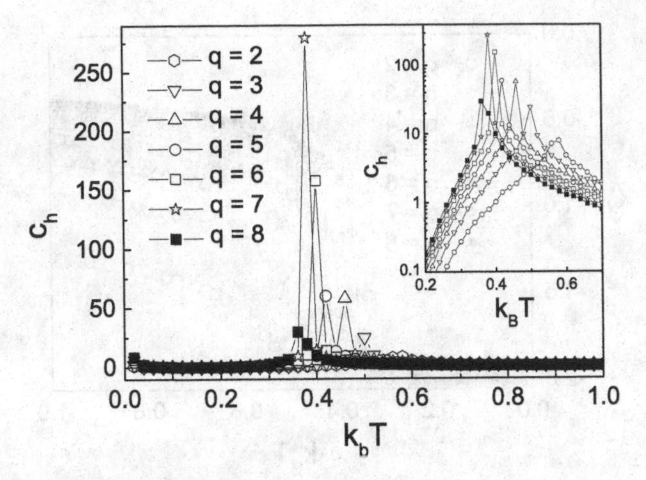

Fig. 18.4 The heat capacity c_h vs temperature $k_B T$ for a q-states POTTS model on a 40×40 square lattice, with $q = 1, 2, \ldots, 8$ and $J = 0.5$. 10^4 measurements have been performed. The *inset* shows the specific heat c_h on a logarithmic scale in the region around the transition temperature

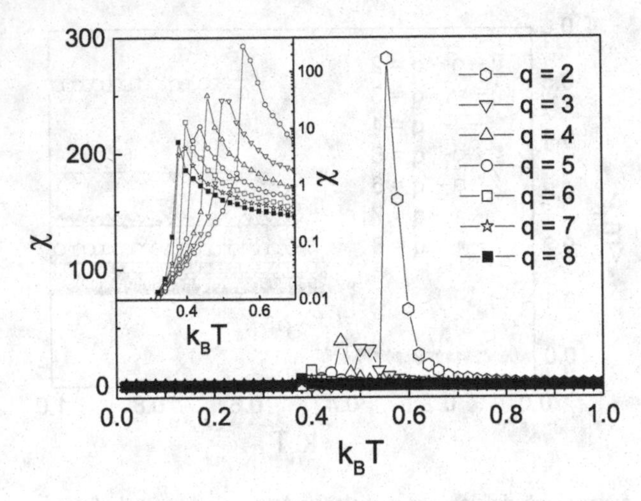

Fig. 18.5 The magnetic susceptibility χ vs temperature $k_B T$ for a q-states POTTS model on a 40×40 square grid, with $q = 1, 2, \ldots, 8$ and $J = 0.5$. 10^4 measurements have been performed. The *inset* shows the magnetic susceptibility χ on a logarithmic scale in the region around the transition temperature

important since the physically relevant property of the POTTS model HAMILTON function (18.16) is its $\mathbb{Z}_q$ symmetry with a degenerate ground state.

A second interesting feature is the observation that the critical temperature T_c also decreases with increasing values of q which becomes particularly transparent from Figs. 18.4 and 18.5. The critical temperatures are quoted in Table 18.1. Finally, we deduce from Fig. 18.2 that the phase transition is smoother for $q = 2$ and becomes discontinuous for large values of q. In particular, the q-states POTTS model

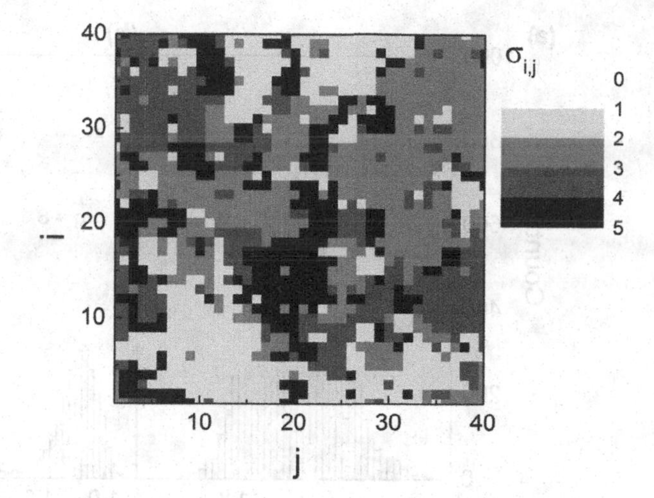

Fig. 18.6 A typical spin configuration $\sigma_{i,j}$ for a q-states POTTS model on a 40×40 square lattice with $q = 4$, $J = 0.5$, and $k_B T = 0.47$

Table 18.1 List of the critical values $\alpha_c = J/(k_B T_c)$ of the q-states POTTS model for $q = 2, 3, \ldots, 8$

q	α_c
2	0.89
3	1.00
4	1.09
5	1.16
6	1.22
7	1.28
8	1.35

exhibits a second order phase transition for $q = 2, 3, 4$ and a first order phase transition for $q > 4$ which is hard to see from Figs. 18.2 and 18.3. However, there is another method to unambiguously identify a first order phase transition. It is referred to as the histogram technique. The mean energies of consecutive measurements near the critical temperature are simply collected in a histogram. If only one peak is observed, the system fluctuates around a single phase, and a second order phase transition was observed. However, the existence of two or more peaks means that the system fluctuates between two or more different phases and, therefore, exhibits a first order phase transition. Figure 18.7 displays two histograms for $q = 2$ ($k_B T = 0.56$) and $q = 8$ ($k_B T = 0.37$) from 10^4 measurements to prove our case.

One possible realization of the $q = 3$ states POTTS model was discussed by M. KARDAR and A.N. BERKER [13]. They studied the over saturated adsorption of Krypton atoms on a graphite surface. A detailed analysis of this system revealed that three energetically degenerate sublattices are formed. Furthermore, the authors demonstrated that the thermodynamic properties of this system can be explained by a $q = 3$ states POTTS model. For a more detailed discussion we refer to the original paper. More applications of the POTTS model were discussed in the review by F.Y. WU [14].

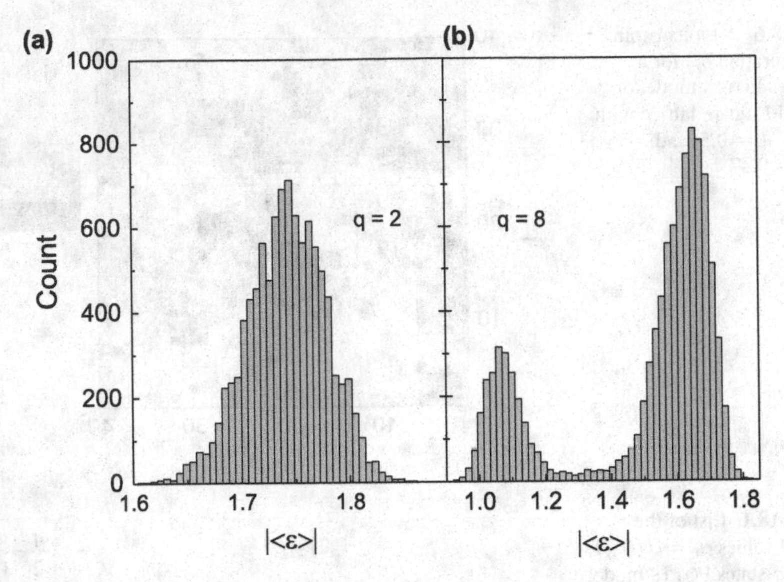

Fig. 18.7 (**a**) Histogram of 10^4 measurements of the absolute value of the mean free energy per particle $|\langle\varepsilon\rangle|$ for a q-states POTTS model on a 40×40 square lattice at temperature $k_B T = 0.56$ with $q = 2$. We observe one single peak which indicates that the system exhibits a second order phase transition. (**b**) The same as (**a**) but for $q = 8$ and temperature $k_B T = 0.37$. We observe two well separated peaks, thus the system exhibits a first order phase transition

The attentive reader may have noticed that our results do not carry error-bars. We neglected error-bars for a clearer illustration. A short discussion of methods used to calculate numerical errors was presented in Sect. 15.2 for the ISING model and they can easily be adapted for the POTTS model. More advanced techniques will be introduced in the next chapter.

18.4 Advanced Algorithms for the POTTS Model

We discuss briefly some advanced techniques for the POTTS model. Although these algorithms are applicable for arbitrary q we restrict our discussion to the case $q = 2$, the ISING model, for reasons of simplicity. Let us briefly motivate the need for more advanced methods: For large values of N we observe the formation of spin domains[3] for temperatures $T \approx T_c$. In such a case the specific METROPOLIS algorithm used so far is disadvantageous because single spin flips will only affect the boundaries of these domains (*critical slowing down*). It is therefore necessary to perform many sweeps in order to produce configurations which are entirely different. It might

[3]This are regions in which all spins point in the same direction, the so-called WEISS domains [15].

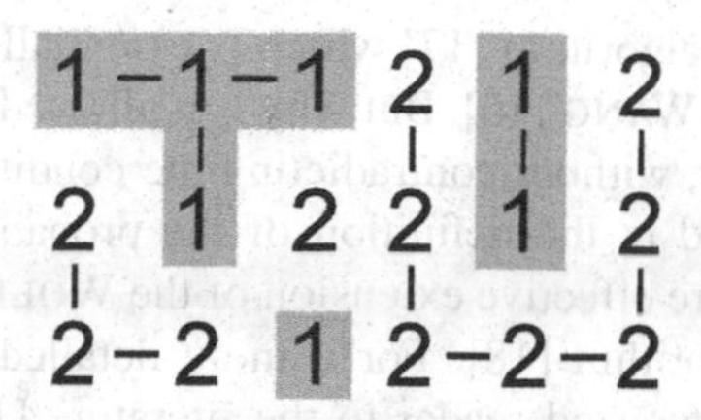

Fig. 18.8 Schematic illustration of the identification of clusters according to the SWENDSEN-WANG algorithm. *1* and *2* denote two different spin orientations, bonds are denoted by *solid lines* and all bonded spins form clusters

therefore be a better approach to flip a whole spin *cluster* at once. Such algorithms are referred to as *cluster* algorithms. The main problem is the identification of clusters as well as the assignment of a probability to the flip of a particular cluster.

As a first example we shall discuss the SWENDSEN-WANG algorithm [16]. The algorithm is executed in the following steps:

1. Identify all *links* between two neighboring identical spins.
2. Define a *bond* between two linked spins with probability
$$P = 1 - \exp(-2\beta J) , \qquad (18.19)$$
with $\beta = 1/(k_B T)$.
3. Identify all *clusters* which are built from spins connected by bonds, see Fig. 18.8.
4. Flip every cluster with probability $1/2$.
5. Delete the bonds and restart the iteration for the next spin configuration.

We note the following properties of the SWENDSEN-WANG algorithm:

- The algorithm is ergodic because every spin forms a cluster on its own with a non-vanishing probability according to Eq. (18.19).
- The algorithm fulfills detailed balance for the BOLTZMANN distribution and thus reproduces the correct stationary distribution.

Since the algorithm breaks domain walls or flips whole clusters, this algorithm can be regarded to be very efficient from a numerical point of view. However, it outperforms the single spin METROPOLIS algorithm only for temperatures near the critical temperature because only then spin domains dominate the observables.

A simpler version of this algorithm consists of the following four steps:

1. Randomly pick a lattice site.
2. Find all neighbors with the same spin and form bonds with probability (18.19).
3. Move to the boundary of the cluster and repeat step 2, i.e. the cluster *grows*.
4. If no new bond is formed, flip the cluster with probability 1.

In this simplified version the identification of clusters is not necessary because each cluster is built dynamically during the simulation. Such a formulation of a cluster algorithm is the WOLFF algorithm [17] which is essentially a generalization of the work of SWENDSEN and WANG [16]. But why is it allowed to accept every step, i.e. flip every formed cluster, without contradicting the condition of detailed balance? The explanation is found in the definition of the probability of bond-formation, Eq. (18.19). An even more effective extension of the WOLFF algorithm to quantum systems is the *loop* algorithm [18]. For a more detailed discussion of all these methods we refer the interested reader to the literature [19] and to the particular papers cited here.

Before proceeding to the next chapter, let us briefly mention that there is also an entirely different approach to improve the METROPOLIS algorithm for quantum systems, the so called *worm* algorithms [18]. However, a detailed discussion of such algorithms is beyond the scope of this book.

Summary

The dominant topic of this chapter was importance sampling, a method to improve Monte Carlo methods by reducing the variance. In this method some hard to sample pdf was approximated as closely as possible by another, easy to sample pdf and one concentrated on intervals which particularly matter for an as accurate as possible estimate of, for instance, an expectation value of some property $f(x)$. In this sense MARKOV-chain Monte Carlo techniques corresponded to importance sampling as long as detailed balance was obeyed. In this particular case the MARKOV-chain was known to approach the equilibrium distribution which must not necessarily be known in detail. The METROPOLIS algorithm with its symmetric acceptance probability was one possible realization of MARKOV-chains which obeyed detailed balance. Another method was the METROPOLIS-HASTINGS algorithm with its asymmetric acceptance probability. It also obeyed detailed balance and improved the variance over the 'classical' METROPOLIS algorithm. The second part of this chapter was dedicated to the simulation of the q-state POTTS model, an extension of the ISING model. The POTTS model had the feature that it developed a second order phase transition for $q \leq 4$ and a first order phase transition for $q > 4$. Moreover, the transition temperature was q-dependent. The numerical simulation of the physics of this model proved to be able to pick up on all these particular features. Finally, some advanced algorithms developed for a more precise handling of various properties of spin-models, particularly around the phase transition, have been presented without going into great detail.

Problems

1. Modify the program designed to solve the Ising model (see Problems in Chap. 15) in such a way that the physics of the q-states Potts model can be simulated for arbitrary values of q. Try to reproduce the figures presented within this chapter. In order to investigate the order of the phase transition, plot the internal energy per particle $\langle \varepsilon \rangle$ for $T \approx T_c$ in a histogram for different measurements.

   The critical temperatures listed in Table 18.1 for $q = 2, 3, \ldots, 8$ can be used to validate your code.
2. Include a non-zero external field h and study its influence on the physics of the q-states Potts model for different values of q.

References

1. Norris, J.R.: Markov Chains. Cambridge Series in Statistical and Probabilistic. Cambridge University Press, Cambridge (1998)
2. Kendall, W.S., Liang, F., Wang, J.S.: Markov Chain Monte Carlo: Innovations and Applications. Lecture Notes Series, vol. 7. Institute for Mathematical Sciences, National University of Singapore. World Scientific, Singapore (2005)
3. Modica, G., Poggiolini, L.: A First Course in Probability and Markov Chains. Wiley, New York (2012)
4. Graham, C.: Markov Chains: Analytic and Monte Carlo Computations. Wiley, New York (2014)
5. Potts, R.B.: Some generalized order-disorder transformations. Math. Proc. Camb. Philos. Soc. **48**, 106–109 (1952). doi:10.1017/S0305004100027419
6. Doucet, A., de Freitas, N., Gordon, N. (eds.): Sequential Monte Carlo Methods in Practice. Information Science and Statistics. Springer, Berlin/Heidelberg (2001)
7. Press, W.H., Teukolsky, S.A., Vetterling, W.T., Flannery, B.P.: Numerical Recipes in C++, 2nd edn. Cambridge University Press, Cambridge (2002)
8. Kalos, M.H., Whitlock, P.A.: Monte Carlo Methods, 2nd edn. Wiley, New York (2008)
9. von der Linden, W., Dose, V., von Toussaint, U.: Bayesian Probability Theory. Cambridge University Press, Cambridge (2014)
10. Berg, B.A.: Markov Chain Monte Carlo Simulations and Their Statistical Analysis. World Scientific, Singapore (2004)
11. German, S.: Stochastic relaxation, gibbs distributions, and the bayesian restoration of images. IEEE Trans. Pattern Anal. Mach. Intell. **6**, 721–741 (1984). doi:10.1109/TPAMI.1984.4767596
12. Neal, R.M.: Slice sampling. Ann. Stat. **31**, 705–767 (2003). doi:10.1214/aos/1056562461
13. Kardar, M., Berker, A.N.: Commensurate-incommensurate phase diagrams for overlayers from a helical potts model. Phys. Rev. Lett. **48**, 1552–1555 (1982). doi:10.1103/PhysRevLett.48.1552
14. Wu, F.Y.: The potts model. Rev. Mod. Phys. **54**, 235–268 (1982). doi:10.1103/RevModPhys.54.235
15. White, R.M.: Quantum Theory of Magnetism, 3rd edn. Springer Series in Solid-State Sciences. Springer, Berlin/Heidelberg (2007)
16. Swendsen, R.H., Wang, J.S.: Nonuniversal critical dynamics in monte carlo simulations. Phys. Rev. Lett. **58**, 86–88 (1987). doi:10.1103/PhysRevLett.58.86

17. Wolff, U.: Collective monte carlo updating for spin systems. Phys. Rev. Lett. **62**, 361–364 (1989). doi:10.1103/PhysRevLett.62.361
18. Evertz, H.G.: The loop algorithm. Adv. Phys. **52**, 1–66 (2003). doi:10.1080/0001873021000049195
19. Newman, M.E.J., Barkema, G.T.: Monte Carlo Methods in Statistical Physics. Clarendon Press, Oxford (1999)

Chapter 19
Data Analysis

19.1 Introduction

It is the aim of this chapter to present some of the most important techniques of statistical data analysis which is of interest for experimental as well as theoretical sciences. In particular, the superstition that numerically generated data sets do not need to be analyzed with statistical methods is certainly not justified if the data was generated by Monte Carlo methods. Some simple methods of statistical analysis have already been discussed in previous chapters. For instance, in Chap. 12 we discussed simple quality tests for random number generators, in Chap. 15 we calculated the errors associated with the observables of the ISING model. Here, these simple methods will be summarized and some more advanced techniques will be introduced on a basic level. For a more advanced discussion of this topic we refer the interested reader to Refs. [1–5].

19.2 Calculation of Errors

We repeat briefly the basics of simple estimators which we made use of previously. We approximate the expectation value $\langle x \rangle$ of some variable x

$$\langle x \rangle = \int \mathrm{d}x \, x p(x) , \tag{19.1}$$

where $p(x)$ is a pdf, by its arithmetic mean

$$\langle x \rangle \approx \overline{x} = \frac{1}{N} \sum_{i=1}^{N} x_i , \tag{19.2}$$

B.A. Stickler, E. Schachinger, *Basic Concepts in Computational Physics*,
DOI 10.1007/978-3-319-27265-8_19

where the numbers x_i follow the distribution $p(x)$. It is of conceptual importance to distinguish between the expectation value $\langle x \rangle$ which is a c-number, while the estimator $\overline{x}$ is a random number fluctuating around $\langle x \rangle$. The error of approximating $\langle x \rangle$ by $\overline{x}$ can be estimated by calculating the variance

$$\mathrm{var}\,(\overline{x}) = \frac{\mathrm{var}\,(x)}{N} = \frac{\langle x^2 \rangle - \langle x \rangle^2}{N}, \tag{19.3}$$

if the random numbers x_i are uncorrelated (see Appendix E). In case of correlated data the treatment becomes more involved and this will be discussed in Sect. 19.3. The expectation values $\langle x^2 \rangle$ and $\langle x \rangle$ in Eq. (19.3) may again be replaced by the corresponding estimators $\overline{x^2}$ and $\overline{x}$ in order to obtain a reasonable estimate of the variance var $(\overline{x})$. In particular, we approximate

$$\langle x^2 \rangle \approx \overline{x^2} = \frac{1}{N} \sum_{i=1}^{N} x_i^2 . \tag{19.4}$$

This approximation has already been applied in our investigation of the ISING model, Chap. 15. When dealing with MARKOV-chain Monte Carlo simulations, the result (19.3) can be interpreted in a rather trivial way: Repeating the simulation under identical conditions results in roughly 68 % of all simulations to yield a mean value $\overline{x} \in [\overline{x} - \sigma_{\overline{x}}, \overline{x} + \sigma_{\overline{x}}]$, where $\sigma_{\overline{x}} = \sqrt{\mathrm{var}\,(\overline{x})}$ is the standard error.

We consider now the, in the meanwhile, quite familiar situation in which the underlying pdf $p(x)$ of a sequence of random numbers $\{x_i\}$ is unknown. In such a case one cannot simply use a particular estimator without some knowledge of the particular form of $p(x)$. A common way to proceed is the *poor person's assumption*: The underlying distribution is symmetric. This assumption has its origin in the central limit theorem (see Appendix, Sect. E.8). However, some intuitive checks may be required if fatal misconceptions are to be avoided. Is the data set reasonably large one can retrieve essential information from collecting the data points in form of a histogram or, if the index i refers to time instances, by plotting a time sequence.

We can deduce a first idea about the form of the underlying pdf from a histogram. For instance, if the data set displays only one peak, as in Fig. 19.1, quantities like the mean or the variance could be useful. But if there are two (or more) separate peaks, as in Fig. 19.2, it does not necessarily make sense to calculate the mean or variance by summing over all the data points. Such a situation can, for instance, occur in statistical spin models, with two phases, as we observed it in the q-state POTTS model, Fig. 18.7a, b.

Time series, in which the data points x_i are plotted as a function of discrete time instances t_i, can also reveal important information about the properties of the data set. For instance, systematic trends, outliers, or hints for correlations may be observed.

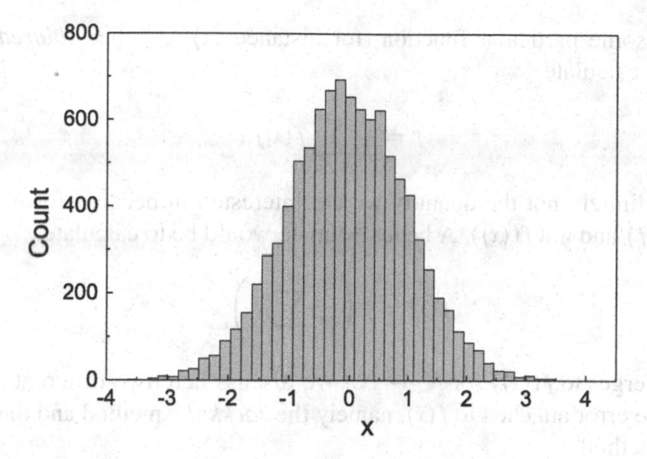

Fig. 19.1 Histogram generated by random sampling of a Gaussian of mean zero and variance one

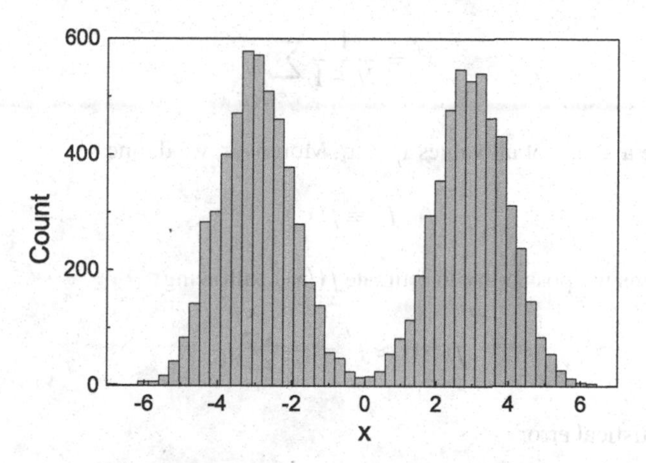

Fig. 19.2 Histogram generated by random sampling of two Gaussians of mean zero and variance one, displaced by $+3$ and -3, respectively

Let us turn our attention to some more advanced estimator techniques. So far we discussed the sample mean and sample variance as candidates for unbiased estimators.[1] In a more general context the calculation of observables from data sets might be more complex. In the following we assume a data set of N data points $(x_1, x_2, \ldots, x_N)$. Basically, we would like to estimate a quantity of the form $f(\langle x \rangle)$

[1] Since mean and variance are calculated from the same data points, they are usually not unbiased. Therefore a common choice is the so called *bias corrected variance* $\text{var}\,(\overline{x})_B$ which is given by $\text{var}\,(\overline{x})_B = \frac{N}{N-1}\text{var}\,(\overline{x})$ where N is the number of data points. A more detailed discussion can be found in any textbook on statistics [6–9].

where f is some particular function (for instance $\langle x\rangle^2$). A bad (*biased*) estimate would be to calculate

$$\bar{f} = \frac{1}{N}\sum f(x_i)\,, \tag{19.5}$$

which is definitely not the quantity we are interested in because for $N \to \infty$ we have $\bar{f} \to \langle f\rangle$ and not $f(\langle x\rangle)$. A better estimate would be to calculate

$$f(\bar{x}) = f\left(\frac{1}{N}\sum x_i\right)\,, \tag{19.6}$$

which converges to $f(\langle x\rangle)$ for $N \to \infty$. We discuss here two different methods to calculate the error attached to $f(\bar{x})$, namely the *Jackknife* method and the *statistical bootstrap* method.

We define Jackknife averages

$$x_i^J = \frac{1}{N-1}\sum_{j\neq i} x_j\,, \tag{19.7}$$

and x_i^J is the average of all values $x_j \neq x_i$. Moreover, we define

$$f_i^J \equiv f(x_i^J)\,, \tag{19.8}$$

and this opens the possibility to estimate $f(\langle x\rangle)$ following

$$f(\langle x\rangle) \approx \bar{f}^J = \frac{1}{N}\sum_i f_i^J\,, \tag{19.9}$$

with the statistical error

$$\sigma^2_{\bar{f}^J} = (N-1)\left[\overline{(f^J)^2} - (\bar{f}^J)^2\right]\,, \tag{19.10}$$

which can be written as

$$\sigma^2_{\bar{f}^J} = \frac{N-1}{N}\sum_i (f_i^J - \bar{f}^J)^2\,, \tag{19.11}$$

for uncorrelated f_i^J (see Appendix E).

In the case of the statistical bootstrap we consider again a set of N data-points $\{x_i\}$. We randomly choose N elements from this data set *without removal* which constitutes the set $\{x_j^{(i)}\}$ and calculate for these N points the observable $f_i = f(1/N\sum_j x_j^{(i)})$. This procedure is repeated M-times and we get

$$f(\langle x\rangle) \approx \bar{f}_{BS} = \frac{1}{M}\sum_i f_i\,, \tag{19.12}$$

and

$$\sigma^2_{\overline{f}_{BS}} = \frac{1}{M} \sum_i \left(f_i - \overline{f}_{BS}\right)^2 \ . \tag{19.13}$$

This method was applied in Chap. 15 to determine estimates for the error-bars of the various observables as a function of temperature in Fig. 15.6. The methods discussed here can, of course, also be employed to derive estimates for the errors attached to the various observables studied in the POTTS model, Chap. 18.

Let us close this section with a short comment on systematic errors. As already highlighted within Chap. 1 one also has to be aware of possible systematic errors. Like in experimental data, these errors are more easily overlooked in numerical data since they are rather hard to identify. In general, there is no method available to investigate systematic errors. For instance, in the simulation of the ISING model, the main source of errors was that the MARKOV-chain was not allowed to completely equilibrate which would have been equivalent to running the simulation forever. The introduction of the concept of an auto-correlation time will, at least, allow for a systematic investigation of this fundamental problem.

19.3 Auto-Correlations

The situation becomes more involved whenever the random numbers of the sequence $\{x_i\}$ are correlated, i.e. $\text{cov}\left(x_i, x_j\right) \neq 0$ for $i \neq j$ [see Appendix, Eq. (E.16)], where the elements of the series $\{x_i\}$ are successive members of a time series. Hence, existing covariances between elements x_i and x_j account for auto-correlations of a certain observable between different time steps. We rewrite Eq. (19.3):

$$\begin{aligned}
\text{var}\left(\overline{x}\right) &= \left\langle \overline{x}^2 \right\rangle - \left\langle \overline{x} \right\rangle^2 \\
&= \frac{1}{N^2} \sum_{i,j=1}^{N} \left\langle x_i x_j \right\rangle - \frac{1}{N^2} \sum_{i,j=1}^{N} \left\langle x_i \right\rangle \left\langle x_j \right\rangle \\
&= \frac{1}{N^2} \sum_{i=1}^{N} \left(\left\langle x_i^2 \right\rangle - \left\langle x_i \right\rangle^2 \right) \\
&\quad + \frac{1}{N^2} \sum_{i \neq j} \left(\left\langle x_i x_j \right\rangle - \left\langle x_i \right\rangle \left\langle x_j \right\rangle \right) \ .
\end{aligned} \tag{19.14}$$

The first term on the right-hand side of Eq. (19.14) is identified as var $(x_i)/N$ which is assumed to be identical for all i, i.e. var $(x_i) \equiv$ var (x). Furthermore, we rewrite the sum

$$\sum_{i \neq j} \cdot = 2 \sum_{i=1}^{N} \sum_{j=i+1}^{N} \cdot ,$$

and obtain

$$\text{var}\,(\overline{x}) = \frac{1}{N} \left[\text{var}\,(x) + \frac{2}{N} \sum_{i=1}^{N} \sum_{j=i+1}^{N} \text{cov}\,(x_i, x_j) \right] . \tag{19.15}$$

Let us assume time translational invariance:

$$\text{cov}\,(x_i, x_j) \equiv C(j-i) , \quad \text{for } j > i. \tag{19.16}$$

We apply this relation to Eq. (19.15) and obtain

$$\begin{aligned} \text{var}\,(\overline{x}) &= \frac{1}{N} \left[\text{var}\,(x) + \frac{2}{N} \sum_{i=1}^{N} \sum_{j=i+1}^{N} C(j-i) \right] \\ &= \frac{1}{N} \left[\text{var}\,(x) + \frac{2}{N} \sum_{k=1}^{N} C(k)\,(N-k) \right] \\ &= \frac{1}{N} \left[\text{var}\,(x) + 2 \sum_{k=1}^{N} C(k) \left(1 - \frac{k}{N} \right) \right] , \end{aligned} \tag{19.17}$$

which can be reformulated as:

$$\text{var}\,(\overline{x}) = \frac{2\text{var}\,(x)\, \hat{\tau}_x^i}{N} . \tag{19.18}$$

We introduced here the (proper) *integrated auto-correlation time* $\hat{\tau}_x^i$

$$\hat{\tau}_x^i = \frac{1}{2} + \sum_{k=1}^{N} A(k) \left(1 - \frac{k}{N} \right) , \tag{19.19}$$

and the normalized auto-correlation function

$$A(k) = \frac{C(k)}{C(0)} = \frac{\text{cov}\,(x_i, x_{i+k})}{\text{var}\,(x_i)} . \tag{19.20}$$

In most cases we are interested in the limit $N \to \infty$ of Eq. (19.19):

$$\tau_x^i = \lim_{N\to\infty} \hat{\tau}_x^i = \frac{1}{2} + \sum_{k=1}^{\infty} A(k) . \tag{19.21}$$

The form of the auto-correlation function $A(x)$ can be approximated using the results of Sect. 16.4. There, we observed that the stationary distribution π was the left-eigenvector of the transition matrix P with eigenvalue 1, Eq. (16.73). Let $\{\varphi_\ell\}$ denote the set of all left-eigenvectors of the matrix P with eigenvalues λ_ℓ, i.e. $\varphi_\ell P = \lambda_\ell \varphi_\ell$.[2] Then some arbitrary state $q(0)$ can be expressed in this basis as:

$$q(0) = \sum_i \alpha_i \varphi_i . \tag{19.22}$$

After n consecutive time-steps we arrive at state $q(n)$

$$q(n) = q(0)P^n = \sum_i \alpha_i \varphi_i P^n = \sum_i \alpha_i \lambda_i^n \varphi_i , \tag{19.23}$$

which follows from Eq. (16.62). We denote the observable we want to calculate by $O(n)$ and expand it according to Ref. [10]

$$O(n) = \sum_i [q(n)]_i o_i = \sum_i \alpha_i \lambda_i^n o_i , \tag{19.24}$$

where o_i stands for the expectation value of O in the i-th eigenstate φ_i. For large n the value of $O(n)$ will be dominated by the largest eigenvalue of P, say λ_0, and we denote this value by $O(\infty) = \alpha_0 o_0$. This allows us to rewrite Eq. (19.24) as

$$O(n) = O(\infty) + \sum_{i\neq 0} \alpha_i o_i \lambda_i^n . \tag{19.25}$$

Let $\lambda_1 \in \mathbb{R}$ be the second largest eigenvalue and let us define the *exponential auto-correlation time* τ_x^e via

$$\tau_x^e = -\frac{1}{\log(\lambda_1)}, \tag{19.26}$$

[2] Note that since P is a stochastic matrix, it follows that $|\lambda_\ell| \leq 1$ for all ℓ. Furthermore, it can be shown that the largest eigenvalue of a stochastic matrix is equal to 1.

and the value of $O(n)$ can, for large values of n, be approximated by

$$O(n) \approx O(\infty) + \beta \exp\left(-\frac{n}{\tau_x^e}\right), \tag{19.27}$$

where β is some constant. Hence, the auto-correlation obeys

$$C(n) \propto [O(0) - O(\infty)]\,[O(n) - O(\infty)] \propto \beta \exp\left(-\frac{n}{\tau_x^e}\right), \tag{19.28}$$

and we can simply set for the auto-correlation function $A(k)$

$$A(k) = \gamma \exp\left(-\frac{k}{\tau_x^e}\right), \tag{19.29}$$

where γ is some constant. We use this result in the expression for the integrated auto-correlation time (19.21) and arrive at:

$$\begin{aligned} \tau_x^i &= \frac{1}{2} + \gamma \sum_{k=1}^{\infty} \left[\exp\left(-\frac{1}{\tau_x^e}\right)\right]^k \\ &= \frac{1}{2} + \gamma \frac{\exp\left(-\frac{1}{\tau_x^e}\right)}{1 - \exp\left(-\frac{1}{\tau_x^e}\right)} . \end{aligned} \tag{19.30}$$

For $\tau_x^e \gg 1$ the exponential function can be expanded into a TAYLOR series. Keeping terms up to first order results in:

$$\tau_x^i = \frac{1}{2} + \gamma \frac{1 - \frac{1}{\tau_x^e}}{\frac{1}{\tau_x^e}} = \frac{1}{2} + \gamma\left(\tau_x^e - 1\right) \propto \gamma \tau_x^e . \tag{19.31}$$

However, we note that in general relation (19.31) is only a poor approximation because usually the exponential auto-correlation time is very different from the integrated auto-correlation time.

Let us briefly discuss our results. A comparison between Eqs. (19.3) and (19.18) reveals that due to correlations in the time series, the number of effective (or useful) data points N_{eff} can be determined from

$$N_{\text{eff}} = \frac{N}{2\tau_x^i} . \tag{19.32}$$

In the limit $\tau_x^e \to 0$ we obtain $\tau_x^i = 1/2$ and, thus, recover Eq. (19.3). The effective number of measurements is the relevant quantity whenever the error of a Monte Carlo integration is calculated.

In another approach, one can determine the exponential auto-correlation time τ_x^e and use it to estimate the number of steps that should be neglected between two successive measurements. This can be achieved by fitting the auto-correlation $A(k)$ with an exponential function. (A brief introduction to least squares fits can be found in Appendix H.) We note that in one and the same system the auto-correlation times may be very different for different observables.

19.4 The Histogram Technique

The histogram technique is a method which allows to approximate the expectation value of some observable for temperatures near a given temperature T_0 without performing further MARKOV-chain Monte Carlo simulations. The basic idea is easily sketched. Suppose the observable O is solely a function of energy E. We perform a MARKOV-chain Monte Carlo simulation for a given temperature T_0 and measure the energy E several times. The resulting measurements are sorted in a histogram with bin width ΔE as was demonstrated in Sect. 18.3. If $n(E)$ denotes the number of configurations measured within the interval $(E, E + \Delta E)$, then the probability that some energy is measured to lay within the interval $(E, E + \Delta E)$ is given by

$$P_H(E, T_0) = \frac{n(E)}{M} , \tag{19.33}$$

where the index H refers to histogram and $M = \sum_E n(E)$ is the number of measurements. However, we note that this probability can also be expressed by the BOLTZMANN distribution

$$P(E, T) = \frac{N(E) \exp\left(-\frac{E}{k_B T}\right)}{\sum_E N(E) \exp\left(-\frac{E}{k_B T}\right)} , \tag{19.34}$$

where $N(E)$ denotes the number of micro-states within the interval $(E, E + \Delta E)$. $N(E)$ is independent of the temperature T and relation (19.34) is valid for all temperatures T. In particular, for $T = T_0$

$$P_H(E, T_0) = P(E, T_0) , \tag{19.35}$$

which immediately yields

$$N(E) = \alpha n(E) \exp\left(\frac{E}{k_B T_0}\right) , \tag{19.36}$$

where α is some constant and we emphasize that $n(E)$ was measured at T_0. Inserting Eq. (19.36) into (19.34) yields

$$P(E,T) = \frac{n(E)\exp\left[-\left(\frac{1}{k_B T} - \frac{1}{k_B T_0}\right)E\right]}{\sum_E n(E)\exp\left[-\left(\frac{1}{k_B T} - \frac{1}{k_B T_0}\right)E\right]}, \quad (19.37)$$

for arbitrary T. The expectation value $\langle O \rangle_T$ of the observable O at some temperature T can now be determined from

$$\langle O \rangle_T = \sum_E O(E)P(E,T) = \frac{\sum_E O(E)n(E)\exp\left[-\left(\frac{1}{k_B T} - \frac{1}{k_B T_0}\right)E\right]}{\sum_E n(E)\exp\left[-\left(\frac{1}{k_B T} - \frac{1}{k_B T_0}\right)E\right]}. \quad (19.38)$$

This result implies, that it is not necessary to run an additional MARKOV-chain Monte Carlo simulation in an attempt to compute the expectation value $\langle O \rangle_T$ for temperature T if T is in the vicinity of T_0. However, if T deviates strongly from T_0, the above procedure (19.38) does not provide a good approximation because the relevant configurations at T may have been very improbable at T_0 and may, therefore, not have been reproduced sufficiently often in the original MARKOV-chain Monte Carlo simulation.

Summary

Data analysis is an important but often neglected part of natural sciences and in particular of numerical simulations. It consists mainly of consistency checks and error analysis. This chapter concentrated in a first step on error analysis. It discussed the most common methods to arrive at an estimate of the error involved whenever expectation values of some property are analyzed. These went beyond all those methods which have already been discussed in some detail throughout this book. In a second step auto-correlations have been discussed. They should be part of consistency checks and give valuable information about possible systematic errors. The auto-correlation analysis was of particular importance whenever the quality of the sequence of random numbers was crucial to a particular simulation. (Experiments in which the events are expected to be random, like radioactive decay, fall also into this category.) Nevertheless, this method proved to be very useful in MARKOV-chain Monte Carlo simulations as it allowed to define and determine an auto-correlation time which could serve as a measure of the number of sweeps which have to be neglected between two consecutive measurements. Finally, the histogram technique was introduced as a method of data interpolation. It allowed, in addition to applications which have already been presented within this book, to derive the expectation value of some property at some 'temperature' T from the already known expectation value of this same property at some other temperature T_0 if $T \sim T_0$ and if the equilibrium distribution was known.

Problems

1. Calculate the auto-correlation function for random numbers generated by the two linear congruential generators discussed in Sect. 12.2. Check also the random number generator provided by your system. Discuss the results.
2. POTTS model: Calculate the error attached to the specific heat c_h and the susceptibility χ using the Jackknife method for all values of $q = 1, \ldots, 8$. Plot the corresponding diagrams and discuss the results. Determine the exponential and integrated correlation time.

References

1. Gaul, W., Opitz, O., Schader, M. (eds.): Data Analysis. Springer, Berlin/Heidelberg (2000)
2. Sivia, D., Skilling, J.: Data Analysis, 2nd edn. Oxford University Press, Oxford (2006)
3. Adèr, H.J., Mellenbergh, G.J., Hand, D.J. (eds.): Advising on Research Methods: A Consultant's Companion, chap. 14, 15. Johannes van Kessel, Huizen (2008)
4. Brandt, S.: Data Analysis. Springer, Berlin/Heidelberg (2014)
5. von der Linden, W., Dose, V., von Toussaint, U.: Bayesian Probability Theory. Cambridge University Press, Cambridge (2014)
6. Iversen, G.P., Gergen, I.: Statistics. Springer Undergraduate Textbooks in Statistics. Springer, Berlin/Heidelberg (1997)
7. Wilcox, R.R.: Basic Statistics. Oxford University Press, New York (2009)
8. Monahan, J.F.: Numerical Methods of Statistics. Cambridge Series in Statistical and Probabilistic Mathematics. Cambridge University Press, Cambridge (2011)
9. Wood, S.: Core Statistics. Institute of Mathematical Statistics Textbooks. Cambridge University Press, Cambridge (2015)
10. Sokal, A.D.: Monte Carlo Methods in Statistical Mechanics: Foundations and New Algorithms. Department of Physics. New York University, New York (1996). www.stat.unc.edu/faculty/cji/Sokal.pdf

Chapter 20
Stochastic Optimization

20.1 Introduction

Suppose $x \in \mathbb{S}$ is some vector in an n-dimensional search space $\mathbb{S}$ and let $\mathbb{H} : \mathbb{S} \to \mathbb{R}$ be a mapping from the search space $\mathbb{S}$ onto the real axis $\mathbb{R}$. The function $\mathbb{H}$ plays a particular role and is usually referred to as the *cost function*. A minimization problem can be defined in a very compact form:

Find $x_0 \in \mathbb{S}$, such that $\mathbb{H}(x_0)$ is the *global* minimum of the cost function $\mathbb{H}$.

In analogue, a maximization problem with cost function $\mathbb{H}$ defines a minimization problem with cost function $\mathbb{G} = -\mathbb{H}$. The class of both problems is referred to as the class of optimization problems [1–3] and only minimization problems will be discussed here.

The reader might be aware that there are numerous applications in physics and related sciences. We list a few in order to remind ourselves of their fundamental importance:

- The set of linear equations $Ax = b$ is often regarded as the minimization problem: $\mathbb{H}(x) = \|Ax - b\|^2$ which can be beneficial for high dimensional problems.
- The quantum mechanical ground state energy E_0 is given by

$$E_0 = \min_{\Psi} \frac{\langle \Psi \,|\, H \,|\, \Psi \rangle}{\langle \Psi \,|\, \Psi \rangle}, \tag{20.1}$$

  where $|\Psi\rangle$ denotes the wave function and H is the Hamiltonian of the system.
- High dimensional and highly non-linear least squares fits. (More details can be found in Appendix H.)
- The equilibrium crystal structure of solids is obtained by minimization of the free energy.
- Protein folding is described by minimization of the forces in a *molecular dynamics* problem.

B.A. Stickler, E. Schachinger, *Basic Concepts in Computational Physics*,
DOI 10.1007/978-3-319-27265-8_20

Whenever the cost function is at least once differentiable, methods of deterministic optimization can be applied [4]. (Two simple deterministic optimization methods are presented in Appendix I.) On the other hand, if $\mathbb{H}$ is not differentiable or too complex, due to a huge search space $\mathbb{S}$ or many local minima, methods of stochastic optimization [5] can be employed. The term stochastic optimization is used for methods which contain at least one step which is based on random number generation. Let us briefly give some examples of problems for which deterministic methods fail:

- The *Traveling Salesperson Problem* [6, 7]: A traveling salesperson has to visit L cities in a tour as short as possible under the constraint that he/she has to return to the starting point in the end. Each city has to be visited only once, hence the cities have to be ordered in such a way that the travel length becomes a global minimum. In particular, the cost function

$$\mathbb{H}(\{i\}) = \sum_{\ell=1}^{L} |x_{i_{\ell+1}} - x_{i_\ell}| \,, \tag{20.2}$$

has to be minimized. Here $\{i\}$ denotes a certain configuration of cities and we set $i_{L+1} = i_1$. Obviously, we cannot calculate the first derivative of $\mathbb{H}$ with respect to $\{i\}$, set it zero, and solve the problem in the classical way. On the other hand, a brute force approach of calculating $\mathbb{H}(\{i\})$ for all possible arrangements $\{i\}$ is not possible since we have $L!$ different possible routes. Since for one particular choice all L starting points and both travel directions yield the same result, we have to calculate $L!/(2L) = (L-1)!/2$ different configurations $\{i\}$. We would have about 10^{155} different choices for $L = 100$ cities! This clearly makes such an approach intractable.
- The arrangement of timetables under certain constraints. In particular, the design of timetables in schools, universities or at airports. This problem is also referred to as the *Nurse Scheduling Problem* [8].
- The ISING *spin glass* [9]: In contrast to the classical ISING model, the ISING spin glass is characterized by nearest neighbor interactions J_{ij} which are, in the most simple case, chosen to be $J_{ij} = +1$ and $J_{ij} = -1$ with the same probability. In this case the ground state below the critical temperature is not simply given by a configuration in which all spins point in the same direction. Of course, the ground state configuration in such a case can be highly degenerate. The fact that such a model can be simulated using MARKOV-chain Monte Carlo methods as they have been discussed within Chaps. 15 and 18 gives us some idea of how one may employ stochastic methods to solve optimization problems.
- The *N-Queens Problem* [10]: Place N queens on a $N \times N$ chessboard in such a way that no two queens attack each other. In particular, this means that two queens are not allowed to share the same row, the same column, and the same diagonal. It can be shown that the problem possesses solutions for $N \geq 4$. One defines a function $\mathbb{H}(\{n\})$ which counts the number of attacks in a certain configuration

$\{n\}$. For instance, for $N = 4$, the configuration

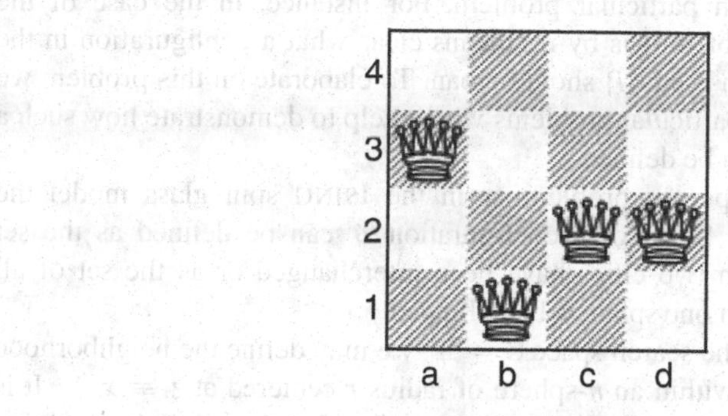

has $\mathbb{H}(\{n\}) = 2$. On the contrary, the configuration

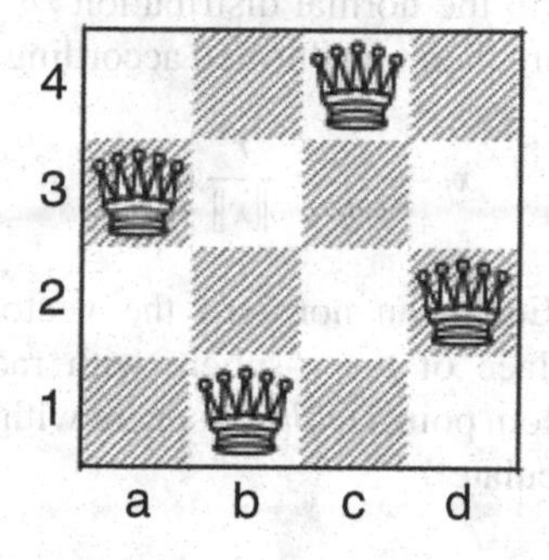

solves the 4-queens problem and $\mathbb{H}(\{n\}) = 0$.

We concentrate here on some of the most basic methods of stochastic optimization: the method of *hill climbing*, the method of *simulated annealing*, and *genetic algorithms*. Ideas on which several more advanced techniques are based will be sketched in Sect. 20.5.

20.2 Hill Climbing

The method of *hill climbing* [11] is probably one of the most simple methods of stochastic optimization. Given a cost function $\mathbb{H}(x)$, we execute the following steps:

1. Choose an initial position x_0.
2. Randomly pick a new x_n from the *neighborhood* of x_{n-1}.
3. Keep x_n if $\mathbb{H}(x_n) \leq \mathbb{H}(x_{n-1})$.
4. Terminate the search if no new x_n can be found in the neighborhood of x_{n-1}.

We note that the algorithm requires a *neighborhood* relation. This relation is to be defined for each particular problem. For instance, in the case of the traveling salesperson problem it is by no means clear what a configuration in the *neighborhood* of a certain route $\{i\}$ should mean. To elaborate on this problem we concentrate here on two particular problems which help to demonstrate how such a neighborhood relation can be defined.

In the traveling salesperson problem or in the ISING spin glass model the neighborhood of a route $\{i\}$ or of a configuration $\mathscr{C}$ can be defined as the set of all routes $\{i\}$ in which two cities have been interchanged or as the set of all configurations $\mathscr{C}$ in which one spin has been flipped.

On the other hand, if the search space $\mathbb{S} = \mathbb{R}^n$ we may define the neighborhood as the number of points within an n-sphere of radius r centered at $z \equiv x_{n-1}$. It is rather simple to sample points from an n-sphere centered at the origin by applying the method of G. MARSAGLIA [12]: For an n-dimensional vector we sample all components $x_1, \ldots, x_n$ from the normal distribution $\mathscr{N}(0, 1)$ with mean zero and variance one. The points are then transformed according to

$$x_j \to x'_j = \frac{r}{\|x\|} x_j + z_j \,, \tag{20.3}$$

where $\|x\|$ denotes the Euclidean norm of the vector x. The points given by Eq. (20.3) lie on the surface of the n-sphere with radius r. In order to obtain uniformly distributed random points *within* a sphere with radius r we draw a random number $u \in [0, 1]$ and calculate

$$x_j \to x'_j = u^{\frac{1}{n}} x_j \,, \tag{20.4}$$

where the factor $1/n$ in the exponent of u ensures that the points are uniformly distributed.

Let us briefly summarize the most important properties of the method of hill climbing:

- The way the algorithm is defined it will terminate in a *local* minimum and not in the global minimum. A classical remedy is the restart of the algorithm from various different initial positions. Information gathered from previous runs can help to make a good choice for the initial positions of restarts.
- It depends highly on the choice of initial conditions if and how the global minimum is found. This situation is very similar to the application of deterministic methods of optimization (see Appendix I). Sometimes it may even be of advantage to accept points which result in a slight increase of the cost function's value just to escape a local minimum.
- For most problems this method is very expensive from a computational point of view.

We apply the method of hill climbing to the N-queens problem for $N = 8$. The algorithm is executed in the following way: In the initial configuration the queens are set randomly on the chessboard and we place only one queen in each row and column. It is then checked whether or not two queens attack each other. If they do, a new configuration is generated by picking two queens at random and by changing their respective positions. This is repeated until a configuration arises in which none of the queens attacks another. Such an algorithm resembles a random walk in a parameter space which spans all possible configurations under the constraint that only one queen is placed in each row and column. The iteration is terminated as soon as no queen is attacked by any other queen. It is rather obvious that this strategy is not very fast, however, one possible solution to the problem for $N = 8$ can easily be found within a few iteration steps:

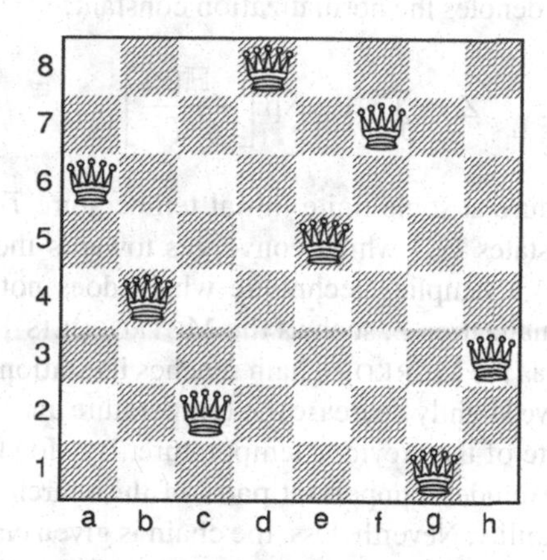

However, for large values of N hill climbing is definitely not a recommendable method to solve the N-queens problem.

20.3 Simulated Annealing

Let us turn our attention to *simulated annealing* [7, 13, 14]. The name of this algorithm stems from the annealing process in metallurgy in which a metal is first heated and then slowly cooled in order to reduce the amount of defects in the material. The reasoning behind this method can quite easily be reconstructed with the help of the ISING model which we discussed in detail in Chap. 15. There we learned from thermodynamics that the equilibrium distribution of possible configurations $P(\mathscr{C}, T)$ at a certain temperature T is a BOLTZMANN distribution

$$P(\mathscr{C}, T) = \frac{1}{Z} \exp\left[-\frac{H(\mathscr{C})}{k_B T}\right], \tag{20.5}$$

where $H(\mathscr{C})$ is the HAMILTON function of the system. In particular, we expect that the system is in its ground state (let us assume a non-degenerate ground-state for the time being) with probability one in the limit $T \to 0$, provided that we cooled sufficiently slowly so that the system had enough time to equilibrate. This can be used to solve the optimization problem: We take the cost function $\mathbb{H}(x)$ and define the probability for the realization of a particular state (configuration) in the search space $x_0 \in \mathbb{S}$ by

$$P(x_0, T) = \frac{1}{Z} \exp\left[-\frac{\mathbb{H}(x_0)}{T}\right], \tag{20.6}$$

where T is some external parameter, which we refer to as *temperature* for reasons of convenience, and Z denotes the normalization constant:

$$Z = \sum\!\!\!\!\!\!\int_{x\in\mathbb{S}} \mathrm{d}x \exp\left[-\frac{\mathbb{H}(x)}{T}\right]. \tag{20.7}$$

We start the procedure at some finite initial temperature $T_0 \neq 0$ and construct a MARKOV-chain of states $\{x_n\}$ which converges towards the distribution (20.6). We choose, of course, a sampling technique which does not require the explicit knowledge of the normalization Z, such as the METROPOLIS-HASTINGS algorithm of Sect. 18.2. As soon as the MARKOV-chain reaches its stationary distribution for a given temperature T, we slightly decrease the temperature and restart the MARKOV-chain with the last state of the previous temperature. By slowly *cooling* the search MARKOV-chain, we exclude unimportant parts of the search space by decreasing their acceptance probability. Nevertheless, the chain is given enough time to explore the whole remaining search space at each temperature. This procedure is commonly referred to as the classical version of *simulated annealing*.

It is of advantage to start with an initial temperature which allows to cover the largest part of possible states in the search space $\mathbb{S}$. Thus, the acceptance probability for a new state in the MARKOV-chain is almost equal to one for all $x \in \mathbb{S}$. If this were not the case, some regions of the search space might be excluded from our search routine right away due to an unlucky choice of the initial configuration. In particular, the result might be a state in the neighborhood of the initial state of the MARKOV-chain and it is, therefore, most likely a local minimum rather than the global minimum.

We note that the algorithm consists of the following essential ingredients: (i) a proposal probability for new states x within the search space $\mathbb{S}$, (ii) an acceptance probability $P_a(x \to x')$ for a proposed x' from a previous state x, and (iii) a cooling strategy $T = T(t)$, where t is time. Let us briefly elaborate on these points.

(i) Proposal Probability

The question of how to generate new states x from a previous state x' within the search space $\mathbb{S}$ has already been answered in the case of hill climbing, Sect. 20.2 by defining the neighborhood of a state in search space. The corresponding proposal probability will be denoted by $P_p(x \to x')$.

(ii) Acceptance of Probability

The acceptance probability has to be chosen in such a way that the sequence of generated states constitutes a MARKOV-chain which converges toward the distribution (20.6). Hence, detailed balance has to be imposed and the implications of this requirement have been discussed extensively in Chap. 18. Note that the proposal probability has to be included into the definition of the acceptance probability as was outlined in Sect. 18.2.

One particular choice of a METROPOLIS-HASTINGS acceptance probability

$$P_a(x \to x', T) = \min\left(1, \frac{P(x', T)}{P(x, T)} \frac{P_p(x' \to x)}{P_p(x \to x')}\right) , \tag{20.8}$$

appears to be quite natural for several reasons:

- It is very general and can, thus, also handle asymmetric proposal probabilities.
- In the symmetric case $P_p(x \to x') = P_p(x' \to x)$ and $\mathbb{H}(x') \leq \mathbb{H}(x)$ we get

$$\frac{P(x', T)}{P(x, T)} = \exp\left\{\frac{1}{T}\left[\mathbb{H}(x) - \mathbb{H}(x')\right]\right\} \geq 1 , \tag{20.9}$$

  according to our choice (20.6) and the state x' is accepted with probability one. On the other hand, for $\mathbb{H}(x') > \mathbb{H}(x)$, x' may still be accepted with some finite probability $P_a(x \to x', T)$ which offers an opportunity to escape a local minimum.

(iii) Cooling Strategy

The design of a proper cooling strategy includes both, the choice of an appropriate initial temperature T_0 as well as the formulation of a mathematical rule which defines $T_{n+1} = f(T_n)$ where $T_{n+1} < T_n$.

First of all we discuss the choice of the initial temperature. A common choice is to choose it in such a way that at least 80 % of all generated states are accepted. The simplest procedure to determine this temperature starts with some arbitrary value $T_0 > 0$ and generates N states. If the number of rejected states N_r is greater than $0.2\,N$, then the temperature T_0 is doubled and the number of rejected states is measured again.

Another more sophisticated choice is based on the following idea: The best choice would be $T_0 \to \infty$ because then the acceptance probability would be one for all possible states independent of $\mathbb{H}(x)$. This corresponds to a random walk in search space $\mathbb{S}$ and we calculate the mean value $\langle \mathbb{H} \rangle_\infty$ and the variance $\text{var}(\mathbb{H})_\infty$. Thus, the function values $\mathbb{H}$ fluctuate between $[\langle \mathbb{H} \rangle_\infty - \sqrt{\text{var}(\mathbb{H})_\infty}, \langle \mathbb{H} \rangle_\infty + \sqrt{\text{var}(\mathbb{H})_\infty}]$. We consider now the expectation value $\langle \mathbb{H} \rangle_{T_0}$ for large values of T_0. We define the small parameter $\epsilon = 1/T_0 \ll 1$ and find with $p(x, \epsilon) = P(x, T)$

$$\begin{aligned} \langle \mathbb{H} \rangle_\epsilon &= \int \mathrm{d}x\, p(x, \epsilon) \mathbb{H}(x) \\ &= \langle \mathbb{H} \rangle_0 - \epsilon \left[\langle \mathbb{H}^2 \rangle_0 - \langle \mathbb{H} \rangle_0^2 \right] . \end{aligned} \tag{20.10}$$

Re-substituting $T_0 = 1/\epsilon$ results, finally, in:

$$\langle \mathbb{H} \rangle_{T_0} \approx \langle \mathbb{H} \rangle_\infty - \frac{\text{var}(\mathbb{H})_\infty}{T_0} . \tag{20.11}$$

The initial temperature T_0 is now chosen in such a way that the expectation value $\langle \mathbb{H} \rangle_{T_0}$ borders the infinite temperature fluctuations from below and we set consequently

$$\langle \mathbb{H} \rangle_{T_0} = \langle \mathbb{H} \rangle_\infty - \sqrt{\text{var}(\mathbb{H})_\infty} , \tag{20.12}$$

with the implication that

$$T_0 = \sqrt{\text{var}(\mathbb{H})_\infty} . \tag{20.13}$$

We are now in a position to investigate appropriate cooling strategies: The *geometric cooling schedule*

$$T_n = T_0 q^n , \tag{20.14}$$

with $0 \ll q < 1$ is very often used. However, particular cost functions $\mathbb{H}(x)$ may develop several *phase transitions* in the course of the cooling process. Naturally, the expectation value $\langle \mathbb{H} \rangle$ changes rapidly in the region $T \approx T_c$, with T_c the temperature at which the phase transition occurs. It is, therefore, certainly of advantage to take such a possibility into account and to design the cooling strategy accordingly.

Hence, a more appropriate strategy is to use temperature changes which cause only slightly modified acceptance probabilities. In particular, we demand that

$$\frac{1}{1+\delta} < \frac{P(x, T_n)}{P(x, T_{n+1})} < 1 + \delta , \tag{20.15}$$

with $0 < \delta \ll 1$. Assuming a BOLTZMANN type distribution for $P(x, T_n)$, we obtain

$$\exp\left[-\mathbb{H}(x)\left(\frac{1}{T_n} - \frac{1}{T_{n+1}}\right)\right] < 1 + \delta \,, \tag{20.16}$$

or

$$T_{n+1} > \frac{T_n}{1 + \frac{T_n}{\mathbb{H}(x)}\ln(1+\delta)} \,. \tag{20.17}$$

Hence, we can choose

$$T_{n+1} \approx \frac{T_n}{1 + \frac{T_n}{3\sqrt{\mathrm{var}(\mathbb{H})_{T_n}}}\ln(1+\delta)} \,, \tag{20.18}$$

where we replaced $\mathbb{H}(x) \approx 3\sqrt{\mathrm{var}\,(\mathbb{H})_{T_n}}$. This choice is plausible if one recognizes that we can replace $\mathbb{H}(x) \to \mathbb{H}(x) - \mathbb{H}_{\min}$ in the above calculations, where $\mathbb{H}_{\min}$ represents the (unknown) minimum of $\mathbb{H}(x)$. This cooling schedule is known as the AARTS *schedule*.

Finally, we have to discuss how to terminate the algorithm. Typically, there are several choices and we present briefly the most popular ones. The obvious choice is to terminate the algorithm as soon as the acceptance ratio is below some predefined threshold value. A more sophisticated choice is to terminate the algorithm whenever the mean value $\langle\mathbb{H}\rangle$ reaches some constant value. A quite different and more formal approach would be to initially define a maximum number of iterations or to set the final temperature T_f to some reasonable value. Nevertheless, the termination condition has to be defined for each particular problem individually.

Before presenting an example, we note some further results associated with cooling strategies. It was demonstrated by S. KIRKPATRIK et al. [15] that the optimal cooling strategy for a BOLTZMANN type distribution is of the form

$$T_n \propto \frac{1}{\ln(n)} \,, \tag{20.19}$$

where n labels the temperature steps. In this case the global minimum is found with probability one. However, the convergence is rather slow. In addition, several extensions of classical simulated annealing have been suggested in the literature. For instance, *fast simulated annealing* uses a CAUCHY distribution

$$P(x, T) = \frac{T}{(x^2 + T^2)^{\frac{d+1}{2}}} \tag{20.20}$$

instead of a BOLTZMANN distribution. Here d is the dimension of the search space $\mathbb{S}$. The optimal cooling strategy for such a distribution function is of the form

$$T_n \propto \frac{1}{n} \,, \tag{20.21}$$

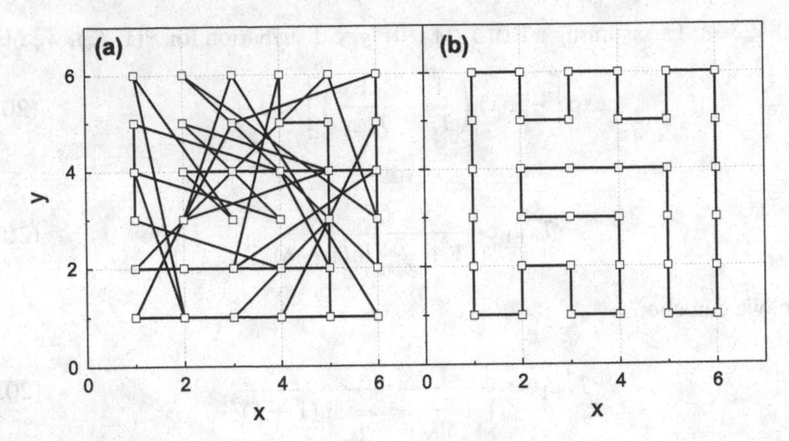

Fig. 20.1 (a) Initial route of the traveling salesperson for 36 cities on a regular grid. (b) One of many optimal routes of the traveling salesperson for 36 cities on a regular grid

which signifies a considerable increase in convergence speed in comparison to Eq. (20.19). Another generalization is referred to as *generalized simulated annealing* and is based on the TSALLIS distribution which depends on an external parameter ϵ:

$$P_\epsilon(x,T) = \frac{1}{Z}\left[1 + \frac{\epsilon \mathbb{H}(x)}{k_B T}\right]^{-\frac{1}{\epsilon}} . \tag{20.22}$$

It can be demonstrated that P_ϵ converges toward the BOLTZMANN distribution for $\epsilon \to 0$. We mention in passing that the concept of the TSALLIS distribution is closely intertwined with the definition of the TSALLIS entropy and the formulation of non-extensive thermodynamics by C. TSALLIS [16].

As a first illustrative example we discuss the *traveling salesperson problem* for $N = 36$ cities on a regular grid because in this case the optimal route is easily identified. We calculate the initial temperature from Eq. (20.13) and employ the geometric cooling schedule (20.14) with $q = 0.99$ together with a termination criterion of the form

$$\langle \mathbb{H} \rangle_{T_n} - \langle \mathbb{H} \rangle_{T_{n-1}} < \eta \, , \tag{20.23}$$

where η is the required accuracy. Figure 20.1a presents one route for the initial temperature and Fig. 20.1b displays one of many optimal routes after convergence has been reached. This case will be called the *first scenario*. In the *second scenario* we place 36 cities in four equally spaced clusters. Results for the optimal route are presented in Fig. 20.2b.

The possibility of phase transitions to occur during the cooling process has already been mentioned. In a genuine physical system the question whether a phase transition is possible at all or if it is of first or second order is solely determined by

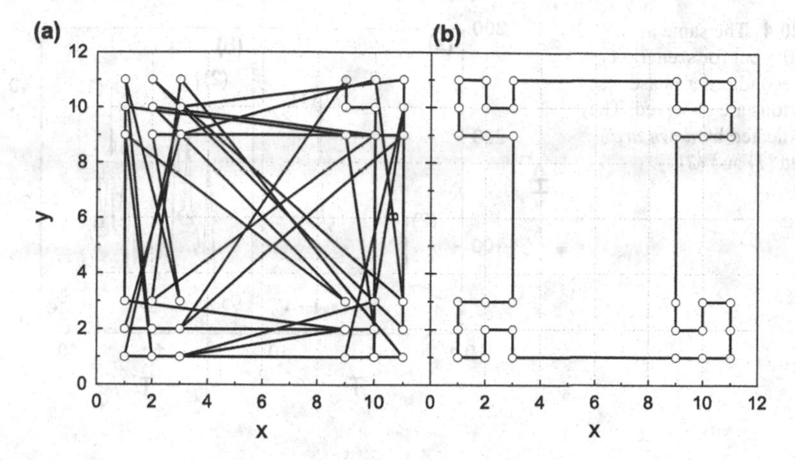

Fig. 20.2 (a) Initial route of the traveling salesperson for 36 cities placed in four equally spaced clusters. (b) One of many optimal routes of the traveling salesperson for 36 cities placed in four equally spaced clusters

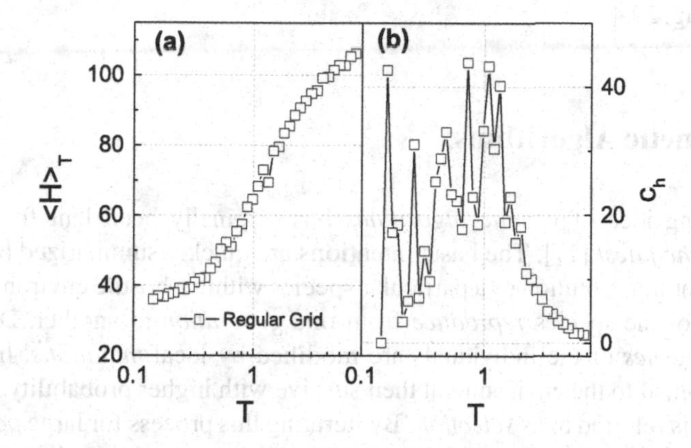

Fig. 20.3 (a) The expectation value $\langle \mathbb{H} \rangle_T$ and (b) the 'specific heat' c_h vs temperature T for scenario one

the HAMILTON function $H(x)$ of the system. As an intriguing example we refer to the q-states POTTS model of Sect. 18.3 where a second order phase transition was observed for $q \leq 4$ and a first order phase transition for $q > 4$. In analogy, the order of a 'phase transition' during the iteration process toward the global minimum in simulated annealing is completely determined by the particular properties of the cost function $\mathbb{H}(x)$. We want to study such a possibility and determine the expectation values $\langle \mathbb{H} \rangle_T$ and the 'specific heat' c_h as functions of temperature T for the two scenarios of the traveling salesperson problem. Figure 20.3 presents the results for scenario one and Fig. 20.4 those for scenario two. The second scenario develops two second order phase transitions while in the first scenario only one second order phase transition can be observed. The first phase transition of the second scenario

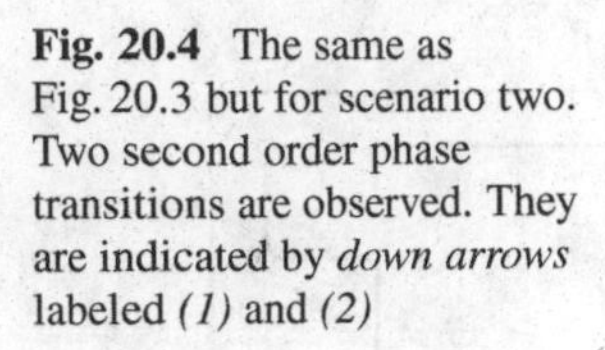

Fig. 20.4 The same as Fig. 20.3 but for scenario two. Two second order phase transitions are observed. They are indicated by *down arrows* labeled *(1)* and *(2)*

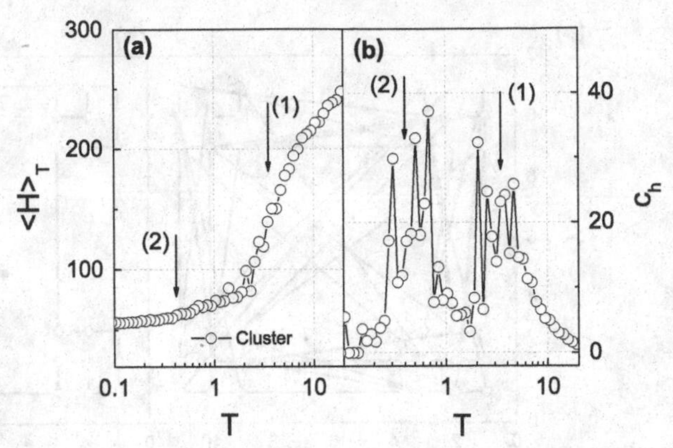

(at $T \approx 3.5$) can be related to the optimization of the clusters' sequence while in the second phase transition (at $T \approx 0.42$) the sequence of cities within the clusters becomes finalized. These two transitions are indicated by down arrows labeled (1) and (2) in Fig. 20.4.

20.4 Genetic Algorithms

The sparkling idea of *genetic algorithms* has originally been lent from natures *survival of the fittest* [17]. The basic intentions are quickly summarized by remembering the natural evolution of a particular species within a hostile environment: The individuals of the species *reproduce* from one *generation* to another. During this process the *genes* of the individuals are modified by local *mutations*. Individuals best accustomed to the environment then survive with higher probability. This very last process is referred to as *selection*. By iterating this process for large *populations* the individuals of the whole species will adjust their properties to the environment *on average*,[1] and, thus, the individuals will be better equipped for survival within the hostile environment. A large population is compulsory in order to obtain a huge variety in the *phenotype* of the individuals. Algorithms based on such a scheme are referred to as *genetic algorithms*.

We are not going into the details of the implementation of genetic algorithms because this is beyond the scope of this book. However, the ideas sketched above will be applied to the problem of the traveling salesperson passing through m-cities just to illustrate the method. Let $s = (s_1, \dots, s_m) \in \mathbb{N}^m$ denote a list of m integers, which obey $s_i \leq i$. For instance, for $m = 10$, s might be given by

[1]Note that in the real world the environment (in particular the natural enemies of a species) develop as well. Moreover, we do not consider any communication within a species, like the formation of *societies*, *learning*, and related processes.

Table 20.1 Sample tour to illustrate the recovery of the order of cities within a genetic algorithm. Elements indicated by [x] are 'selected' elements which are added to the column Tour

$\hat{s}$	1	2	3	4	5	6	7	8	9	10		Tour
9	1	2	3	4	5	6	7	8	[9]	10	→	9
4	1	2	3	[4]	5	6	7	8	10		→	4
3	1	2	[3]	5	6	7	8	10			→	3
3	1	2	[5]	6	7	8	10				→	5
5	1	2	6	7	[8]	10					→	8
1	[1]	2	6	7	10						→	1
4	2	6	7	[10]							→	10
2	2	[6]	7								→	6
2	2	[7]									→	7
1	[2]										→	2

$$s = (1, 2, 2, 4, 1, 5, 3, 3, 4, 9) . \tag{20.24}$$

The order of cities is then recovered by setting $\hat{s} = (s_m, \ldots, s_1)$ and performing the steps illustrated in Table 20.1.

In words: The vector $\hat{s}$ labels the elements taken from the list $(1, 2, \ldots, m)$ *with* removal. The resulting list *Tour* specifies the optimum sequence of the cities. The genetic algorithm is executed in the following steps:

- Define M initial individuals.
- *Mutation*: for each individual we introduce a single random local modification with probability p_{mut}.
- *Reproduction*: We produce M additional individuals by pairwise combining the *parents*. This is performed by

  (a) Pick two individuals at random.
  (b) Draw a random integer $r \in [1, m-1]$ and replace the first r genes of the first individual by the first r genes of the second individual and vice versa.

  In this way, we obtain $2M$ individuals.
- *Selection*: The M individuals with the highest *fitness* which corresponds to the lowest value of the cost function survive.

The above steps are repeated until the desired number of generations has been achieved.

In Fig. 20.5 we show the optimal path for the traveling salesperson problem discussed in the previous section, but now for $N = 30$ cities. It was obtained with the genetic algorithm described here. The number of individuals was chosen to be $M = 5000$ and the number of generations to be $G = 5000$.

Some remarks are appropriate: First of all we note that there are many different permutations of how a genetic algorithm can be realized. In particular, it is the problem which determines the most convenient form to implement the essential

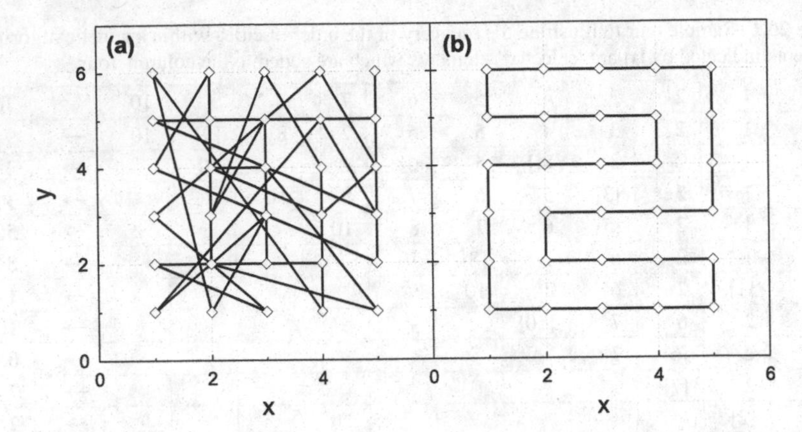

Fig. 20.5 (**a**) The random route of one individual out of the population of 5000. (**b**) One of many optimal routes of the traveling salesperson for $N = 30$ cities as obtained by a genetic algorithm

ingredients: mutation, reproduction, and selection. However, particular care is required in formulating the algorithm in such a way that it does not produce individuals which are *too* similar. In such a case the algorithm is very likely to terminate in a local minimum.

Another remark comments on how to treat optimization problems with continuous variables x. Here it might be advantageous to represent the variable x in its binary form because it makes the reproduction step particularly simple.

20.5 Some Further Methods

We briefly list some alternative stochastic optimization techniques without going into detail. Two famous alternatives which are closely related to simulated annealing are:

- *Threshold Accepting Algorithms*: The new configuration x' is accepted with probability one if $\mathbb{H}(x') \leq \mathbb{H}(x) + T$. During the simulation the *temperature* or *threshold level* T is continuously decreased. The above choice of an acceptance probability is very effective to allow for an escape from local minima.
- *Deluge Algorithms*: These algorithms are very similar to threshold accepting algorithms. We present it in the original formulation which is suited to find the global maximum of a function $\mathbb{G}(x)$. The global minimum of $\mathbb{H}(x)$ can be found by searching the maximum of $\mathbb{G}(x) = -\mathbb{H}(x)$. One accepts a new state x' with probability one if $\mathbb{G}(x') > T$, where T is continuously increased during the simulation. Hence, the whole *landscape* of $\mathbb{G}(x)$ is *flooded* with increasing T until only the summits of $\mathbb{G}(x)$ are left. Finally, only the biggest *mountain* will reach out of the *water* and the global maximum has been found.

Two famous ideas which are closely related to genetic algorithms are:

- *Grouping Genetic Algorithms*: The idea is to put the individuals of the population in distinct *groups*. These groups may for instance be formed by comparing the genes or grouping individuals with similar cost function values. All members of a group have one part of the genes in common, and all operators acting on genes act on the whole group. Such an approach can significantly improve the convergence rate of a classical genetic algorithm.
- *Ant Colony Optimization*: The idea is, again, borrowed from nature, in particular from an ant colony searching the optimum path between two or more fixed or variable points. In a real world an ant travels from one point to another randomly, leaving a trail of pheromone on its traveled path. Following ants are very likely to follow the pheromone trail, however, some random nature remains. The key point is that with time the pheromone trail starts to evaporate, hence its impact on the path of following ants is reduced if the path is not traveled frequently or often enough so that the pheromones evaporated. In this way one prevents the algorithm to get stuck in a local minimum and the global minimum may be found by sending out artificial ants.

There are many further methods available in the literature (see, for instance, Refs. [2, 18]) to which we refer the interested reader.

Summary

The local maximum/minimum of some cost function $\mathbb{H}(x)$ within a search space $\mathbb{S}$ can be determined using stochastic methods, thus establishing a particular class of algorithms known as *Stochastic Optimization*. The most straightforward method was the algorithm of *hill climbing* which resembled a controlled random walk within a restricted search space $\mathbb{S}$ called neighborhood. Because of this feature hill climbing will find in general local minima within this neighborhood and the global minimum has to be found under variation of initial conditions. This made this method too expensive for more complex problems from a computational point of view. To move from a random walk formulation to a formulation on the basis of MARKOV-chain Monte Carlo was the logical next step. The method of choice was named *simulated annealing*. It used the METROPOLIS-HASTINGS algorithm to generate new configurations within a search space $\mathbb{S}$ from a temperature dependent equilibrium distribution. A cooling strategy was used to slowly restrict the search space to the neighborhood of the global minimum. This global minimum was always found, albeit rather slowly. We mentioned some flavors of this basic algorithm which either differed in the definition of the acceptance probability or in the cooling strategy. A completely different class of algorithms was established with the so-called *genetic algorithms*. They were adapted from nature's concept of the survival of the fittest. They were based on the notions of: (i) Mutation, a single random local modification of a certain probability. (ii) Reproduction, additional 'individuals' were generated by pairwise combining parents. (iii) Selection: Individuals with the lowest value of the cost function survived and mutation started again. Genetic algorithms established a very versatile class of solvers to cover a huge body of optimization problems.

Problems

Solve the traveling salesperson problem for $N = 20$ cities on a regular grid with the help of simulated annealing. As a cooling schedule, use the geometric cooling as explained in Sect. 20.3. Determine the initial temperature by demanding an acceptance rate of 90 % and terminate the algorithm if the mean value of the cost function $\langle \mathbb{H} \rangle$ remains unchanged for at least 10 successive temperatures. Calculate the expectation value $\langle \mathbb{H} \rangle_T$ for different temperatures and identify the transition temperature. In a second step produce a list of 20 cities which are randomly distributed on a two-dimensional grid. Optimize this problem as well. Note that you should produce the list of cities only *once* in order to obtain comparable and reproducible results.

References

1. Panos, M.P., Resende, M.G.C. (eds.): Handbook of Applied Optimization. Oxford University Press, New York (2002)
2. Hartmann, A.H., Rieger, H.: Optimization Algorithms in Physics. Wiley – VCH, Berlin (2002)
3. Locatelli, M., Schoen, F.: Global Optimization. MPS-SIAM Series on Optimization. Cambridge University Press, Cambridge (2013)
4. Scholz, D.: Deterministic Global Optimization. Springer, Berlin/Heidelberg (2012)
5. Schneider, J.J., Kirkpatrick, S. (eds.): Stochastic Optimization. Springer, Berlin/Heidelberg (2006)
6. Lawler, E.L., Lenstra, K.K., Rinnooy Kan, A.H.G., Shmoys, D.B.: The Traveling Salesman Problem: A Guided Tour of Combinatorial Optimization. Wiley, New York (1985)
7. Press, W.H., Teukolsky, S.A., Vetterling, W.T., Flannery, B.P.: Numerical Recipes in C++, 2nd edn. Cambridge University Press, Cambridge (2002)
8. Kundu, S., Acharyya, S.: Stochastic local search approaches in solving the nurse scheduling problem. In: Chaki, N., Cortesi, A. (eds.) Computer Information Systems – Analysis and Technologies, Communications in Computer and Information Science, vol. 245, pp. 202–211. Springer, Berlin/Heidelberg (2011)
9. Marinari, E., Parisi, G., Ritort, F.: On the 3d Ising spin glass. J. Phys. A: Math. Gen. **27**, 2687 (1994). doi:10.1088/0305-4470/27/8/008
10. Watkins, J.J.: Across the Board: The Mathematics of Chessboard Problems. Princeton University Press, Princeton (2012)
11. Skiena, S.S.: The Algorithm Design Manual. Springer, Berlin/Heidelberg (2008)
12. Marsaglia, G.: Choosing a point from the surface of a sphere. Ann. Math. Stat. **43**, 645–646 (1972). doi:10.1214/aoms/1177692644
13. Bertsimas, D., Tsitsiklis, J.: Simulated annealing. Stat. Sci. **8**, 10–15 (1993)

14. Salamon, P., Sibani, P., Frost, R.: Facts, Conjectures, and Improvements for Simulated Annealing. Cambridge University Press, Cambridge (2002)
15. Kirkpatrik, S., Gellat, C.D., Jr., Vecchi, M.P.: Simulated annealing. Science **220**, 671 (1983)
16. Tsallis, C.: Introduction to Nonextensive Statistical Mechanics. Springer, Berlin/Heidelberg (2009)
17. Man, K.F., Tang, K.S., Kwong, S.: Genetic Algorithms. Springer, Berlin/Heidelberg (1999)
18. Bäck, T.: Evolutionary Algorithms in Theory and Practice. Oxford University Press, New York (1996)

Appendix A
The Two-Body Problem

Consider two mass points with positions $r_i(t) \in \mathbb{R}^3$, $i = 1, 2$ and masses m_i, $i = 1, 2$. It is assumed that the point masses interact through a central potential $U = U(|r_1(t) - r_2(t)|)$ and that external forces are neglected. Thus, the system is closed. The explicit notation of time t is now omitted for the sake of a more compact presentation. Furthermore, we introduce with $p_i \in \mathbb{R}^3$, $i = 1, 2$ the point mass' momentum and the LAGRANGE function [1–5] of the system takes on the form

$$L(r_1, r_2, p_1, p_2) = \frac{p_1^2}{2m_1} + \frac{p_2^2}{2m_2} - U(|r_1 - r_2|) . \tag{A.1}$$

The moments p_i are replaced by

$$p_i = m_i \dot{r}_i, \quad i = 1, 2 , \tag{A.2}$$

and this yields for the LAGRANGE function (A.1)

$$L(r_1, r_2, \dot{r}_1, \dot{r}_2) = \frac{m_1}{2}\dot{r}_1^2 + \frac{m_2}{2}\dot{r}_2^2 - U(|r_1 - r_2|) , \tag{A.3}$$

where $\dot{r}_i$ denotes the time derivative of r_i. We note the following symmetries: the LAGRANGE function is (i) translational invariant, (ii) rotational invariant, and (iii) time invariant. We know from classical mechanics that each symmetry of the LAGRANGE function corresponds to a constant of motion (a quantity that is conserved throughout the motion) and, thus, results in a reduction of the dimensionality of the 12-dimensional phase space.

Let us demonstrate these symmetries: In order to prove translational invariance, we transform to center of mass coordinates which are defined as

$$R = \frac{m_1 r_1 + m_2 r_2}{m_1 + m_2} \qquad \text{and} \qquad r = r_2 - r_1 . \tag{A.4}$$

B.A. Stickler, E. Schachinger, *Basic Concepts in Computational Physics*,
DOI 10.1007/978-3-319-27265-8

It is easily verified that we can express the original coordinates r_1 and r_2 with the help of (A.4) as

$$r_1 = R + \frac{m_2}{m_1 + m_2} r \qquad \text{and} \qquad r_2 = R - \frac{m_1}{m_1 + m_2} r \,. \tag{A.5}$$

The LAGRANGE function (A.3) is rewritten in these new coordinates (A.4) and this yields

$$\begin{aligned} L(r, R, \dot{r}, \dot{R}) &= \frac{M}{2}\dot{R}^2 + \frac{m}{2}\dot{r}^2 - U(|r|) \\ &\equiv L(r, \dot{r}, \dot{R}) \,, \end{aligned} \tag{A.6}$$

where we introduced the total mass M and the *reduced* mass m:

$$M = m_1 + m_2 \qquad \text{and} \qquad m = \frac{m_1 m_2}{m_1 + m_2} \,. \tag{A.7}$$

Obviously, the center of mass coordinate R plays in Eq. (A.6) the role of a *cyclic* coordinate: It does not appear explicitly in the LAGRANGE function. This means that the system is translational invariant and we can deduce from LAGRANGE's equations that

$$\frac{\mathrm{d}}{\mathrm{d}t}\frac{\partial}{\partial \dot{R}} L = \frac{\partial}{\partial R} L = 0 \,, \tag{A.8}$$

and the center of mass momentum is conserved. Hence, we obtain that

$$\frac{\partial}{\partial \dot{R}} L = M\dot{R} = \text{const} \,, \tag{A.9}$$

with the solution

$$R(t) = At + B \,, \tag{A.10}$$

where $A, B \in \mathbb{R}^3$ are constants determined by the initial conditions of the problem. As a result, the center of mass moves along a straight line with constant velocity. We collect all results and reformulate the LAGRANGE function (A.6) as

$$\begin{aligned} L(r, \dot{r}) &= \frac{M}{2}A^2 + \frac{m}{2}\dot{r}^2 - U(|r|) \\ &\equiv \tilde{L}(r, \dot{r}) + \text{const} \,. \end{aligned} \tag{A.11}$$

Hence, the problem was reduced to a one-body problem with the LAGRANGE function $\tilde{L}(r, \dot{r})$. In what follows the tilde is omitted and the LAGRANGE function

$$L(r, \dot{r}) = \frac{m}{2}\dot{r}^2 - U(|r|) , \tag{A.12}$$

is now studied instead of Eq. (A.11). It is an effective one-body LAGRANGE function.

In the next step the effect of rotational invariance is investigated. Equation (A.12) resembles the LAGRANGE function of a particle of mass m which is located at position r and moves in the field of a central force $F \in \mathbb{R}^3$. This force points to the center of the coordinate system (or points from the center of the coordinate system to the particle). This situation is clearly invariant under a rotation of the coordinate system since $U = U(|r|)$ depends only on the modulus of r. Consequently, r is parallel to F for all $t \geq 0$. In such a case the vector of angular momentum $\ell \in \mathbb{R}^3$ is conserved, since

$$\frac{\mathrm{d}}{\mathrm{d}t}\ell = \mathscr{M} = r \times F = 0 , \quad \to \ell = \text{const} , \tag{A.13}$$

where $\mathscr{M}$ is the *torque*. This allows us to arbitrarily rotate our coordinate system. We take advantage if this property and rotate it in such a way that

$$\ell = |\ell| e_z , \tag{A.14}$$

where e_z is the unit vector in z-direction. Moreover, since the angular momentum ℓ is given by

$$\ell = m r \times \dot{r} = \text{const} , \tag{A.15}$$

and because $\ell \| e_z$ we conclude that $r \perp e_z$. This allows us to set $z = 0$ which means that the whole motion of the point mass can be described in the $x - y$ plane. Rotational invariance led us to the conservation of angular momentum and this made the reduction from a three-dimensional problem to a two dimensional problem possible. The particular form (A.12) of the LAGRANGE function suggests the introduction of polar coordinates (ρ, φ):

$$L(\rho, \dot{\rho}, \dot{\varphi}) = \frac{m}{2}\left(\dot{\rho}^2 + \rho^2\dot{\varphi}^2\right) - U(\rho). \tag{A.16}$$

We solve now LAGRANGE's equations (A.6) on the basis of Eq. (A.16): The first step deals with the differential equation for the radius ρ

$$\frac{\mathrm{d}}{\mathrm{d}t}\frac{\partial}{\partial \dot{\rho}}L = m\ddot{\rho} = \frac{\partial}{\partial \rho}L = m\rho\dot{\varphi}^2 - \frac{\partial}{\partial \rho}U(\rho) , \tag{A.17}$$

thus

$$m\ddot{\rho} - m\rho\dot{\varphi}^2 + \frac{\mathrm{d}}{\mathrm{d}\rho}U(\rho) = 0 \,. \tag{A.18}$$

The differential equation for the angle φ follows from

$$\frac{\mathrm{d}}{\mathrm{d}t}\frac{\partial}{\partial\dot{\varphi}}L = \frac{\mathrm{d}}{\mathrm{d}t}m\rho^2\dot{\varphi} = \frac{\partial}{\partial\varphi}L = 0 \,, \tag{A.19}$$

which corresponds to

$$\frac{\mathrm{d}}{\mathrm{d}t}\left(m\rho^2\dot{\varphi}\right) = 0 \,. \tag{A.20}$$

Equation (A.20) is trivially fulfilled since according to Eq. (A.15)

$$m\rho^2\dot{\varphi} = |\ell| = \mathrm{const} \,. \tag{A.21}$$

However, we solve Eq. (A.21) for $\dot{\varphi}$

$$\dot{\varphi} = \frac{|\ell|}{m\rho^2} \,, \tag{A.22}$$

plug (A.22) into (A.18), and obtain

$$m\ddot{\rho} - \frac{|\ell|^2}{m\rho^3} + \frac{\mathrm{d}}{\mathrm{d}\rho}U(\rho) = 0 \,. \tag{A.23}$$

We make use of the time invariance of the LAGRANGE function (A.16). This equation does not explicitly depend on time t and we have

$$\frac{\partial}{\partial t}L = 0 \,. \tag{A.24}$$

This implies conservation of energy, as can easily be demonstrated. For this purpose, we regard the total time derivative of the LAGRANGE function L

$$\frac{\mathrm{d}}{\mathrm{d}t}L = \dot{\rho}\frac{\partial}{\partial\rho}L + \ddot{\rho}\frac{\partial}{\partial\dot{\rho}}L + \ddot{\varphi}\frac{\partial}{\partial\dot{\varphi}}L + \frac{\partial}{\partial t}L \,, \tag{A.25}$$

and solve for $\frac{\partial}{\partial t}L$

$$\frac{\mathrm{d}}{\mathrm{d}t}\left(\dot{\rho}\frac{\partial}{\partial\dot{\rho}}L + \dot{\varphi}\frac{\partial}{\partial\dot{\varphi}}L - L\right) = -\frac{\partial}{\partial t}L = 0 \,. \tag{A.26}$$

Consequently

$$\dot{\rho}\frac{\partial}{\partial\dot{\rho}}L + \dot{\varphi}\frac{\partial}{\partial\dot{\varphi}}L - L = \text{const}\,, \tag{A.27}$$

which states the conservation of energy. We evaluate this expression with the help of Eq. (A.16). We obtain

$$\begin{aligned}\dot{\rho}\frac{\partial}{\partial\dot{\rho}}L + \dot{\varphi}\frac{\partial}{\partial\dot{\varphi}}L - L &= \frac{m}{2}\left(\dot{\rho}^2 + \rho^2\dot{\varphi}^2\right) + U(\rho)\\ &= \frac{m}{2}\dot{\rho}^2 + \frac{|\ell|^2}{2m\rho^2} + U(\rho)\\ &= E\,.\end{aligned} \tag{A.28}$$

Here we employed, in the second step, relation (A.22). In summary, time invariance resulted in:

$$\frac{m}{2}\dot{\rho}^2 + \frac{|\ell|^2}{2m\rho^2} + U(\rho) = E\,. \tag{A.29}$$

This is a first order differential equation in ρ.

The necessary step required for a solution of the two-body problem can now be outlined: (i) Calculate $R(t)$ according to Eq. (A.10), (ii) solve Eq. (A.29) in order to obtain $\rho(t)$, (iii) plug $\rho(t)$ into Eq. (A.22) and solve for $\varphi(t)$, (iv) since $z(t) = 0$, the original vectors $r_1(t)$, $r_2(t)$ can be constructed from $\rho(t)$ and $\varphi(t)$. All integration constants are uniquely determined by the initial conditions of the problem at hand.

From Eq. (A.29) we obtain

$$\dot{\rho} = \pm\sqrt{\frac{2}{m}\left(E - U(\rho) - \frac{|\ell|^2}{2m\rho^2}\right)}\,, \tag{A.30}$$

which results in an implicit equation for ρ

$$t = t_0 + \int_{\rho_0}^{\rho} \mathrm{d}\rho' \frac{m\rho'}{\sqrt{2m\rho'^2\left[E - U(\rho')\right] - |\ell|^2}}\,, \tag{A.31}$$

where we defined $\rho_0 \equiv \rho(t_0)$, t_0 is some initial time, and we neglected the negative root. Equation (A.31) defines t as a function of ρ, $t = t(\rho)$, which has to be inverted to, finally, obtain the required solution $\rho = \rho(t)$. Whether Eq. (A.31) can be solved analytically depends on the particular form of the potential $U(\rho)$. If Eq. (A.31) cannot be solved analytically one has to employ numerical approximations.

Finally, the angle φ can be expressed as a function of the radius ρ, i.e. $\varphi = \varphi(\rho)$. We get from Eqs. (A.22) and (A.30)

$$\frac{\mathrm{d}\varphi}{\mathrm{d}\rho} = \frac{\mathrm{d}\varphi}{\mathrm{d}t}\frac{\mathrm{d}t}{\mathrm{d}\rho} = \pm\frac{|\ell|}{m\rho^2}\left[\frac{2}{m}\left(E - U(\rho) - \frac{|\ell|^2}{2m\rho^2}\right)\right]^{-\frac{1}{2}} , \tag{A.32}$$

integrate over ρ, and find the desired relation

$$\varphi = \varphi_0 \pm |\ell| \int_{\rho_0}^{\rho} \frac{\mathrm{d}\rho'}{\rho'\sqrt{2m\rho'^2\left[E - U(\rho')\right] - |\ell|^2}} , \tag{A.33}$$

where $\varphi_0 \equiv \varphi(t_0)$.

Appendix B
Solving Non-linear Equations: The NEWTON Method

We give a brief introduction into the solution of non-linear equations with the help of NEWTON's method.[1] We regard a differentiable function $F(x)$ and we would like to find the solution of the equation

$$F(x) = 0 . \tag{B.1}$$

The simplest approach is to transform the equation into an equation of the form

$$x = f(x) , \tag{B.2}$$

which is always possible. This equation could be solved iteratively by simply repeating

$$x_{t+1} = f(x_t) , \tag{B.3}$$

where we start with some initial value x_0. If this method converges, one can approximate the solution arbitrarily close, however, convergence is not guaranteed and will in fact depend on the transformation from Eqs. (B.1) to (B.2). A more advanced technique is the so called NEWTON method [6, 7]. It is based on the definition of $f(x)$ as

$$f(x) = x - \frac{F(x)}{F'(x)} , \tag{B.4}$$

which allows the iteration

$$x_{t+1} = x_t - \frac{F(x_t)}{F'(x_t)} . \tag{B.5}$$

[1]This method is also referred to as the NEWTON-RAPHSON method.

B.A. Stickler, E. Schachinger, *Basic Concepts in Computational Physics*,
DOI 10.1007/978-3-319-27265-8

Here $F'(x)$ denotes the derivative of $F(x)$ with respect to x. The convergence behavior of the iteration (B.5) highly depends on the form of the function $F(x)$ and on the choice of the starting point x_0. The routine can be regarded as converged if $|x_{t+1} - x_t| < \epsilon$, where ϵ is the accuracy required.

If $F(x)$ is not differentiable one can use the *regula falsi* or employ *stochastic methods* which are discussed in the second part of this book. The iteration of the method known as *regula falsi* is [6, 7]

$$x_{t+1} = x_t - F(x_t)\frac{x_t - x_{t-1}}{F(x_t) - F(x_{t-1})} \,. \tag{B.6}$$

A more detailed discussion on methods to solve transcendental equations numerically can be found in any textbook on numerical methods, see for instance Refs. [8, 9]. We shall also briefly introduce the case of a non-linear system of equations of the form (B.1) where $F(x) \in \mathbb{R}^N$ and $x \in \mathbb{R}^N$. In this case the iteration scheme is given by

$$x_{t+1} = x_t - J^{-1}(x_t)F(x_t) \,, \tag{B.7}$$

where

$$J(x) = \nabla_x F(x) = \begin{pmatrix} \frac{\partial F_1(x)}{\partial x_1} & \frac{\partial F_1(x)}{\partial x_2} & \cdots & \frac{\partial F_1(x)}{\partial x_N} \\ \frac{\partial F_2(x)}{\partial x_1} & \frac{\partial F_2(x)}{\partial x_2} & \cdots & \frac{\partial F_2(x)}{\partial x_N} \\ \vdots & \vdots & \ddots & \vdots \\ \frac{\partial F_N(x)}{\partial x_1} & \frac{\partial F_N(x)}{\partial x_2} & \cdots & \frac{\partial F_N(x)}{\partial x_N} \end{pmatrix} . \tag{B.8}$$

is the JACOBI matrix of $F(x)$. We can also make use of the methods discussed in Chap. 2 to calculate numerically the derivatives in Eqs. (B.5) or (B.8).

Appendix C
Numerical Solution of Linear Systems of Equations

We discuss briefly two of the most important methods to solve non-homogeneous systems of linear equations applying numerical methods. We consider a system of n equations of the form

$$\begin{aligned}
a_{11}x_1 + a_{12}x_2 + \ldots + a_{1n}x_n &= b_1 \;, \\
a_{21}x_1 + a_{22}x_2 + \ldots + a_{2n}x_n &= b_2 \;, \\
\vdots \quad & \quad \vdots \\
a_{n1}x_1 + a_{n2}x_2 + \ldots + a_{nn}x_n &= b_n \;,
\end{aligned} \tag{C.1}$$

which is usually transformed into a matrix equation,

$$Ax = b \;. \tag{C.2}$$

The coefficients of the matrix $A = \{a_{ij}\}$ as well as the vector $b = \{b_i\}$ are assumed to be real valued and, furthermore, if

$$\sum_{i=1}^{n} |b_i| \neq 0 \;, \tag{C.3}$$

the problem (C.2) is referred to as non-homogeneous (inhomogeneous). The solution of non-homogeneous linear systems of equations is one of the central problems in numerical analysis, since numerous numerical methods, such as the finite difference approach to a boundary value problem, see Chap. 8, can be reduced to such a problem.

B.A. Stickler, E. Schachinger, *Basic Concepts in Computational Physics*,
DOI 10.1007/978-3-319-27265-8

The solution of (C.2) is well defined as long as the matrix A is non-singular, i.e. as long as

$$\det(A) \neq 0 . \tag{C.4}$$

Then the unique solution of (C.2) can be written as

$$x = A^{-1} b . \tag{C.5}$$

However, the inversion of matrix A is very complex for $n \geq 4$ and one would prefer methods which are computationally more effective. Basically, one distinguishes between *direct* and *iterative* methods. Since a complete discussion of this huge topic would be too extensive, we will mainly focus on two methods.

In contrast to iterative procedures, direct procedures do not contain any methodological errors and can, therefore, be regarded as *exact*. However, these methods are often computationally very extensive and rounding errors are in many cases not negligible. As an example we will discuss the LU decomposition. On the other hand, many iterative methods are fast and rounding errors can be controlled easily. However, it is not guaranteed that an iterative procedure converges, even in cases where the system of equations is known to have unique solutions. Moreover, the result is an approximate solution. As an illustration for an iterative procedure we will discuss the GAUSS-SEIDEL method.

C.1 The *LU* Decomposition

The LU decomposition [6, 10] is essentially a numerical realization of GAUSSIAN elimination which is based on a fundamental property of linear systems of equations (C.2). This property states the system (C.2) to remain unchanged when a linear combination of rows is added to one particular row. This property is then employed in order to obtain a matrix in *triangular* form. It was demonstrated by DOOLITTLE and CROUT [6, 10, 11] that the GAUSSIAN elimination can be formulated as a decomposition of the matrix A into two matrices L and U:

$$A = LU . \tag{C.6}$$

Here, U is an *upper triangular* matrix and L is a *lower triangular* matrix. In particular, U is of the form

$$U = \begin{pmatrix} u_{11} & u_{12} & \dots & u_{1n} \\ 0 & u_{22} & \dots & u_{2n} \\ \vdots & & & \vdots \\ 0 & 0 & \dots & u_{nn} \end{pmatrix} , \tag{C.7}$$

and L is of the form

$$L = \begin{pmatrix} 1 & 0 & & \dots 0 \\ m_{21} & 1 & 0 & \dots 0 \\ m_{31} & m_{32} & 1 & \dots 0 \\ \vdots & & & \vdots \\ m_{n1} & m_{n2} & m_{n3} & \dots 1 \end{pmatrix} . \tag{C.8}$$

The factorization (C.6) is referred to as *LU decomposition*. The corresponding procedure can be easily identified by equating the elements in (C.6). One can show that the following operations yield the desired result: For $j = 1, 2, \dots, n$ one computes

$$u_{ij} = a_{ij} - \sum_{k=1}^{i-1} m_{ik} u_{kj} \qquad i = 1, 2, \dots, j , \tag{C.9}$$

$$m_{ij} = \frac{1}{u_{jj}} \left(a_{ij} - \sum_{k=1}^{j-1} m_{ik} u_{kj} \right) \quad i = j+1, j+2, \dots, n , \tag{C.10}$$

with the requirement that $u_{jj} \neq 0$. Note that in this notation we used the convention that the contribution of the sum is equal to zero if the upper boundary is less than the lower boundary. We rewrite Eq. (C.2) with the help of the LU decomposition (C.6)

$$Ax = LUx = b , \tag{C.11}$$

and by defining $y = Ux$, we retrieve a system of equations for the variable y:

$$Ly = b . \tag{C.12}$$

The particular form of L allows to solve the system (C.12) immediately by *forward substitution*. We find the solution

$$y_i = b_i - \sum_{k=1}^{i-1} m_{ik} y_k , \qquad i = 1, 2, \dots, n , \tag{C.13}$$

and the equation

$$Ux = y , \tag{C.14}$$

remains. It can solved by *backward substitution*:

$$x_i = \frac{1}{u_{ii}} \left(y_i - \sum_{k=i+1}^{n} u_{ik} x_k \right) , \quad i = n, n-1, \dots, 1 . \tag{C.15}$$

We note that this method can also be employed to invert the matrix A. The strategy is based on the relation

$$AX = I\,, \tag{C.16}$$

where $X = A^{-1}$ is to be determined and I is the n-dimensional identity. Equation (C.16) is equivalent to the following system of equations:

$$\begin{aligned} Ax_1 &= \begin{pmatrix} 1 \\ 0 \\ \vdots \\ 0 \end{pmatrix}, \\ Ax_2 &= \begin{pmatrix} 0 \\ 1 \\ 0 \\ \vdots \\ 0 \end{pmatrix}, \\ \vdots\ & \quad\ \vdots \\ Ax_n &= \begin{pmatrix} 0 \\ \vdots \\ 0 \\ 1 \end{pmatrix}, \end{aligned} \tag{C.17}$$

where the vectors x_i are the rows of the unknown matrix X, i.e. $X = (x_1, x_2, \ldots, x_n)$. The n equations of the system (C.17) can be solved with the help of the LU decomposition.

Furthermore, one can easily calculate the determinant of A using the LU decomposition. We note that

$$\det(A) = \det(LU) = \det(L)\det(U) = \det(U)\,, \tag{C.18}$$

since L and U are triangular matrices, the determinants are equal to the product of the diagonal elements, which yields $\det(L) = 1$. Hence we have

$$\det(A) = \det(U) = \prod_{i=1}^{n} u_{ii}\,. \tag{C.19}$$

In conclusion we remark that there are many specialized methods which have been designed particularly for matrices of specific forms, such as tridiagonal matrices, symmetric matrices, block-matrices, ... Such matrices commonly appear in physics applications. For instance, we remember that the matrix we encountered in Sect. 8.2 within the context of a finite difference approximation of boundary value problems, was tridiagonal. These specialized methods are usually the first choice if one has a matrix of such a specific form because they are much faster and more stable than methods developed for matrices of more general form. Since a full treatment of these methods is beyond the scope of this book, we refer the interested reader to books on numerical linear algebra, for instance Refs. [10, 11].

C.2 The GAUSS-SEIDEL Method

The GAUSS-SEIDEL method is an iterative procedure to approximate the solution of non-homogeneous systems of linear equations [6, 12]. The advantage of an iterative procedure, in contrast to a direct approach, is that its formulation is in general much simpler. However, one might have problems with the convergence of the method, even in cases where a solution exists and is unique. We note that the GAUSS-SEIDEL method is of particular interest whenever one has to deal with *sparse* coefficient matrices.[1] This requirement is not too restrictive since most of the matrices encountered in physical applications are indeed sparse. As an example we remember the matrices arising in the context of a finite difference approach to boundary value problems, Sect. 8.2.

Again, we use Eq. (C.1) as a starting point for our discussion. It is a requirement of the GAUSS-SEIDEL method that *all* diagonal elements of A are non-zero. We then solve each row of (C.1) for x_i. This creates the following hierarchy

$$
\begin{aligned}
x_1 &= -\frac{1}{a_{11}}\left(a_{12}x_2 + a_{13}x_3 + \ldots + a_{1n}x_n - f_1\right) , \\
x_2 &= -\frac{1}{a_{22}}\left(a_{21}x_1 + a_{23}x_3 + \ldots + a_{2n}x_n - f_2\right) , \\
&\vdots \\
x_n &= -\frac{1}{a_{nn}}\left(a_{n1}x_1 + a_{n2}x_2 + \ldots + a_{n,n-1}x_{n-1} - f_n\right) ,
\end{aligned}
\tag{C.20}
$$

or in general for $i = 1, \ldots, n$

$$
x_i = -\frac{1}{a_{ii}}\left(\sum_{\substack{j=1 \\ j\neq i}}^{n} a_{ij}x_j - f_i\right) . \tag{C.21}
$$

[1] A matrix A is referred to as sparse, when the matrix is populated primarily by zeros.

We note that Eq. (C.21) can be rewritten as a matrix equation

$$x = Cx + b, \tag{C.22}$$

where we defined the matrix $C = \{c_{ij}\}$ via

$$c_{ij} = \begin{cases} -\dfrac{a_{ij}}{a_{ii}} & i \neq j\,, \\ 0 & i = j\,, \end{cases} \tag{C.23}$$

and the vector $b = \{b_i\}$ as

$$b_i = \frac{f_i}{a_{ii}}\,. \tag{C.24}$$

We recognize that Eq. (C.21) can be transformed into an iterative form with the help of a trivial manipulation

$$x_i = x_i - \left[x_i + \frac{1}{a_{ii}} \left(\sum_{\substack{j=1 \\ j\neq i}}^{n} a_{ij}x_j - f_i \right) \right]\,, \tag{C.25}$$

or

$$x_i^{(t+1)} = x_i^{(t)} - \Delta x_i^{(t)}\,, \tag{C.26}$$

where

$$\Delta x_i^{(t)} = x_i^{(t)} + \frac{1}{a_{ii}} \left(\sum_{j=1}^{i-1} a_{ij}x_j^{(t+1)} + \sum_{j=i+1}^{n} a_{ij}x_j^{(t)} - f_i \right)\,. \tag{C.27}$$

Equation (C.26) in combination with (C.27) produces a sequence of vectors

$$x^{(0)} \to x^{(1)} \to x^{(2)} \to \ldots \to x^{(m)}\,, \tag{C.28}$$

where $x^{(0)}$ is referred to as the initialization vector or trial vector. One can prove that if this sequence converges, it approaches the exact solution x arbitrarily close:

$$\lim_{t\to\infty} x^{(t)} = x\,. \tag{C.29}$$

We remark that if the terms $x_i^{(t+1)}$ on the right hand side of Eq. (C.27) are replaced by $x_i^{(t)}$ the method is referred to as the JACOBI method.

To terminate the GAUSS-SEIDEL method, we need an exit condition: One should terminate the iteration whenever:

- The approximate solution $x^{(t)}$ obeys the required accuracy ϵ or $\tilde{\epsilon}$, for instance

$$\max\left(|x_i^{(t)} - x_i^{(t-1)}|\right) \leq \epsilon \,, \tag{C.30}$$

where ϵ is the absolute error, or

$$\max\left(\frac{|x_i^{(t)} - x_i^{(t-1)}|}{|x_i^{(t)}|}\right) \leq \tilde{\epsilon} \,, \tag{C.31}$$

where $\tilde{\epsilon}$ is the relative error.
- When a maximum number of iterations is reached. This condition may be interpreted as an emergency exit which ensures that the iteration terminates even if the process is not convergent or has still not converged.

Let us discuss one final, however, crucial point of this section: In many cases the convergence of the GAUSS-SEIDEL method can be significantly improved by including a *relaxation parameter* ω to the iterative process. In this case the update routine (C.26) takes on the form

$$x_i^{(t+1)} = x_i^{(t)} - \omega \Delta x_i^{(t)} \,. \tag{C.32}$$

If the relaxation parameter ω obeys $\omega > 1$ one speaks of *over-relaxation*, if $\omega < 1$ of *under-relaxation* and if $\omega = 1$ the regular GAUSS-SEIDEL method is recovered. An appropriate choice of the relaxation parameter may fasten the convergence of the method significantly. The best result will certainly be obtained if the *ideal* value of ω, ω_i were known. Unfortunately, it is impossible to determine ω_i prior to the iteration in the general case. We remark the following properties:

- The method (C.32) is only convergent for $0 < \omega \leq 2$.
- If the matrix C is positive definite and $0 < \omega < 2$, the GAUSS-SEIDEL method converges for any choice of $x^{(0)}$ (OSTROWSKI-REICH theorem, [13]).
- In many cases, $1 \leq \omega_i \leq 2$. We note that this inequality holds only under particular restrictions for the matrix C [see Eq. (C.23)]. However, we note without going into detail, that these restrictions are almost always fulfilled when one is confronted with applications in physics.
- If C is positive definite and tridiagonal, the ideal value ω_i can be calculated using

$$\omega_i = \frac{2}{1 + \sqrt{1 - \lambda^2}} \,, \tag{C.33}$$

where λ is the largest eigenvalue of C, Eq. (C.23).

- Since the calculation of λ is in many cases quite complex, one could employ the following idea: It is possible to prove that

$$\lim_{t\to\infty} \frac{|\Delta x^{(t+1)}|}{|\Delta x^{(t)}|} \to \lambda^2 . \tag{C.34}$$

Hence, one may start with $\omega = 1$, perform t_0 ($20 < t_0 < 100$) iterations and then approximate ω_i with the help of Eq. (C.33) and

$$\lambda^2 \approx \frac{|\Delta x^{(t_0)}|}{|\Delta x^{(t_0-1)}|} . \tag{C.35}$$

The iteration is then continued with the approximated value of ω_i until convergence is reached.

In conclusion we remark that numerous numerical libraries contain sophisticated routines to solve linear systems of equations. In many cases it is, thus, advisable to rely on such routines.

Appendix D
Fast Fourier Transform

Integral transforms are indispensable in modern mathematics and natural science because they can be employed to simplify complex mathematical problems. In this Appendix we will discuss the FOURIER transform as one prominent representative of integral transforms in general. Loosely speaking, the FOURIER transform is the unambiguous decomposition of a function $f(x)$ into its frequency components. Its applications range from the harmonic analysis of periodic signals to the solution of differential equations and the description of wave phenomena in classical mechanics [2, 4], electrodynamics [14–16], quantum mechanics [17–19], and many more. Here, we briefly discuss its numerical implementation, the *fast* FOURIER *transform* (FFT) and its applications in Computational Physics.

We start by recalling the concept of FOURIER series: It is asserted by FOURIER's theorem that every square-integrable, d-periodic function $f(x), f(x+d) = f(x)$, can be (uniquely) represented as[1]

$$f(x) = \sum_{n \in \mathbb{Z}} \hat{f}_n \exp\left(i\frac{2\pi nx}{d}\right), \tag{D.1}$$

where the complex coefficients $\hat{f}_n \in \mathbb{C}$ are related to $f(x)$ by the inverse transform

$$\hat{f}_n = \frac{1}{d}\int_0^d \mathrm{d}x f(x) \exp\left(-i\frac{2\pi nx}{d}\right). \tag{D.2}$$

[1]In other words, the plane waves $\exp(in2\pi x/d)$ with period d form a complete, orthonormal basis in the space of d-periodic, square integrable functions with the scalar product (10.10). We remark that this also applies to functions which are defined on a compact interval of length d [20].

B.A. Stickler, E. Schachinger, *Basic Concepts in Computational Physics*,
DOI 10.1007/978-3-319-27265-8

The representation (D.1) of $f(x)$ is referred to as the FOURIER series of $f(x)$ and the coefficient (D.2) is the FOURIER coefficient of order n. Equation (D.1) is an unambiguous expansion of the function $f(x)$ into contributions which oscillate with an integer multiple of the frequency $2\pi/d$. There are numerous important properties, examples and applications of FOURIER series for which we refer to the literature [21–24].

The concept of FOURIER series can be generalized to the idea of the FOURIER transform of a square integrable function $f(x)$ by formally letting $d \to \infty$ [23]. The FOURIER transform relates the function $f(x)$ to its transform $\hat{f}(k)$, $k \in \mathbb{R}$, via[2]

$$f(x) = \int_{-\infty}^{\infty} \mathrm{d}k \hat{f}(k) \exp(ikx), \tag{D.3}$$

and the inverse transform is obtained as

$$\hat{f}(k) = \frac{1}{2\pi} \int_{-\infty}^{\infty} \mathrm{d}x f(x) \exp(-ikx). \tag{D.4}$$

The transform (D.3) and its inverse (D.4) can be used to considerably simplify mathematical problems. For instance, a linear differential equation for the function $f(x)$ is mapped onto a linear algebraic equation for $\hat{f}(k)$. The solution of the differential equation is then obtained by back-transforming the solution $\hat{f}(k)$ of the algebraic equation. Again, we refer to the literature for further applications and the various properties of the transforms (D.3) and (D.4). Instead, let us concentrate on the question of how to compute the FOURIER transform (D.2) numerically.

It appears to be reasonable to start with the concepts developed in Chap. 3.[3] For this purpose, we assume that the function $f(x)$ is solely known on a grid of N equidistant grid-points x_ℓ, $\ell = 0, \ldots, N-1$. In addition, we note that it is sufficient to limit our discussion to 2π-periodic functions. Thus, we can choose our grid-points to be $x_\ell = x_0 + \ell h$ where $x_0 = 0$ and $h = 2\pi/N$, so that $x_{N-1} = 2\pi(1-1/N)$.

Approximating the integral (D.2) with the help of the forward rectangular rule, Chap. 3, yields

$$\hat{f}_n = \frac{1}{N} \sum_{\ell=0}^{N-1} f_\ell \exp\left(-\frac{2\pi n \ell}{N}\right) + \mathcal{O}(h^2). \tag{D.5}$$

It follows from this equation that the coefficients $\hat{f}_n$ are periodic in n with period N due to the finite number of grid-points. Hence, the maximal number of distinct

[2]We work here with the asymmetric definition of the FOURIER transform. For other definitions, the pre-factors have to be adapted consistently.

[3]If $f(x)$ is not periodic we have to truncate the integral (D.4) and restrict the integration to a suitable finite interval so that the problem again reduces to the evaluation of Eq. (D.2).

coefficients is equal to the number of grid-points. The inversion of Eq. (D.5) follows directly from Eq. (D.1) and reads[4]

$$f_\ell = \sum_{n=0}^{N-1} \hat{f}_n \exp\left(i\frac{2\pi n\ell}{N}\right). \tag{D.6}$$

The transforms (D.5) and (D.6) are referred to as the *discrete* FOURIER *transform* (DFT) and its inverse, respectively. We cast these relations into matrix form by defining vectors $F = (f_0, \ldots, f_{N-1})^T$ and $\hat{F} = (\hat{f}_0, \ldots, \hat{f}_{N-1})^T$ together with the matrix W of elements:

$$W_{nm} = \omega_N^{nm}. \tag{D.7}$$

Here, $\omega_N = \exp\left(\frac{2\pi i}{N}\right)$ denotes the N-th root of unity. The transformation matrix W is known as the FOURIER matrix or DFT matrix and it is easy to prove that its inverse W^{-1} has the elements

$$\left(W^{-1}\right)_{nm} = \omega_N^{-nm}. \tag{D.8}$$

All this allows to rewrite Eqs. (D.5) and (D.6) in compact form:

$$\hat{F} = \frac{1}{N} W^{-1} F, \qquad \text{and} \qquad F = W\hat{F}. \tag{D.9}$$

Thus, we reduced the problem of numerically implementing the FOURIER transform (D.2) to the task of multiplying the $N \times N$ complex matrix W with the N-element vector F. This means that we have to perform N^2 complex multiplications and $N(N-1)$ complex additions. However, the symmetry $W_{nm} = W_{mn}$ already suggests that there is further room for improvement. In fact, there are methods that do much better and these algorithms are known as *fast* FOURIER *transform* (FFT) algorithms.

We limit our presentation to the version proposed by COOLEY and TUKEY [6, 25, 26] which is, with some variations, the most common algorithm. In its simplest form it is based on the observation that one can always split the FOURIER transform (D.5) into an even and an odd part

$$\hat{f}_n = \frac{1}{N}\sum_{\ell=0}^{N/2} f_{2\ell}\,\omega_N^{2n\ell} + \frac{\omega_N^n}{N}\sum_{\ell=0}^{N/2} f_{2\ell+1}\,\omega_N^{2n\ell}, \tag{D.10}$$

[4]It follows directly from the summation rule of the geometric series that

$$\sum_{m=0}^{N-1} \exp\left(i\frac{2\pi mn}{N}\right) = N\delta_{n0}.$$

provided that N is even.[5] Since $\omega_N^{2k} = \omega_{N/2}^n$, we can interpret Eq. (D.10) as the linear combination of two FOURIER transforms of length $N/2$. Denoting the FOURIER coefficients of the function values $f_{2\ell}$ on even grid-points by $\hat{A}_n$ and of the values $f_{2\ell+1}$ on odd grid-points by $\hat{B}_n$, we obtain for $n = 1, \ldots, N/2$

$$\hat{f}_n = \hat{A}_n + \omega_N^n \hat{B}_n. \tag{D.11}$$

We now make use of the property that the FOURIER coefficients are periodic, i.e. $\hat{A}_{n+N/2} = \hat{A}_n$, $\hat{B}_{n+N/2} = \hat{B}_n$, and that $\omega_N^{n+N/2} = -\omega_N^n$. Thus, we can calculate the remaining coefficients $\hat{f}_n$, $n = N/2 + 1, \ldots, N$ with the help of:

$$\hat{f}_{n+N/2} = \hat{A}_n - \omega_N^n \hat{B}_n. \tag{D.12}$$

Because of Eqs. (D.11) and (D.12) the N FOURIER coefficients can be computed as a linear combination of two FOURIER transforms of size $N/2$. The recursive application of the very same scheme to $\hat{A}_n$ and $\hat{B}_n$ constitutes the core of the FFT algorithm in its simplest variation [6]. It is also the efficiency of this algorithm that makes the FOURIER transform an attractive tool for numerical calculations [27]. In fact, there are several problems in this book where an algorithm based on FFT could have been evoked. Let us discuss two examples in more detail in order to illustrate this.

(i) In Chap. 9.3 we could have solved the stationary inhomogeneous heat equation for its FOURIER coefficients followed by the back-transform.[6] Denoting by $\hat{T}_n$ the FOURIER coefficients of the temperature $T(x)$ and by $\hat{\Gamma}_n$ the FOURIER coefficients of the heat source/drain, we obtain from Eq. (9.20)

$$\hat{T}_n = -\frac{\hat{\Gamma}_n}{(n\omega)^2}, \tag{D.13}$$

where $\omega = 2\pi/L$. Performing the inverse FFT on Eq. (D.13) and adding the homogeneous solution (9.4) immediately gives the required temperature profile $T(x)$. In a similar fashion, FFT could have been used for solving the partial differential equations discussed in Chap. 11. From the examples in this chapter, the time-dependent SCHRÖDINGER equation serves as our second application.

(ii) The Hamiltonian of a free point particle (for simplicity in one dimension) is diagonal in momentum space, $H = P^2/2m$, with the momentum operator (10.6). Given the position-space representation of the initial state $\psi(x, t)$, we can then

[5]This is not a limitation because we can always choose N to be even.

[6]Here we use the fact that the plane waves $\exp(in2\pi x/d)$ form a complete, orthonormal basis of the functions defined on a compact interval of length L [20].

compute the time evolved wave packet $\psi(x, t + \Delta t)$ for arbitrary $\Delta t > 0$ according to Eq. (10.17) as

$$\psi(x, t + \Delta t) = \frac{1}{2\pi\hbar} \int \mathrm{d}p \exp\left(-\frac{i\Delta t}{\hbar}\frac{p^2}{2m} + \frac{i}{\hbar}px\right) \hat{\psi}(p, t), \tag{D.14}$$

where $\hat{\psi}(p, t)$ is the momentum space representation of the initial state $\psi(x, t)$, i.e. its FOURIER transform with $k = p/\hbar$. Hence, the time evolution of the free wave packet is readily computed numerically with the help of the FFT and its inverse.

It is now certainly interesting to investigate whether or not a similar approach can be applied to solve Eq. (10.1) in the presence of a potential $V(x)$ which is diagonal in position space. Although this can be achieved by solving the full stationary eigenvalue problem (10.9) followed by the application of the eigenvector expansion (10.17), we present here a more efficient but approximate solution valid for small time steps Δt. In order to see this, we transform Eq. (D.14) into a slightly more compact form. Denoting by $\mathscr{F}$ the FOURIER transform operator, $\hat{\psi}(p) = \mathscr{F}\psi(x)$, we can write Eq. (D.14) as

$$\psi(x, t + \Delta t) = \mathscr{F}^{-1} U_{\Delta t} \mathscr{F} \psi(x, t), \tag{D.15}$$

where $U_{\Delta t} = \exp\left(-i\Delta t p^2/2\hbar m\right)$ is the unitary time evolution operator for the time interval Δt.[7] The correct result can not be obtained by multiplying Eq. (D.15) with the position-space time evolution of the potential $V_{\Delta t} = \exp(-i\Delta t V(x)/\hbar)$ because the operators V and P do not commute. However, by applying the BAKER-CAMPBELL-HAUSDORFF formula[8] [17, 19, 28], we can approximate the time evolution $\psi(x, t) \to \psi(x, t + \Delta t)$ for a small time step Δt by:

$$\psi(x, t + \Delta t) = V_{\Delta t} \mathscr{F}^{-1} U_{\Delta t} \mathscr{F} \psi(x, t) + \mathscr{O}(\Delta t^2). \tag{D.16}$$

An even better approximation is obtained by the symmetrized form

$$\psi(x, t + \Delta t) = \mathscr{F}^{-1} U_{\Delta t/2} \mathscr{F} V_{\Delta t} \mathscr{F}^{-1} U_{\Delta t/2} \mathscr{F} \psi(x, t) + \mathscr{O}(\Delta t^3). \tag{D.17}$$

This method, known as the *split operator technique* [29], is a frequently used method to numerically solve time dependent problems in quantum mechanics with the help of FFT.

[7] U_t is the momentum space representation of the free unitary time evolution operator $U = \exp(-itP^2/2\hbar m)$.

[8] The BAKER-CAMPBELL-HAUSDORFF formula states how the exponential function $\exp(X + Y)$ of two non-commuting operators X and Y can be expanded in terms of products of exponentials of their commutators [17, 28].

Appendix E
Basics of Probability Theory

E.1 Classical Definition

It is the aim of this Appendix to summarize the most important definitions and results from basic probability theory as required within this book. For a more in depth presentation we refer to the literature [30–34].

The classical probability $P(A)$ for an event A is defined by the number of favorable results n, divided by the number of possible results m,

$$P(A) = \frac{n}{m} \,. \tag{E.1}$$

For two events A and B we can deduce the following rules[1]

$$P(A \vee B) = P(A) + P(B) - P(A \wedge B) \,, \tag{E.2a}$$

$$P(Z) = 0 \quad \text{impossible event; } Z \ldots \text{zero element} \,, \tag{E.2b}$$

$$P(I) = 1 \quad \text{certain event; } I \ldots \text{identity element} \,, \tag{E.2c}$$

$$0 \leq P(A) \leq 1 \,, \tag{E.2d}$$

$$P(A|B) = \frac{P(A \wedge B)}{P(B)} \,, \tag{E.2e}$$

[1]Here we use the symbols $\vee$ and $\wedge$ to denote the Boolean operators OR and AND, respectively.

B.A. Stickler, E. Schachinger, *Basic Concepts in Computational Physics*,
DOI 10.1007/978-3-319-27265-8

where $P(A|B)$ is the probability for the event A under the constraint that event B is true. Moreover, if $\overline{A}$ is the complementary event[2] to A we have

$$P(\overline{A}) = 1 - P(A)\ . \tag{E.3}$$

The statistical definition of the probability for an event A is given by:

$$P(A) = \lim_{m\to\infty} \frac{n}{m}\ . \tag{E.4}$$

E.2 Random Variables and Moments

A random variable is a functional which assigns to an event ω a real number x from the set of possible outcomes Ω: $x = X(\omega)$ [31].[3] Roughly speaking it is a variable whose value is assigned to the observation of some random process. The mean value of a discrete random variable X is defined by

$$\langle X\rangle = \sum_{\omega\in\Omega} X(\omega)P_\omega\ , \tag{E.5}$$

where P_ω is the probability for the event ω. For instance, in case of a dice-throw $X(\omega) \equiv n = 1, 2, \ldots, 6$.

We restrict ourselves now to discrete random variables and, thus, x can only take on discrete values. Furthermore, we introduce the function of random variables $Y = f(X)$ and define quite generally its mean value:

$$\langle f(X)\rangle \equiv \langle f\rangle = \sum_i f(x_i)P_i\ . \tag{E.6}$$

Note that

$$\langle 1\rangle \equiv \sum_i P_i = 1\ . \tag{E.7}$$

Moments of order k of a random variable X are defined by

$$m_k := \left\langle X^k\right\rangle\ , \tag{E.8}$$

[2] This means that $A \vee \overline{A} = I$ and $A \wedge \overline{A} = Z$.

[3] A more exact formulation will follow in the course of this Appendix.

and central moments are introduced via the relation

$$\mu_k := \left\langle (\Delta X)^k \right\rangle = \left\langle (X - \langle X \rangle)^k \right\rangle . \tag{E.9}$$

Of particular interest is the second central moment, the *variance*:

$$\mathrm{var}\,(X) := \left\langle (X - \langle X \rangle)^2 \right\rangle = \left\langle X^2 \right\rangle - \langle X \rangle^2 . \tag{E.10}$$

Finally, the *standard deviation* σ is defined as the square root of the variance:

$$\sigma := \mathrm{std}(X) = \sqrt{\mathrm{var}\,(X)} . \tag{E.11}$$

We study now a discrete set of observations x_i where $i = 1, \ldots, N$. Then the sample mean value is given by

$$\overline{x} = \frac{1}{N} \sum_i x_i , \tag{E.12}$$

and the error (standard deviation) of $\overline{x}$ (*standard error*) can be determined from:

$$\mathrm{var}\,(\overline{x}) = \mathrm{var}\left(\frac{1}{N} \sum_i x_i \right) = \frac{\sigma^2}{N} . \tag{E.13}$$

We assumed here the x_i to be uncorrelated with the consequence that $\mathrm{cov}\,(x_i, x_j) = \mathrm{var}\,(x_i)\, \delta_{ij}$ [defined in Eq. (E.16)]. Therefore,

$$\text{standard error} = \sigma_{\overline{x}} = \frac{\sigma}{\sqrt{N}} , \tag{E.14}$$

where σ is the standard deviation of the observations as defined above.

In the case of multiple random variables we can proceed as above. For instance, the expectation value of a function of two random variables is given by

$$\langle f(X, Y) \rangle := \sum_{i,j} f(x_i, y_j) P_{ij} , \tag{E.15}$$

and the covariance between two random variables:

$$\mathrm{cov}\,(X, Y) := \langle (X - \langle X \rangle)(Y - \langle Y \rangle) \rangle = \langle XY \rangle - \langle X \rangle \langle Y \rangle . \tag{E.16}$$

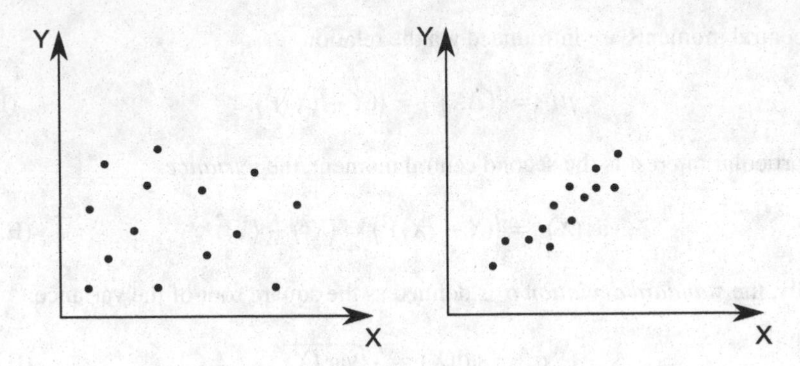

Fig. E.1 Uncorrelated (*left panel*) and positively correlated (*right panel*) variables X and Y

The value of the covariance together with its sign determines important properties of the random variables X and Y in their relation to each other:

$$\mathrm{cov}\,(X,Y) = \begin{cases} > 0 & \text{for } Y - \langle Y \rangle > 0 \Rightarrow X - \langle X \rangle > 0\,, \\ & \qquad \text{(positive linear correlation)} \\ < 0 & \text{for } Y - \langle Y \rangle > 0 \Rightarrow X - \langle X \rangle < 0\,, \\ < 0 & \text{for } X - \langle X \rangle > 0 \Rightarrow Y - \langle Y \rangle < 0\,, \\ & \qquad \text{(negative linear correlation)} \\ = 0 & \text{no linear dependence between } X \text{ and } Y\,. \end{cases} \tag{E.17}$$

Random variables whose covariance is zero are called *uncorrelated*. [This property was used in the derivation of Eq. (E.13).] To give an example, Fig. E.1 compares schematically uncorrelated and positively correlated random variables X and Y.

E.3 Binomial Distribution and Limit Theorems

The binomial distribution is given by

$$P(k|n,p) = \binom{n}{k} p^k (1-p)^{n-k}\,, \tag{E.18}$$

where $\binom{n}{k}$ is the binomial coefficient

$$\binom{n}{k} = \frac{n!}{k!(n-k)!}\,. \tag{E.19}$$

For large values of n STIRLING's approximation can be applied to calculate an estimate of $n!$:

$$n! = n^{n+\frac{1}{2}} e^{-n} \sqrt{2\pi} \left[1 + \mathcal{O}(n^{-1})\right] . \tag{E.20}$$

Furthermore, it is easy to prove that the mean value and the variance of the binomial distribution are given by

$$\langle k \rangle = np , \tag{E.21}$$

$$\text{var}\,(k) = np(1-p) . \tag{E.22}$$

The DE MOIVRE-LAPLACE theorem states that for var$(k) \gg 1$

$$P(k|n,p) \approx g(k|k_0,\sigma) = \frac{1}{\sqrt{2\pi\sigma^2}} \exp\left[-\frac{(k-k_0)^2}{2\sigma^2}\right] , \tag{E.23}$$

where $k_0 = \langle k \rangle$ and $\sigma = \sqrt{\text{var}\,(k)}$. We can also deduce that

$$P(k = np|n,p) = \frac{1}{\sqrt{2\pi np(1-p)}} \longrightarrow 0 , \tag{E.24}$$

for $n \to \infty$. From this, BERNOULLI's law of large numbers follows

$$P(|k/n - p| < \epsilon|n,p) \to 1 \quad \forall \epsilon > 0 . \tag{E.25}$$

E.4 POISSON Distribution and Counting Experiments

If the mean expectation value μ is independent of the number of experiments n, i.e. $np = \mu \equiv$ const, it follows from Eq. (E.18) that

$$\lim_{n\to\infty} P\left(k \middle| n, p = \frac{\mu}{n}\right) = \exp(-\mu)\frac{\mu^k}{k!} =: P(k|\mu) . \tag{E.26}$$

The distribution $P(k|\mu)$ is called POISSON distribution. We obtain for the POISSON distribution:

$$\langle k \rangle = \mu , \tag{E.27}$$

$$\text{var}\,(k) = \mu . \tag{E.28}$$

It is important to note that counting experiments, as for instance radioactive decay, follow the Poisson statistics. A typical counting experiment observes within the time interval t (in average) μ events. This time interval is now divided into n subintervals with $\Delta t = t/n$. If the counting events can be assumed to be independent, the process follows a binomial distribution and we have $\mu = np$. This is equivalent to $p = \mu/n$. We return to the case of radioactive decay: We count μ signals within one minute which are uniformly distributed over the time interval. The experiment is now reduced to a time interval of one second and the probability of detecting a signal consequently reduces to $\mu/60$. For $p \ll 1$ but $np \gg 1$ the binomial distribution $P(k|n,p)$ can be approximated by $P(k|\mu)$ and we can use for large values of μ

$$P(k|\mu) = \frac{1}{\sqrt{2\pi\sigma^2}} \exp\left[-\frac{(k-\mu)^2}{2\sigma^2}\right] , \tag{E.29}$$

with

$$\sigma = \sqrt{\mu} \approx \sqrt{k} . \tag{E.30}$$

In most experimentally relevant cases is μ unknown and is approximated by:

$$\mu = k \pm \sqrt{k} . \tag{E.31}$$

E.5 Continuous Variables

We define the *cumulative distribution function* (cdf) [31, 32], $F(x)$, of a continuous variable x by[4]

$$F(x) := P(X \leq x|\mathscr{B}) , \tag{E.32}$$

where $\mathscr{B}$ is a generalized condition (*condition complex*). Moreover, we define the *probability density function* (pdf), $p(x)$ by

$$p(x) = \frac{\mathrm{d}}{\mathrm{d}x} F(x) . \tag{E.33}$$

It follows that

$$p(x)\mathrm{d}x = [F(x+\mathrm{d}x) - F(x)] \stackrel{!}{=} P(x \leq X \leq x + \mathrm{d}x|\mathscr{B}) . \tag{E.34}$$

[4]For convenience we use here the notation $F(x)$ for the cumulative distribution function in contrast to the notation $P(x)$ used throughout the second part of this book.

Hence,

$$F(x) = \int_{-\infty}^{x} \mathrm{d}x' p(x') \,. \tag{E.35}$$

Note that the pdf is normalized

$$\int \mathrm{d}x' p(x') = F(\infty) = P(X \leq \infty|\mathscr{B}) = 1 \,, \tag{E.36}$$

and non-negative

$$p(x) \geq 0 \,. \tag{E.37}$$

E.6 BAYES' Theorem

We regard a set of discrete events A_i under the generalized condition $\mathscr{B}$. Then we have the normalization condition

$$\sum_i P(A_i|\mathscr{B}) = 1 \,, \tag{E.38}$$

and the marginalization rule

$$P(B|\mathscr{B}) = \sum_i P(B|A_i, \mathscr{B}) P(A_i|\mathscr{B}) \,. \tag{E.39}$$

BAYES' theorem [33, 35] for discrete variables follows from Eq. (E.2e) since $P(A \wedge B) = P(B \wedge A)$:

$$P(A|B, \mathscr{B}) = \frac{P(B|A, \mathscr{B}) P(A|\mathscr{B})}{P(B|\mathscr{B})} \,. \tag{E.40}$$

In case of continuous variables the above equations modify accordingly. The marginalization and BAYES' theorem for pdfs are then given by

$$P(B|\mathscr{B}) = \int \mathrm{d}x P(B|x, \mathscr{B}) p(x|\mathscr{B}) \,, \tag{E.41}$$

and

$$p(y|x, \mathscr{B}) = \frac{p(x|y, \mathscr{B}) p(y|\mathscr{B})}{p(x|\mathscr{B})} \,. \tag{E.42}$$

E.7 Normal Distribution

The normal distribution (GAUSS distribution) is defined by the pdf:

$$p(x) = \mathcal{N}(x|x_0, \sigma) = \frac{1}{\sqrt{2\pi\sigma^2}} \exp\left[-\frac{(x-x_0)^2}{2\sigma^2}\right] . \tag{E.43}$$

The corresponding cdf

$$\begin{aligned} F(x) &= \frac{1}{\sqrt{2\pi\sigma^2}} \int_{-\infty}^{x} \mathrm{d}x' \exp\left[-\frac{(x'-x_0)^2}{2\sigma^2}\right] \\ &= \Phi\left(\frac{x-x_0}{\sigma}\right) = \frac{1}{2} + \frac{1}{2}\mathrm{erf}\left(\frac{x-x_0}{\sqrt{2\sigma^2}}\right) , \end{aligned} \tag{E.44}$$

follows. Here $\Phi(x)$ is given by

$$\Phi(x) = \frac{1}{\sqrt{2\pi}} \int_{-\infty}^{x} \mathrm{d}x' e^{-x'^2/2} , \tag{E.45}$$

and $\mathrm{erf}(x)$ is the error function [36, 37]:

$$\mathrm{erf}(x) = \frac{2}{\sqrt{\pi}} \int_{0}^{x} \mathrm{d}x' e^{-x'^2} . \tag{E.46}$$

Furthermore, we obtain

$$\langle x \rangle = x_0 , \tag{E.47}$$

$$\mathrm{var}\,(x) = \sigma^2 . \tag{E.48}$$

E.8 Central Limit Theorem

Let S denote a random variable defined by

$$S = \sum_{i=1}^{N} c_i X_i , \tag{E.49}$$

where the X_i are independent and identically distributed random numbers with mean μ and variance σ^2 and

$$\lim_{N\to\infty} \frac{1}{N} \sum_{i=1}^{N} c_i^k = \mathrm{const} , \quad \forall k \in \mathbb{Z} . \tag{E.50}$$

Then,

$$p(S|N,\mathscr{B}) \approx \mathscr{N}[S|\langle S\rangle, \mathrm{var}(S)], \tag{E.51}$$

with

$$\langle S\rangle = \mu \sum_{i=1}^{N} c_i, \tag{E.52}$$

and

$$\mathrm{var}(S) = \sigma^2 \sum_{i=1}^{N} c_i^2, \tag{E.53}$$

for large values of N. The theorem of DE MOIVRE-LAPLACE is a special case of the central limit theorem, with the result that the X_i are binomial distributed.

E.9 Characteristic Function

The characteristic function $G(k)$ of a stochastic variable X is defined by [31, 32]

$$G(k) = \left\langle e^{ikX}\right\rangle = \int_I \mathrm{d}x e^{ikx} p(x), \tag{E.54}$$

where I denotes the range of the pdf $p(x)$. It follows that

$$G(0) = 1 \qquad \text{and} \qquad |G(k)| \le 1. \tag{E.55}$$

Expanding Eq. (E.54) in a Taylor series with respect to k yields

$$G(k) = \sum_m \frac{(ik)^m}{m!} \int_I \mathrm{d}x\, x^m p(x) \equiv \sum_m \frac{(ik)^m}{m!} \langle X^m\rangle. \tag{E.56}$$

Hence, the characteristic function is a *moment generating function.*

E.10 The Correlation Coefficient

We shall briefly define and discuss the correlation coefficient. Two random variables X and Y form a random vector $Z = (X, Y)$ which follows the pdf $p(Z) = p(X, Y)$ with the normalization

$$\int \mathrm{d}x\mathrm{d}y\, p(x, y) = 1. \tag{E.57}$$

The correlation coefficient r is now defined as

$$r = \frac{\mathrm{cov}\,(X, Y)}{\sqrt{\mathrm{var}\,(X)\,\mathrm{var}\,(Y)}} \,, \tag{E.58}$$

where $\mathrm{cov}\,(X, Y)$ is the covariance (E.16) of X and Y while $\mathrm{var}\,(\cdot)$ denotes the variance (E.10) of the respective argument. It follows from the CAUCHY-SCHWARZ inequality that $0 \le r^2 \le 1$ and, therefore, $-1 \le r \le 1$.[5]

The random variables X and Y are said to be the stronger correlated the bigger r^2 becomes because for statistically independent (uncorrelated) variables we have $p(x, y) = q_1(x)q_2(y)$ with the consequence that $\mathrm{cov}\,(X, Y) = 0$ and, thus, $r = 0$.

The definition of the correlation coefficient is commonly motivated by the problem of linear regression: Suppose we have a set of data points Y associated with data points X. We would like to find a linear function $f(X) = a + bX$ which approximates the data points Y as good as possible. The problem may be stated as

$$\left\langle [Y - f(X)]^2 \right\rangle = \left\langle (Y - a - bX)^2 \right\rangle \to \min \,, \tag{E.60}$$

where a and b are real constants. This corresponds to GAUSS's method of minimizing the square of errors. We have

$$\frac{\partial}{\partial a} \left\langle (Y - a - bX)^2 \right\rangle = -2\,\langle Y - a - bX \rangle = 0 \,, \tag{E.61}$$

and

$$\frac{\partial}{\partial b} \left\langle (Y - a - bX)^2 \right\rangle = -2\,\langle (Y - a - bX)X \rangle = 0 \,. \tag{E.62}$$

Equations (E.61) and (E.62) result in:

$$a + b\,\langle X \rangle = \langle Y \rangle \,, \tag{E.63}$$

$$a\,\langle X \rangle + b\left\langle X^2 \right\rangle = \langle XY \rangle \,. \tag{E.64}$$

Both are easily solved for a and b and one obtains

$$a = \langle Y \rangle - b\,\langle X \rangle \,, \tag{E.65}$$

[5] One defines the scalar product between random variables $(X, Y) = \mathrm{cov}\,(X, Y)$ and therefore $\|X\|^2 = (X, X) = \mathrm{var}\,(X)$. The CAUCHY-SCHWARZ inequality reads

$$|(X, Y)|^2 \le \| X \|^2 \| Y \|^2 \,, \tag{E.59}$$

and therefore $0 \le r^2 \le 1$.

where

$$b = \frac{\langle XY\rangle - \langle X\rangle\langle Y\rangle}{\langle X^2\rangle - \langle X\rangle^2} = \frac{\mathrm{cov}\,(X,Y)}{\mathrm{var}\,(X)} . \tag{E.66}$$

Thus, the linear function $f(X)$ which approximates the data points Y optimally is given by

$$\begin{aligned} f(X) &= \langle Y\rangle - \frac{\mathrm{cov}\,(X,Y)}{\mathrm{var}\,(X)}(X - \langle X\rangle) \\ &= \langle Y\rangle - r\sqrt{\frac{\mathrm{var}\,(Y)}{\mathrm{var}\,(X)}}(X - \langle X\rangle) , \end{aligned} \tag{E.67}$$

and it follows immediately for the squared error:

$$\left\langle [y - F(x)]^2 \right\rangle = \mathrm{var}\,(Y)\,(1 - r^2) . \tag{E.68}$$

Hence, the best result is achieved for $r = \pm 1$ in which case the association of the data points Y with the data points X is really linear while the worst result is found when $r = 0$ (no association what so ever).

E.11 Stable Distributions

A stable distribution is a distribution which reproduces itself [32]. In particular, consider two random variables X_1 and X_2 which are independent copies of the random variable X following the distribution p_X.[6]

The pdf p_X is referred to as a *stable distribution* if for arbitrary constants a and b the random variable $aX_1 + bX_2$ has the same distribution as the random variable $cX + d$ for some positive c and some $d \in \mathbb{R}$.

For this case one can write down the characteristic function analytically. We will give a special case, the so called *symmetric* LÉVY *distributions* [38]:

$$G_\alpha(k) = \exp\left(-\sigma |k|^\alpha\right) . \tag{E.69}$$

Here $\sigma > 0$ and $0 < \alpha \leq 2$. The pdf of such a distribution shows the asymptotic behavior

$$p_\alpha(x) \propto \frac{\alpha}{|x|^{1+\alpha}}, \qquad |x| \to \infty . \tag{E.70}$$

[6] Independent copies of a random variable, are random variables, which are independent and follow the same distribution as the original random variable.

The normal distribution follows from Eq. (E.69) for $\alpha = 2$. Moreover, we observe from Eq. (E.70) that the variance diverges for all $\alpha < 2$. However, the existence of the variance was the criterion for the validity of the central limit theorem formulated in Sect. E.8. We note that stable distributions reproduce themselves and are attractors for sums of independent identical distributed random variables. This is referred to as the *generalized central limit theorem.*

We remark, in conclusion, that for $\alpha = 1$ the CAUCHY distribution results from Eq. (E.69), and note that stable distributions are also referred to as LÉVY *α-stable distributions* [32].

Appendix F
Phase Transitions

F.1 Some Basics

In many systems transitions between different phases can be observed if an external parameter, such as the temperature or the particle density, changes. Familiar examples are the liquid-gaseous phase transition or the ferromagnetic-paramagnetic transition. The two phases exhibit different properties and often develop a different physical structure, like in disorder-to-order transitions. This suggests the introduction of an *order parameter* φ which is zero in one phase and takes on some finite value $\varphi \neq 0$ in the other one.[1] For instance, in the case of paramagnetic-ferromagnetic transitions the magnetization plays the role of the order parameter [43].

In order to classify phase transitions we briefly repeat some basics from statistical mechanics [39–42, 44, 45]. In a canonical ensemble the probability to find the system in micro-state r (as a function of the external parameters temperature T, volume V and number of particles N) is proportional to the BOLTZMANN-factor

$$P_r(T, V, N) = \frac{1}{Z(T, V, N)} \exp\left[-\beta E_r(V, N)\right] . \tag{F.1}$$

Here, $\beta = 1/(k_B T)$, where k_B is the BOLTZMANN constant, T is the temperature, and E_r is the energy of micro-state r. The canonical partition function $Z(T, V, N)$,

$$Z(T, V, N) = \sum_r \exp\left[-\beta E_r(V, N)\right] , \tag{F.2}$$

[1] The choice of the order parameter may not be unique [39–42].

B.A. Stickler, E. Schachinger, *Basic Concepts in Computational Physics*,
DOI 10.1007/978-3-319-27265-8

ensures the normalization of $P_r(T,V,N)$ and determines the free energy $F(T,V,N)$ according to

$$F(T,V,N) = -\frac{1}{\beta} \ln Z(T,V,N) \,. \tag{F.3}$$

The EHRENFEST classification [46] of phase transitions is based on the behavior of F near the transition point: If F is a continuous function of its variables at the transition point and its first derivative with respect to some thermodynamic variable is discontinuous we call it a *first order phase transition*. For instance, transitions from the liquid to the gaseous phase are classified as first order phase transitions because the density, which is proportional to the first derivative of the free energy with respect to the chemical potential, changes discontinuously at the boiling temperature $T = T_B$. We remark the following characteristics of first order phase transitions:

1. The transition involves a *latent heat* ΔQ: The system absorbs or releases energy. A familiar example is the *latent heat of fusion* in the case of melting or freezing.
2. Both phases can coexist at the transition point.
3. A metastable phase can be observed.

In a *second order phase transition* the first derivative of the free energy F with respect to some thermodynamic variable is continuous but the second derivative of F exhibits a discontinuity. For instance, in a *ferromagnetic phase transition* the magnetization (first derivative of F with respect to the external magnetic field B) changes continuously while the magnetic susceptibility χ (the second derivative of F with respect to B) is discontinuous at the CURIE temperature T_c [43].

The modern classification is based on the behavior of the order parameter near the critical point. The order parameter changes discontinuously for first order phase transitions while it changes continuously for second and higher order phase transitions. Second order transitions are typically related to spontaneous symmetry breaking, as for instance in the paramagnetic-ferromagnetic transition. Based on this observation, LANDAU developed a general description of second order phase transitions which we briefly discuss in the following section.

F.2 LANDAU Theory

We regard a second order phase transition characterized by the scalar order parameter φ [47]. Since $\langle\varphi\rangle$ changes continuously at $T = T_c$, it is convenient to define φ in such a way that $\langle\varphi\rangle\,|_{T\geq T_c} = 0$ while $\langle\varphi\rangle\,|_{T<T_c} \neq 0$.

For the free energy $F(T, h, \varphi)$, one chooses

$$F(T, h, \varphi) = F_0(T) + V\left[\frac{a(T-T_c)}{2}\varphi^2 + \frac{b}{4}\varphi^4 - h\varphi\right], \tag{F.4}$$

where a and b are some material constants and h denotes the external field. This ansatz is motivated by the theory of the paramagnetic-ferromagnetic phase transition [43]. Thus, in equilibrium we have

$$\frac{\delta F}{\delta \varphi} = 0, \tag{F.5}$$

which results in

$$a(T-T_c)\varphi + b\varphi^3 = h. \tag{F.6}$$

For $h = 0$ and $T < T_c$ we obtain

$$\langle \varphi_0 \rangle = \sqrt{\frac{a}{b}(T_c - T)} \sim (T_c - T)^{\gamma}, \tag{F.7}$$

where $\gamma = 1/2$ is called the *critical exponent*. For $T \geq T_c$ we have $\langle \varphi_0 \rangle = 0$. We now regard a weak external field h. The order parameter will change

$$\varphi = \langle \varphi_0 \rangle + \delta\varphi. \tag{F.8}$$

Again, we obtain for equilibrium:

$$\frac{\delta F}{\delta \varphi} = a(T-T_c)(\langle \varphi_0 \rangle + \delta\varphi) + b(\langle \varphi_0 \rangle + \delta\varphi)^3 - h = 0. \tag{F.9}$$

Neglecting contributions of order $\mathcal{O}(\delta\varphi^2)$ yields for the susceptibility

$$\chi = \frac{\partial}{\partial h}\langle \varphi \rangle = \frac{\langle \delta\varphi \rangle}{h} \sim |T - T_c|^{\delta}, \tag{F.10}$$

where $\delta = -1$ is a second critical exponent. This is the CURIE-WEISS law [43]. Finally for $T = T_c$ we obtain from Eq. (F.6)

$$\varphi = \left(\frac{h}{b}\right)^{\frac{1}{3}} \sim h^{\frac{1}{\epsilon}}, \tag{F.11}$$

with the third critical exponent ϵ. The LANDAU theory is a mean-field approximation since local fluctuations of the order parameter are neglected.[2] Although the critical exponents obtained with LANDAU's approach deviate from experimental values, the theory is qualitatively correct. We remark that the critical exponents are universal (a property referred to as *universality* [40]) as they depend only on the dimensionality and the symmetry of the interaction.

[2]The extension to space dependent order parameters is referred to as GINZBURG-LANDAU theory [48].

Appendix G
Fractional Integrals and Derivatives in 1D

This section introduces briefly the common definitions and notations associated with fractional calculus in one dimension [49].

The RIEMANN-LIOUVILLE fractional integrals of order $\alpha \in \mathbb{C}$ $[\Re(\alpha) > 0]$, $I^{\alpha}_{a+}f(x)$ and $I^{\alpha}_{b-}f(x)$ on a finite interval $[a, b]$ on the real axis $\mathbb{R}$ are given by

$$I^{\alpha}_{a+}f(x) := \frac{1}{\Gamma(\alpha)} \int_a^x \mathrm{d}x' \frac{f(x')}{(x - x')^{1-\alpha}} \quad \text{for} \quad (x > a,\ \Re(\alpha) > 0), \tag{G.1a}$$

$$I^{\alpha}_{b-}f(x) := \frac{1}{\Gamma(\alpha)} \int_x^b \mathrm{d}x' \frac{f(x')}{(x' - x)^{1-\alpha}} \quad \text{for} \quad (x < b,\ \Re(\alpha) > 0)\,, \tag{G.1b}$$

where $\Gamma(x)$ denotes the Gamma function [36, 37], $\Re(\alpha)$ is the real part of α, and $f(x)$ is a sufficiently well behaved continuous, differentiable function for which the integrals in (G.1) exist. The corresponding RIEMANN-LIOUVILLE fractional derivatives $D^{\alpha}_{a+}f(x)$ and $D^{\alpha}_{b-}f(x)$ of order $\alpha \in \mathbb{C}$ $[\Re(\alpha) \geq 0]$ are defined by

$$\begin{aligned} D^{\alpha}_{a+}f(x) &:= \left(\frac{\mathrm{d}}{\mathrm{d}x}\right)^n (I^{n-\alpha}_{a+}f)(x) \\ &= \frac{1}{\Gamma(n-\alpha)} \left(\frac{\mathrm{d}}{\mathrm{d}x}\right)^n \int_a^x \mathrm{d}x' \frac{f(x')}{(x - x')^{\alpha-n+1}} \quad \text{for} \quad x > a\,, \end{aligned} \tag{G.2a}$$

and

$$\begin{aligned} D^{\alpha}_{b-}f(x) &:= \left(-\frac{\mathrm{d}}{\mathrm{d}x}\right)^n (I^{n-\alpha}_{b-}f)(x) \\ &= \frac{1}{\Gamma(n-\alpha)} \left(-\frac{\mathrm{d}}{\mathrm{d}x}\right)^n \int_x^b \mathrm{d}x' \frac{f(x')}{(x' - x)^{\alpha-n+1}} \quad \text{for } x < b\,, \end{aligned} \tag{G.2b}$$

B.A. Stickler, E. Schachinger, *Basic Concepts in Computational Physics*,
DOI 10.1007/978-3-319-27265-8

with $n = [\Re(\alpha)] + 1$. Here $[\Re(\alpha)]$ denotes the integer part of $\Re(\alpha)$. For $a \to -\infty$ and $b \to +\infty$ the RIEMANN-LIOUVILLE fractional integrals and derivatives are referred to as WEYL fractional integrals and derivatives. In what follows, they will be denoted by $I^\alpha_\pm$ and $D^\alpha_\pm$, respectively.

If $\alpha \in \mathbb{C}$ $[\Re(\alpha) \geq 0]$ and $[a, b] \in \mathbb{R}$ is a finite interval, then the left- and right-sided CAPUTO fractional derivatives ${}^C D^\alpha_{a+} f(x)$ and ${}^C D^\alpha_{b-} f(x)$ are defined by

$$ {}^C D^\alpha_{a+} f(x) = D^\alpha_{a+} f(x) - \sum_{k=0}^{n-1} \frac{f^{(k)}(a)}{\Gamma(k-\alpha+1)} (x-a)^{k-\alpha} , \tag{G.3a} $$

and

$$ {}^C D^\alpha_{b-} f(x) = D^\alpha_{b-} f(x) - \sum_{k=0}^{n-1} \frac{(-1)^k f^{(k)}(b)}{\Gamma(k-\alpha+1)} (b-x)^{k-\alpha} , \tag{G.3b} $$

with

$$ n = \begin{cases} [\Re(\alpha)] + 1 & \alpha \notin \mathbb{N} , \\ \alpha & \alpha \in \mathbb{N}_0 . \end{cases} \tag{G.3c} $$

This is, however, equivalent to

$$ \begin{aligned} {}^C D^\alpha_{a+} f(x) &= \frac{1}{\Gamma(n-\alpha)} \int_a^x \mathrm{d}x' \, \frac{f^{(n)}(x')}{(x-x')^{\alpha-n+1}} \\ &= (I^{n-\alpha}_{a+} D^n f)(x) , \end{aligned} \tag{G.4a} $$

and

$$ \begin{aligned} {}^C D^\alpha_{b-} f(x) &= \frac{(-1)^n}{\Gamma(n-\alpha)} \int_x^b \mathrm{d}x' \, \frac{f^{(n)}(x')}{(x'-x)^{\alpha-n+1}} \\ &= (-1)^n (I^{n-\alpha}_{b-} D^n f)(x) . \end{aligned} \tag{G.4b} $$

The symmetric fractional integrals $I^\alpha_{|x|}$ and derivatives $\mathscr{D}^\alpha_{|x|}$ are referred to as RIESZ fractional integrals or derivatives and are of the form

$$ I^\alpha_{|x|} = \frac{I^\alpha_+ + I^\alpha_-}{2\cos\left(\frac{\alpha\pi}{2}\right)} , \tag{G.5} $$

for $\alpha \in (0, 1)$ and

$$ \mathscr{D}^\alpha_{|x|} = \begin{cases} (-1)^{\frac{n}{2}} \frac{D^\alpha_+ + D^\alpha_-}{2\cos(\alpha\pi/2)} & \text{for} \quad n = [\Re(\alpha)] + 1 \equiv 2k, \ k \in \mathbb{N}_0 , \\ (-1)^{\frac{n-1}{2}} \frac{D^\alpha_+ - D^\alpha_-}{2\sin(\alpha\pi/2)} & \text{for} \quad n = [\Re(\alpha)] + 1 \equiv 2k+1, \ k \in \mathbb{N}_0 . \end{cases} \tag{G.6} $$

Appendix H
Least Squares Fit

H.1 Motivation

In numerous physics applications a set of corresponding data points (x_k, y_k) was measured or calculated and a set of certain parameters $\{\alpha_j\}$ characterizing a function $f(x_k, \{\alpha_j\})$ is to be determined in such a way that

$$\chi^2 = \sum_k c_k \left[y_k - f(x_k, \{\alpha_j\})\right]^2 \to \min . \tag{H.1}$$

This is referred to as a *least squares fit* problem [6, 7]. Here, $c_k \geq 0$ are weights, which indicate the relevance of a certain data point (x_k, y_k) for the fitting routine, and $f(x, \{\alpha_j\})$ is referred to as the *model function*. Besides numerous applications within the context of experimentally obtained data points, we already came across such a problem in our discussion of data analysis in Chap. 19. Here it was of interest to determine the *experimental auto-correlation time* by fitting an exponential function to the measured auto-correlation coefficient $A(k)$, discussed in Sect. 19.3. Hence, we note that in many applications the parameters $\{\alpha_j\}$ can be associated with a physical property of interest.

We distinguish between two different cases: (i) the function $f(x_k, \{\alpha_j\})$ is a linear function of the parameters $\{\alpha_j\}$ and (ii) the function $f(x_k, \{\alpha_j\})$ is not linear in its parameters $\{\alpha_j\}$. It should be emphasized that in both cases the function does not need to be linear in x_k. This section will discuss methods for linear as well as non-linear least squares fits. However, before proceeding some comments on the data points $\{y_k\}$ seem to be required.

B.A. Stickler, E. Schachinger, *Basic Concepts in Computational Physics*,
DOI 10.1007/978-3-319-27265-8

Suppose the points (x_k, y_k) stem from a measurement which has been repeated N-times. In this case for every value x_k we have N different values $\{y_k^j\}$ and we may use the arithmetic mean

$$\overline{y_k} = \frac{1}{N} \sum_j y_k^j , \tag{H.2}$$

instead of y_k in expression (H.1). We may also calculate the variance var (y_k) via[1]

$$\text{var}\,(y_k) = \frac{1}{N} \sum_j (y_k^j - \overline{y_k})^2 . \tag{H.3}$$

If we assume that the data points y_k^j follow a normal distribution with mean $\langle y_k \rangle$ and variance var (y_k) we may proceed in the following way: The weights c_k are chosen as

$$c_k = \frac{1}{\text{var}\,(y_k)} . \tag{H.4}$$

The resulting fit parameters $\{\alpha_j\}$ are then regarded as mean values of parameters where the variances var (α_i) as well as the covariances cov (α_i, α_j) can be obtained from the matrix

$$N_{ij} = \frac{1}{2} \frac{\partial^2 \chi^2}{\partial \alpha_i \partial \alpha_j} , \tag{H.5}$$

via inversion, i.e.

$$C = N^{-1} , \tag{H.6}$$

and

$$C_{ij} = \text{cov}\,(\alpha_i, \alpha_j) . \tag{H.7}$$

The matrix C is commonly referred to as *covariance matrix*.

[1]In many cases one employs the *bias corrected variance* var $(y_k)_B = \frac{N}{N-1}$ var (y_k). For a detailed discussion of the bias corrected variance the interested reader is encouraged to consult a statistics textbook [45, 50–53].

H.2 Linear Least Squares Fit

In this particular case the model function $f(x_k, \{\alpha_j\})$ is defined as

$$f(x_k, \{\alpha_j\}) = \sum_j \alpha_j \varphi_j(x_k) , \tag{H.8}$$

where $\varphi_j(x_k)$ are linear independent basis functions, which do not have to be linear in x_k. The particular case of a *linear regression*, discussed in Sect. E.10, is included. Equation (H.8) specifies the model function $f(x_k, \{\alpha_j\})$ in (H.1) and this yields

$$\chi^2 = \sum_k c_k \left[y_k - \sum_j \alpha_j \varphi_j(x_k) \right]^2 , \tag{H.9}$$

which is supposed to tend to a minimum. We calculate

$$\frac{\partial \chi^2}{\partial \alpha_\ell} = -2 \sum_k c_k \varphi_\ell(x_k) \left[y_k - \sum_j \alpha_j \varphi_j(x_k) \right] \stackrel{!}{=} 0 , \tag{H.10}$$

and arrive at:

$$\sum_j \alpha_j \sum_k c_k \varphi_\ell(x_k) \varphi_j(x_k) = \sum_k c_k y_k \varphi_\ell(x_k) , \quad \forall \ell. \tag{H.11}$$

This equation can be reformulated as the linear equation

$$M\alpha = \beta , \tag{H.12}$$

where the vectors $\alpha = (\alpha_1, \alpha_2, \dots)^T$ and $\beta = (\beta_1, \beta_2, \dots)^T$ with

$$\beta_i = \sum_k c_k y_k \varphi_i(x_k) , \tag{H.13}$$

and the matrix M

$$M_{ij} = \sum_k c_k \varphi_i(x_k) \varphi_j(x_k) , \tag{H.14}$$

have been introduced.

Equation (H.12) can, for instance, be solved with the help of the methods discussed in Appendix C. It is also particularly simple to determine the covariances because we have

$$N_{ij} = \frac{1}{2} \frac{\partial^2 \chi^2}{\partial \alpha_i \partial \alpha_j} = M_{ij} , \tag{H.15}$$

and the covariances follow from Eqs. (H.6) and (H.7).

H.3 Nonlinear Least Squares Fit

Before we discuss the most general case of a completely arbitrary model function $f(x_k, \{\alpha_j\})$ we want to point out that it is in most cases of advantage to linearize the model function if at all possible. For instance, if the model function is an exponential function, it may be linearized by taking the data points $\ln(y_k)$ instead of y_k.

However, if this is not possible there are numerous alternatives to find a solution of the problem. For instance, the GAUSS-NEWTON method can be employed if the model function $f(x_k, \{\alpha_j\})$ and its derivatives with respect to the parameters α_j are known analytically. Another possibility is offered by the application of an deterministic optimization algorithm as they will be introduced in Appendix I. If even this method is not applicable, the methods of stochastic optimization, discussed in Chap. 20, might be an obvious choice.

We describe now the GAUSS-NEWTON method which is essentially a generalization of the NEWTON method presented in Appendix B. The GAUSS-NEWTON method is a method developed to minimize the expression (H.1) iteratively. The derivatives

$$\frac{\partial f(x_k, \{\alpha_j\})}{\partial \alpha_\ell} , \tag{H.16}$$

are assumed to be known analytically. This is an iterative algorithm and, thus, an iteration index is introduced and indicated by a superscript index n like in α_j^n. The algorithm is described by the following steps:

1. Choose a set of initial values $\{\alpha_j^0\}$ for the iteration.
2. Linearize the function $f(x_l, \{\alpha_j^n\})$ and insert the result into Eq. (H.1):

$$\chi^2 \approx \sum_k c_k \left\{ y_k - f(x_k, \{\alpha_j^n\}) - \sum_\ell \left[\frac{\partial f(x_k, \{\alpha_j\})}{\partial \alpha_\ell} \right]_{\{\alpha_j\}=\{\alpha_j^n\}} (\alpha_\ell - \alpha_\ell^n) \right\}^2 . \tag{H.17}$$

We introduce the following abbreviations for a more compact notation:

$$\mathrm{d}f_{k,\ell}^{n} = \left[\frac{\partial f(x_k, \{\alpha_j\})}{\partial \alpha_\ell}\right]_{\{\alpha_j\}=\{\alpha_j^n\}}, \tag{H.18}$$

and

$$f_k^n = f(x_k, \{\alpha_j^n\}). \tag{H.19}$$

3. We have to solve

$$\frac{\partial \chi^2}{\partial \alpha_i} = -2\sum_k c_k \mathrm{d}f_{k,i}^n \left[y_k - f_k - \sum_\ell \mathrm{d}f_{k,\ell}^n(\alpha_\ell - \alpha_\ell^n)\right] \stackrel{!}{=} 0, \tag{H.20}$$

for all parameters $\{\alpha_j\}$. Therefore, we introduce vectors $\alpha = (\alpha_1, \alpha_2, \ldots)^T$, $\beta = (\beta_1, \beta_2, \ldots)^T$ with

$$\beta_i = \sum_k c_k(y_k - f_k^n)\mathrm{d}f_{k,i}^n, \tag{H.21}$$

and the matrix M with elements:

$$M_{ij} = \sum_k c_k \mathrm{d}f_{k,i}^n \mathrm{d}f_{k,j}^n. \tag{H.22}$$

This transforms Eq. (H.20) into a linear system of equations

$$M(\alpha - \alpha^n) = \beta, \tag{H.23}$$

which is solved for $\Delta\alpha^n = \alpha - \alpha^n$. Please note that α^n denotes the vector α after n iterations. The vector α^{n+1} for the next iteration step is guessed from:

$$\alpha^{n+1} = \alpha^n + \Delta\alpha^n. \tag{H.24}$$

4. The iteration is terminated if for all parameters the desired accuracy was achieved. For instance, the condition $|\alpha_j^{n+1} - \alpha_j^n| \le \epsilon$ can be used with ϵ a small parameter. A criterion for the relative error can be formulated in analogue.

Some comments concerning the covariance matrix are in order: It is more complicated in the nonlinear case because we also have to consider the second partial derivatives of the model function $f(x_k, \{\alpha_\ell\})$. However, if these can for some reason be neglected we obtain, again, that $N_{ij} = M_{ij}$, as in Appendix, Sect. H.2. Another, more serious problem is found in the fact that the GAUSS-NEWTON method suffers from severe instability problems. However, a possible remedy was formulated by D. MARQUART [54] who suggested to multiply the diagonal elements of the matrix M with a factor $(1 + \lambda)$ where $\lambda > 0$. A detailed analysis shows that one can choose λ sufficiently large and in such a way that the value of χ_n^2 decreases monotonically, i.e. $\chi_{n+1}^2 \leq \chi_n^2$ for all iteration steps n. However, an increase of λ decreases the convergence rate and more iterations are necessary until the required accuracy was obtained. It is therefore desirable to choose λ values in such a way that the error decreases monotonically but that, at the same time, a convergence rate is maintained which is as large as possible. A possible strategy is to start with some given value of λ and to reduce it after every iteration step by a constant rate. However, if at some point the error χ^2 increases, i.e. $\chi_{n+1}^2 > \chi_n^2$, then λ has again to be increased.

Appendix I
Deterministic Optimization

I.1 Introduction

We use the term *deterministic optimization* to distinguish these particular optimization methods from the stochastic optimization methods discussed in Chap. 20. There are numerous different deterministic methods designed to find the minimum (or maximum) of a given function $f(x)$, where x can be a vector. Roughly speaking, we can distinguish between methods which require the knowledge of the Hessian,[1] methods which need gradients only, and methods which are based on function values only. For instance, if the gradient of a function is known analytically one may exploit NEWTON's method, as it was introduced in Appendix B. Note that such an approach requires the Hessian of the function $f(x)$.

We plan to discuss here in some detail two specific methods, namely the method of *steepest descent* and the method of *conjugate gradients*. Both methods require the knowledge of the gradient of the function, however, the gradient can also be approximated with the help of finite differences (see Chap. 2). A discussion of additional methods is beyond the scope of this book and the interested reader is referred to the available literature [55].

However, before discussing these two methods in more detail, let us briefly consider the quadratic problem which can be solved analytically. In this case the function $f(x)$ can be written as

$$f(x) = \frac{1}{2}x^T Ax - b^T x + c\,, \tag{I.1}$$

[1]The Hessian, or HESSE matrix, $H \in \mathbb{R}^{N\times N}$ of a function $f(x)$, $x \in \mathbb{R}^N$ is the Jacobian of the Jacobian $J(x)$ of $f(x)$ defined in Eq. (B.8). Thus, it is the matrix of second order partial derivatives of a function. It describes the local curvature of a function of many variables.

B.A. Stickler, E. Schachinger, *Basic Concepts in Computational Physics*,
DOI 10.1007/978-3-319-27265-8

where $x \in \mathbb{R}^N, A \in \mathbb{R}^{N\times N}, b \in \mathbb{R}^N$ and $c \in \mathbb{R}$ where we restrict the discussion to real valued functions for reasons of simplicity. We demonstrate now that for symmetric and positive definite matrices A, i.e. $A^T = A$ and $x^T Ax > 0$ for all $x \neq 0$, the minimum of $f(x)$ is given by $x = A^{-1}b$. The gradient of $f(x)$ is readily evaluated and is given by[2]:

$$\nabla f(x) = \frac{1}{2}A^T x + \frac{1}{2}Ax - b . \tag{I.2}$$

This immediately yields the desired result:

$$Ax = b . \tag{I.3}$$

It follows that $x = A^{-1}b$ is a minimum because we assumed A to be positive definite. It is possible to solve the optimization problem even if A is not symmetric by inverting the symmetrized matrix $\left(A + A^T\right)/2$. Finally, the linear equation (I.3) can be solved with the methods discussed in Appendix C.

I.2 Steepest Descent

The most simple gradient based method is the method of *steepest descent* [6]. It is based on the rather straight forward idea of moving in each iteration step into the opposite direction of the gradient, i.e. *downhill*. Hence, we may formulate it mathematically in the following way: Let x_n be the current position of our search for the minimum. Then we choose

$$x_{n+1} = x_n - \alpha_n \nabla f(x_n) , \tag{I.4}$$

where the step-size in direction of the negative gradient, α_n, has to be determined in an additional step. The step-size should be chosen in such a way that we reach the line minimum in direction $\nabla f(x_n)$:

$$\frac{\mathrm{d}}{\mathrm{d}\alpha_n} f[x_{n+1}(\alpha_n)] = -\nabla f(x_{n+1}) \cdot \nabla f(x_n) \stackrel{!}{=} 0 . \tag{I.5}$$

[2] We remember from vector analysis that

$$\nabla_x \left(x^T Ax\right) = \underbrace{\nabla_x \left(x^T A\right)}_{=A} x + \underbrace{\nabla_x \left(x^T A^T\right)}_{=A^T} x = (A + A^T)x .$$

Hence, we observe that for an optimal choice of α_n the search directions are orthogonal. In practice α_n is estimated with the help of a separate minimization technique, such as bisection. This technique has already been used in our discussion of the shooting methods in Chap. 10.

We provide an example which is supposed to make the method more transparent and to help in the discussion of its caveats: We want to determine the global minimum of the function

$$f(x, y) = \cos(2x) + \sin(4y) + \exp(1.5x^2 + 0.7y^2) + 2x \,. \tag{I.6}$$

Its gradient is easily evaluated

$$\frac{\partial f(x, y)}{\partial x} = -2\sin(2x) + 3x\exp(1.5x^2 + 0.7y^2) + 2 \,, \tag{I.7}$$

and

$$\frac{\partial f(x, y)}{\partial y} = 4\cos(4y) + 1.4y\exp(1.5x^2 + 0.7y^2) \,. \tag{I.8}$$

We define the algorithm *steepest descent* with the following steps:

1. Choose some initial values x_0 and y_0.
2. Calculate the gradient $\nabla f(x_n, y_n)$ in iteration step n.
3. Determine α_n in such a way that

$$f[x_{n+1}(\alpha_n), y_{n+1}(\alpha_n)] \to \min \,, \tag{I.9}$$

which is equivalent to

$$g(\alpha_n) := \nabla f[x_{n+1}(\alpha_n), y_{n+1}(\alpha_n)] \cdot \nabla f(x_n, y_n) = 0 \,. \tag{I.10}$$

This is achieved by a bisection technique similar to the one employed in Sect. 10.3,

a. Set $\alpha_n^a = 0$ and chose α_n^b arbitrary.
b. Increase α_n^b until $g(\alpha_n^a)g(\alpha_n^b) < 0$.
c. Define

$$\alpha_n^c = \frac{\alpha_n^a + \alpha_n^b}{2} \,, \tag{I.11}$$

and determine $g(\alpha_n^c)$.

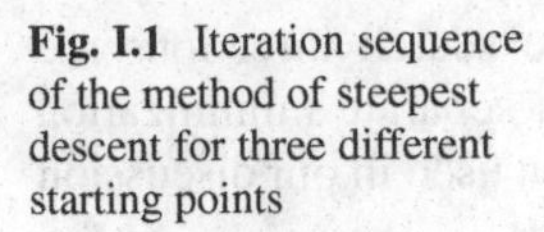
Fig. I.1 Iteration sequence of the method of steepest descent for three different starting points

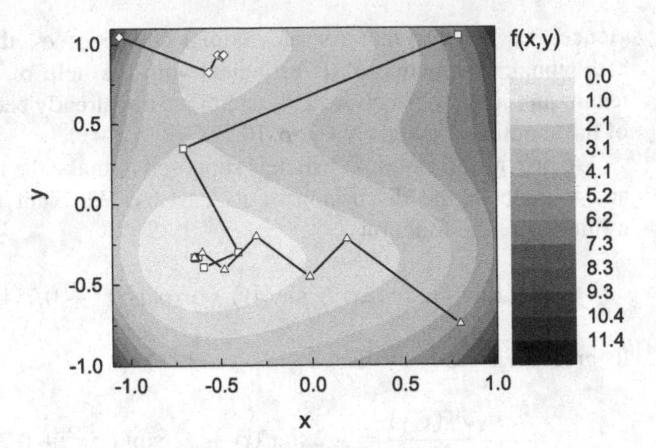

d. If $g(\alpha_n^a)g(\alpha_n^c) < 0$, set $\alpha_n^b = \alpha_n^c$ and return to step c. Otherwise, set $\alpha_n^a = \alpha_n^c$ and return to step c.
e. The bisection is terminated if $|g(\alpha_n^a)| < \epsilon$, with ϵ some required accuracy for the bisection part.

4. Check whether $|f(x_{n+1}, y_{n+1}) - f(x_n, y_n)| \leq \eta$ with η some required accuracy. Return to step 2 for the next iteration step if the algorithm is not converged.

The above algorithm was executed for the function $f(x, y)$ given by Eq. (I.6) for three different starting points, (0.8, −0.75), (0.8, 1.05), and (−1.05, 1.05). The function $f(x, y)$ as well as the iteration sequence towards the minimum for all three starting points is illustrated in Fig. I.1.

We note the following properties of the method: First of all it is a rather slow method due to the orthogonality of subsequent search directions. Moreover, as we observe from Fig. I.1, we can only find the local minimum closest to the starting point and not the global minimum of the function $f(x, y)$. The convergence rate is also highly affected by the choice of the initial position. However, it is a very simple method which works in spaces of arbitrary dimension.

I.3 Conjugate Gradients

The method of *conjugate gradients* [6, 55] is based on the definition of N orthogonal search directions $\{\psi_i\}$ in an N dimensional space. In contrast to steepest descent it is designed in such a way that we take only *one* step in each search direction and the minimum is found after at most N steps, if the function $f(x)$ is of the quadratic form (I.1). In the more general case, however, it will take more steps but will, nevertheless, be much more efficient than the method of steepest descent. Let us formulate the method for a general function $f(x)$.

We approximate the function $f(x)$, with $x \in \mathbb{R}^N$, in the vicinity of the reference point x_n in the n-th iteration step up to second order and name the resulting function $\hat{f}(x)$:

$$\begin{aligned}\hat{f}_n(x) &:= f(x_n) + \nabla f(x_n) \cdot (x - x_n) + \frac{1}{2}(x - x_n) \cdot [\Delta f(x_n)(x - x_n)] \\ &\equiv f(x_n) - b_n^T(x - x_n) + \frac{1}{2}(x - x_n)^T A_n (x - x_n)\,.\end{aligned} \tag{I.12}$$

Here, A_n denotes the Hessian[3] at position x_n and b_n is the negative gradient at x_n. In particular, for a quadratic function $f(x)$ the equality $\hat{f}(x) = f(x)$ holds. We now write the minimum $\hat{x}$ of $f(x)$ as a linear combination of search directions $\{\psi_i\}$ with coefficients λ_i and the initial point x_0:

$$\hat{x} = x_0 + \sum_{i=0}^{M} \lambda_i \psi_i\,. \tag{I.13}$$

Note that in the quadratic case (I.1) this sum will be restricted to $M = N - 1$. At each iteration instance we have the relation

$$x_{n+1} = x_n + \lambda_n \psi_n\,, \tag{I.14}$$

together with the goal

$$x_M \stackrel{!}{=} \hat{x}\,. \tag{I.15}$$

Let us define now a couple of useful quantities. The deviation from the minimum at iteration step $n + 1$, δ_{n+1}, is given by

$$\begin{aligned}\delta_{n+1} &= x_{n+1} - \hat{x} \\ &= x_n + \lambda_n \psi_n - \hat{x} \\ &= \delta_n + \lambda_n \psi_n\,.\end{aligned} \tag{I.16}$$

We define, furthermore, the residual

$$\begin{aligned}r_{n+1} &:= -\nabla \hat{f}_n(x_{n+1}) \\ &= b_n - A_n(x_{n+1} - x_n) \\ &= b_n - \lambda_n A_n \psi_n\,,\end{aligned} \tag{I.17}$$

[3]Note that the Hessian is always symmetric for real valued functions $f(x)$ due to the symmetry of second order derivatives.

where we employed that

$$\frac{1}{2}\nabla\left[(x-x_n)^T A_n(x-x_n)\right] = A_n(x-x_n)\,. \tag{I.18}$$

Finding the minimum of the quadratic approximation $\hat{f}(x)$ of $f(x)$ around x_n is equivalent to the condition

$$r_{n+1} = 0\,. \tag{I.19}$$

In particular, we have to find the product $\lambda_n\psi_n$ in such a way that $r_{n+1} = 0$. Of course, we could invert the Hessian A_n in order to obtain this result. However, this would be too expensive from a computational point of view. The idea is to apply the *ideal* search strategy for quadratic functions to $\hat{f}_n(x)$ in order to obtain x_{n+1}. Hence, the method of conjugate gradients executes packages of N steps, where each package solves the quadratic problem around x_n, until the minimum of the original function $f(x)$ has been found. Therefore, we have to generalize the relations (I.14), (I.16), and (I.17) for iterations within step n.

We have, in particular, for every iteration step n

$$x_{n+1} = x_n + \sum_{\ell=0}^{N-1}\lambda_n^\ell\psi_n^\ell\,, \tag{I.20}$$

together with the definitions

$$x_n^{\ell+1} = x_n^\ell + \lambda_n^\ell\psi_n^\ell\,, \tag{I.21}$$

where $x_{n+1} \equiv x_n^N$. Furthermore, we define the deviation

$$\delta_n^{\ell+1} = x_n^{\ell+1} - x_{n+1} = \delta_n^\ell + \lambda_n^\ell\psi_n^\ell\,, \tag{I.22}$$

and the residual

$$\begin{aligned} r_n^{\ell+1} &= -\nabla\hat{f}_n(x_n^{\ell+1}) \\ &= b_n - A_n(x_n^{\ell+1} - x_n)\,. \end{aligned} \tag{I.23}$$

In contrast to relation (I.17), Eq. (I.23) features the difference $(x_n^{\ell+1} - x_n)$ rather than $(x_{n+1} - x_n)$. We insert the recurrence (I.21) and obtain

$$\begin{aligned} r_n^{\ell+1} &= b_n - A_n(x_n^\ell - x_n) - \lambda_n^\ell A_n\psi_n^\ell \\ &= r_n^\ell - \lambda_n^\ell A_n\psi_n^\ell\,. \end{aligned} \tag{I.24}$$

Hence, in contrast to relation (I.17) Eq. (I.24) defines a recurrence relation. Again, we want to choose the search directions ψ_n^ℓ and the step length λ_n^ℓ in such a way that we find the minimum as quickly as possible. Suppose we already knew the search direction ψ_n^ℓ. The line minimum in this direction is then given by

$$\begin{aligned} \frac{\mathrm{d}}{\mathrm{d}\lambda_n^\ell}\hat{f}_n(x_n^{\ell+1}) &= \nabla\hat{f}(x_n^{\ell+1})\cdot\psi_n^\ell \\ &= -r_n^{\ell+1}\cdot\psi_n^\ell \\ &= -(r_n^\ell - \lambda_n^\ell A_n\psi_n^\ell)^T\psi_n^\ell \\ &= -(r_n^\ell)^T\psi_n^\ell + \lambda_n^\ell(\psi_n^\ell)^T A_n\psi_n^\ell \\ &\stackrel{!}{=} 0\,, \end{aligned} \tag{I.25}$$

and we have

$$\lambda_n^\ell = \frac{(r_n^\ell)^T\psi_n^\ell}{(\psi_n^\ell)^T A_n\psi_n^\ell}\,. \tag{I.26}$$

Hence, the remaining unknown quantities in our algorithm are the search directions ψ_n^ℓ. So far, the only information we obtained is that the search direction ψ_n^ℓ is orthogonal to the residual $r_n^{\ell+1}$, see Eq. (I.25).

However, we also know that

$$\begin{aligned} 0 &= A_n\,(x_{n+1}-x_n) - b_n \\ &= A_n\sum_{\ell=0}^{N-1}\lambda_n^\ell\psi_n^\ell - b_n\,, \end{aligned} \tag{I.27}$$

and therefore

$$0 = \left(\psi_n^k\right)^T A_n\sum_{\ell=0}^{N-1}\lambda_n^\ell\psi_n^\ell - \left(\psi_n^k\right)^T b_n\,, \tag{I.28}$$

for arbitrary k. A sufficient condition to ensure the validity of relation (I.28) is to impose A_n-*orthogonality*:

$$\left\langle\psi_n^k\,\middle|\,\psi_n^\ell\right\rangle_{A_n} \equiv (\psi_n^k)^T A_n\psi_n^\ell = \delta_{k,\ell}\left\langle\psi_n^k\,\middle|\,\psi_n^k\right\rangle_{A_n}\,. \tag{I.29}$$

We note that $\left\langle\psi_n^k\,\middle|\,\psi_n^\ell\right\rangle_A$ constitutes indeed a scalar product since A_n is positive definite in the neighborhood of a minimum.

Let us briefly demonstrate that the choice (I.29) fulfills Eq. (I.28). First of all we note that we obtain from Eq. (I.24)

$$r_n^{\ell+1} = b_n - \sum_{k=0}^{\ell} \lambda_n^k A_n \psi_n^k \,, \tag{I.30}$$

and, therefore, we derive the coefficients λ_n^ℓ from Eq. (I.26) in the convenient form:

$$\lambda_n^\ell = \frac{b_n^T \psi_n^\ell}{\langle \psi_n^\ell \mid \psi_n^\ell \rangle_{A_n}} \,. \tag{I.31}$$

The condition of orthogonality (I.29) is used to rewrite Eq. (I.28) as

$$0 = \lambda_n^k \langle \psi_n^k \mid \psi_n^k \rangle_{A_n} - \left(\psi_n^k\right)^T b_n \,, \tag{I.32}$$

which together with Eq. (I.31) proves the equality (I.28). Hence, the strategy is clear: We choose an initial direction ψ_n^0 and then construct the further directions in such a way that they fulfill A_n-orthogonality (I.29). Before discussing the construction of search directions in more detail we observe the following property:

$$(\psi_n^k)^T r_n^\ell = (\psi_n^k)^T b_n - \sum_{m=0}^{\ell-1} \lambda_n^m \langle \psi_n^k \mid \psi_n^m \rangle_{A_n} = \begin{cases} (\psi_n^k)^T b_n & \text{for } k \geq \ell \,, \\ 0 & \text{else.} \end{cases} \tag{I.33}$$

This means that all search directions ψ_n^k for $k \leq \ell - 1$ are orthogonal to the residual r_n^ℓ, or in other words, all residuals r_n^ℓ are orthogonal (in the classical sense) to all previous search directions.

We shall now briefly outline the resulting update algorithm for search directions: Let $\{\varphi_n^\ell\}$ be a set of linear independent vectors that span our search space for $\hat{f}_n(x)$.[4] We write the search direction ψ_n^k as

$$\psi_n^k = \varphi_n^k + \sum_{\ell=0}^{k-1} \beta_n^{k\ell} \psi_n^\ell \,, \tag{I.34}$$

together with

$$\psi_n^0 = \varphi_n^0 \,. \tag{I.35}$$

[4]In principle these linear independent vectors $\{\varphi_n^\ell\}$ do not need to depend on the index n, i.e. on the actual position x_n. However, we consider here the most general case as will soon become clear.

The expansion coefficients $\beta_n^{\ell k}$ can be determined recursively by imposing A_n-orthogonality for all $\ell < k$:

$$
\begin{aligned}
0 &= \langle \psi_n^k \,|\, \psi_n^\ell \rangle_{A_n} \\
&= \langle \varphi_n^k \,|\, \psi_n^\ell \rangle_{A_n} + \sum_{m=0}^{k-1} \beta_n^{km} \langle \psi_n^m \,|\, \psi_n^\ell \rangle_{A_n} \\
&= \langle \varphi_n^k \,|\, \psi_n^\ell \rangle_{A_n} + \beta_n^{k\ell} \langle \psi_n^\ell \,|\, \psi_n^\ell \rangle_{A_n} \,,
\end{aligned}
\tag{I.36}
$$

and, therefore:

$$
\beta_n^{k\ell} = -\frac{\langle \varphi_n^k \,|\, \psi_n^\ell \rangle_{A_n}}{\langle \psi_n^\ell \,|\, \psi_n^\ell \rangle_{A_n}} \,. \tag{I.37}
$$

This procedure is known as the GRAM-SCHMIDT *conjugation* [6, 12].

Now, the question arises how one should choose the basis vectors φ_n^ℓ and whether or not it is advantageous to choose the φ_n^ℓ as a function of n. A particularly clever choice is to take the residuals, i.e.

$$
\varphi_n^\ell = r_n^\ell \,. \tag{I.38}
$$

In this case we have for $\ell < k$

$$
\begin{aligned}
\beta_n^{k\ell} &= -\frac{\langle r_n^k \,|\, \psi_n^\ell \rangle_{A_n}}{\langle \psi_n^\ell \,|\, \psi_n^\ell \rangle_{A_n}} \\
&= -\frac{(r_n^k)^T A_n \psi_n^\ell}{\langle \psi_n^\ell \,|\, \psi_n^\ell \rangle_{A_n}} \\
&= -\frac{(r_n^k)^T}{\langle \psi_n^\ell \,|\, \psi_n^\ell \rangle_{A_n}} \left[\frac{r_n^\ell - r_n^{\ell+1}}{\lambda_n^\ell} \right] ,
\end{aligned}
\tag{I.39}
$$

where we used recurrence (I.24). We now calculate with the help of Eq. (I.34)

$$
(r_n^k)^T (r_n^\ell) = (r_n^k)^T \psi_n^\ell - (r_n^k)^T \sum_{m=0}^{\ell-1} \beta_n^{\ell m} \psi_n^m = 0 \,, \tag{I.40}
$$

for $\ell < k$ due to the orthogonality of the search direction and the residuals, see Eq. (I.33). Hence, we obtain for all $\ell < k$

$$\begin{aligned} \beta_n^{k\ell} &= \frac{1}{\lambda_n^{k-1}} \frac{(r_n^k)^T r_n^k \delta_{\ell+1,k}}{\langle \psi_n^{k-1} \mid \psi_n^{k-1} \rangle_{A_n}} \\ &= \frac{(r_n^k)^T r_n^k}{(r_n^{k-1})^T r_n^{k-1}} \delta_{\ell,k-1} \, . \end{aligned} \tag{I.41}$$

Hence, the name *conjugated gradients*.

We are now in a position to describe the algorithm for the method of *conjugated gradients*:

1. Choose an initial position x_0.
2. Determine the vector b_n and the matrix A_n for a given position x_n.
3. Perform the following N steps in order to calculate x_{n+1}:

   a. Set

$$\psi_n^0 = r_n^0 = b_n \quad \text{and} \quad \lambda_n^0 = \frac{b_n^T \psi_n^0}{\langle \psi_n^0 \mid \psi_n^0 \rangle_{A_n}} , \tag{I.42}$$

   as well as

$$x_{n+1} = x_n + \lambda_n^0 \psi_n^0 \, . \tag{I.43}$$

   b. Calculate for $k = 1, \ldots, N-1$ the residuals,

$$r_n^k = r_n^{k-1} - \lambda_n^{k-1} A_n \psi_n^{k-1} , \tag{I.44}$$

   the new search directions

$$\psi_n^k = r_n^k + \frac{(r_n^k)^T r_n^k}{(r_n^{k-1})^T r_n^{k-1}} \psi_n^{k-1} , \tag{I.45}$$

   the step lengths

$$\lambda_n^k = \frac{b_n^T \psi^k}{\langle \psi_n^k \mid \psi_n^k \rangle_{A_n}} , \tag{I.46}$$

   and, finally, the modified positions

$$x_{n+1} = x_n + \lambda_n^k \psi_n^k \, . \tag{I.47}$$

4. If $|f(x_{n+1}) - f(x_n)| < \epsilon$, with ϵ some required accuracy, terminate the iteration, otherwise return to step 2. In case of a convex function $f(x)$ terminate also after N steps.

Strictly speaking, this algorithm is only valid for convex functions because we note that one might get into trouble whenever a position is reached at which the Hessian is not positive definite. It is therefore desirable to exclude the Hessian from the algorithm. This can be achieved by an algorithm developed by FLETCHER and REEVES [56]. Based on our previous discussion the generalization is rather obvious: If we do not want to use the Hessian explicitly, we have to determine the step length λ_n^ℓ by minimizing $f(x_n^\ell + \lambda_n^\ell \psi_n^\ell)$ for a given search direction ψ_n^ℓ numerically. The residuals are then taken to be the exact gradient of the function $f(x_n^\ell)$ rather than of $\hat{f}_n(x_n^\ell)$. The next search direction ψ_n^{k+1} is then determined via

$$\psi_n^{k+1} = -\nabla f(x_n^{k+1}) + \frac{\|\nabla f(x_n^{k+1})\|^2}{\|\nabla f(x_n^k)\|^2} \psi_n^k \,. \tag{I.48}$$

Hence, we have the following algorithm (FLETCHER-REEVES *algorithm*):

1. Choose an initial position x_0.
2. Perform the following N steps in order to calculate x_{n+1}:

   a. Set

   $$\psi_n^0 = -\nabla f(x_n) \,. \tag{I.49}$$

   b. Calculate for $k = 0, \ldots, N-1$ λ_n^k by minimizing $f(x_n^k + \lambda_n^k \psi_n^k)$, the new position $x_n^{k+1} = x_n^k + \lambda_n^k \psi_n^k$, and the new search direction via

   $$\psi_n^k = -\nabla f(x_n^{k+1}) + \frac{\|\nabla f(x_n^{k+1})\|^2}{\|\nabla f(x_n^k)\|^2} \psi_n^k \,. \tag{I.50}$$

3. If $|f(x_{n+1}) - f(x_n)| < \epsilon$, with ϵ some required accuracy, terminate the iteration, otherwise return to step 2.

The resulting sequence of steps towards the minimum for the same function and initial conditions as were used for Fig. I.1 is illustrated in Fig. I.2. In comparing Figs. I.1 and I.2 we note immediately that the search strategy developed for the method of conjugate gradients superbly outperforms the search strategy of the method of steepest descent. In particular, if the ratio between the gradient in x and y direction is large, a strategy of orthogonal search directions is disadvantageous. This particular case is illustrated in Fig. I.3 for both, steepest descent and conjugate gradients. Here we investigate the convex function

$$f(x, y) = x^2 + 10y^2 \,, \tag{I.51}$$

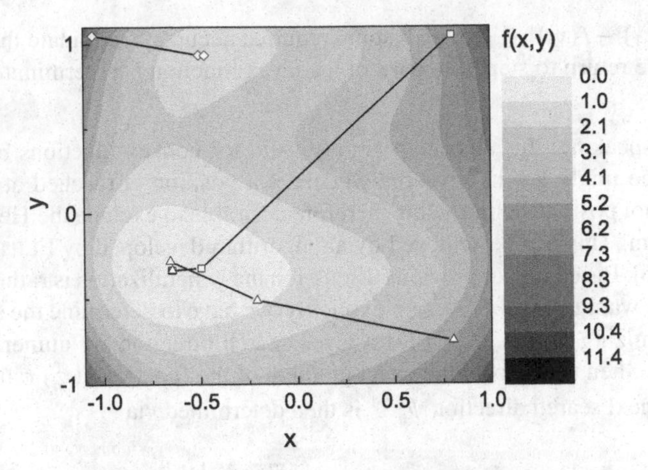

Fig. I.2 Iteration sequence of the method of conjugated gradients for three different starting points

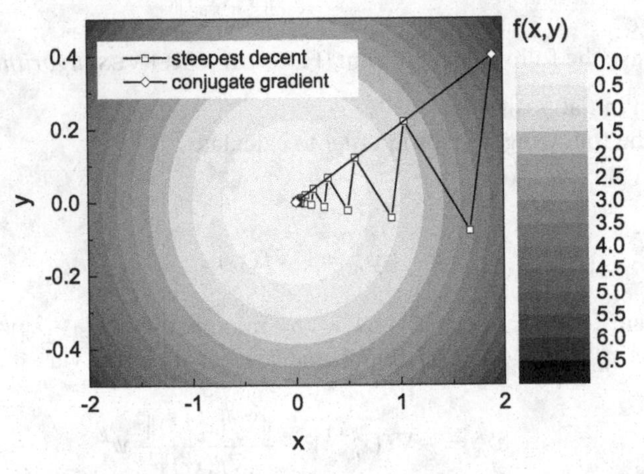

Fig. I.3 Comparison of the iteration sequence between the method of steepest descent and the method of conjugated gradients

together with an initial position $(x_0, y_0) = (1.9, 0.4)$. The resulting sequence of points towards the minimum is illustrated in Fig. I.3. In the case of steepest descent the sequence approaches the minimum rather slowly since subsequent search directions have to be orthogonal to each other in the classical sense. The advantage of conjugate gradients is that A_n-orthonormality accelerates the convergence towards the minimum. In this example we reach it within two steps and a required absolute accuracy of $\eta = 10^{-7}$.

As a final remark we note that also the method of conjugate gradients will only find the *local minimum* closest to the initial position. Hence, the outcome of the method highly depends on the choice of x_0. Moreover, the calculation of the gradients may be very tedious and time-consuming from a numerical point of view.

References

1. Arnol'd, V.I.: Mathematical Methods of Classical Mechanics, 2nd edn. Graduate Texts in Mathematics, vol. 60. Springer, Berlin/Heidelberg (1989)
2. Fetter, A.L., Walecka, J.D.: Theoretical Mechanics of Particles and Continua. Dover, New York (2004)
3. Scheck, F.: Mechanics, 5th edn. Springer, Berlin/Heidelberg (2010)
4. Goldstein, H., Poole, C., Safko, J.: Classical Mechanics, 3rd edn. Addison-Wesley, Menlo Park (2013)
5. Fließbach, T.: Mechanik, 7th edn. Lehrbuch zur Theoretischen Physik I. Springer, Berlin/Heidelberg (2015)
6. Press, W.H., Teukolsky, S.A., Vetterling, W.T., Flannery, B.P.: Numerical Recipes in C++, 2nd edn. Cambridge University Press, Cambridge (2002)
7. Ralston, A., Rabinowitz, P.: A First Course in Numerical Analysis, 2nd edn. Dover, New York (2001)
8. Jacques, I., Judd, C.: Numerical Analysis. Chapman and Hall, London (1987)
9. Burden, R.L., Faires, J.D.: Numerical Analysis. PWS-Kent Publishing Comp., Boston (1993)
10. Stoer, J., Bulirsch, R.: Introduction to Numerical Analysis, 2nd edn. Springer, Berlin/Heidelberg (1993)
11. Westlake, J.R.: A Handbook of Numerical Matrix Inversion and Solution of Linear Equations. Wiley, New York (1968)
12. Hazewinkel, M. (ed.): Encyclopaedia of Mathematics. Springer, Berlin/Heidelberg (1994)
13. Varga, R.S.: Matrix Iterative Analysis, 2nd edn. Springer Series in Computational Mathematics, vol. 27. Springer, Berlin/Heidelberg (2000)
14. Jackson, J.D.: Classical Electrodynamics, 3rd edn. Wiley, New York (1998)
15. Greiner, W.: Classical Electrodynamics. Springer, Berlin/Heidelberg (1998)
16. Griffiths, D.J.: Introduction to Electrodynamics, 4th edn. Addison-Wesley, Menlo Park (2013)
17. Sakurai, J.J.: Modern Quantum Mechanics. Addison-Wesley, Menlo Park (1985)
18. Sakurai, J.J.: Advanced Quantum Mechanics. Addison-Wesley, Menlo Park (1987)
19. Ballentine, L.E.: Quantum Mechanics. World Scientific, Hackensack (1998)
20. Courant, R., Hilbert, D.: Methods of Mathematical Physics, vol. 1. Wiley, New York (1989)
21. Schücker, T.: Distributions, Fourier Transforms and Some of Their Applications to Physics. Lecture Notes in Physics, vol. 37. World Scientific, Hackensack (1991)
22. Kammler, D.W.: A First Course in Fourier Analysis. Cambridge Univerity Press, Cambridge (2008)
23. Hansen, E.W.: Fourier Transforms: Principles and Applications. Wiley, New York (2014)
24. Bernatz, R.: Fourier Series and Numerical Methods for Partial Differential Equations. Wiley, New York (2010)
25. Cooley, J.W., Tukey, J.W.: An algorithm for the machine calculation of complex Fourier series. Math. Comput. **19**, 297–301 (1965). doi:10.1090/S0025-5718-1965-0178586-1
26. Nussbaumer, H.J.: Fast Fourier Transform and Convolution Algorithms. Springer Series in Information Sciences, vol. 2. Springer, Berlin/Heidelberg (1981)
27. Frigo, M., Johnson, S.G.: FFTW, The Fastest Fourier Transform in the West,Vers. 3.3.4 (2015). http://www.fftw.org
28. Bakhturin, Y.A.: Campell-Hausdorff formula. In: Hazewinkel, M. (ed.) Encyclopaedia of Mathematics. Springer, Berlin/Heidelberg (1994)
29. McLachlan, R.I., Quispel, G.R.W.: Splitting methods. Acta Numer. **11**, 341–434 (2002). doi:10.1017/S0962492902000053
30. Rotar, V.: Probability Theory. World Scientific, Singapore (1998)
31. Papoulis, A., Pillai, S.: Probability, Random Variables and Stochastic Processes. McGraw Hill, New York (2001)
32. Breuer, H.P., Petruccione, F.: Open Quantum Systems, chap. 1. Clarendon Press, Oxford (2010)

33. von der Linden, W., Dose, V., von Toussaint, U.: Bayesian Probability Theory. Cambridge University Press, Cambridge (2014)
34. Klenke, A.: Probability Theory. Universitext. Springer, Berlin/Heidelberg (2014)
35. Lee, P.M.: Bayesian Statistics: An Introduction. Wiley, New York (2012)
36. Abramovitz, M., Stegun, I.A. (eds.): Handbook of Mathemathical Functions. Dover, New York (1965)
37. Olver, F.W.J., Lozier, D.W., Boisvert, R.F., Clark, C.W.: NIST Handbook of Mathematical Functions. Cambridge University Press, Cambridge (2010)
38. Applebaum, D.: Lévy Processes and Stochastic Calculus. Cambridge Studies in Advanced Mathematics. Cambridge University Press, Cambridge (2006)
39. Yeomans, J.M.: Statistical Mechanics of Phase Transitions. Clarendon Press, Oxford (1992)
40. Cardy, J.: Scaling and Renormalization in Statistical Physics. Cambridge Lecture Notes in Physics. Cambridge University Press, Cambridge (1996)
41. Fließbach, T.: Statistische Physik. Lehrbuch zur Theoretischen Physik IV. Springer, Berlin/Heidelberg (2010)
42. Pathria, R.K., Beale, P.D.: Statistical Mechanics, 3rd edn. Academic, San Diego (2011)
43. White, R.M.: Quantum Theory of Magnetism, 3rd edn. Springer Series in Solid-State Sciences. Springer, Berlin/Heidelberg (2007)
44. Mandl, F.: Statistical Physics, 2nd edn. Wiley, New York (1988)
45. Keener, R.W.: Theoretical Statistics. Springer, Berlin/Heidelberg (2010)
46. Jaeger, G.: The Ehrenfest classification of phase transitions: introduction and evolution. Arch. Hist. Exact Sci. 51–81 (1998). doi:10.1007/s004070050021
47. Landau, L.D., Lifshitz, E.M.: Course of Theoretical Physics, vol. 5: Statistical Physics. Pergamon Press, London (1963)
48. Ter Haar, D.: Collected Papers of L. D. Landau, p. 546. Pergamon Press, London (1965)
49. Kilbas, A.A., Srivastava, H.M., Trujillo, J.J.: Theory and Applications of Fractal Differential Equations. Elsevier, Amsterdam (2006)
50. Iversen, G.P., Gergen, I.: Statistics. Springer Undergraduate Textbooks in Statistics. Springer, Berlin/Heidelberg (1997)
51. Wilcox, R.R.: Basic Statistics. Oxford University Press, New York (2009)
52. Monahan, J.F.: Numerical Methods of Statistics. Cambridge Series in Statistical and Probabilistic Mathematics. Cambridge University Press, Cambridge (2011)
53. Wood, S.: Core Statistics. Institute of Mathematical Statistics Textbooks. Cambridge University Press, Cambridge (2015)
54. Marquart, D.: An algorithm for least-squares estimation of nonlinear parameters. J. Soc. Ind. Appl. Math. **11**, 431–441 (1963). doi:10.1137/0111030
55. Avriel, M.: Nonlinear Programming: Analysis and Methods. Dover, New York (2003)
56. Fletcher, R., Reeves, C.M.: Function minimization by conjugate gradients. Comput. J. **7**, 149–154 (1964). doi:10.1093/comjnl/7.2.149

Index

B.A. Stickler, E. Schachinger, *Basic Concepts in Computational Physics*,
DOI 10.1007/978-3-319-27265-8

O

P

S

T

V

W

Printed in the United States
By Bookmasters